Palgrave Studies in the History of Science and Technology

Designed to bridge the gap between the history of science and the history of technology, this series publishes the best new work by promising and accomplished authors in both areas. In particular, it offers historical perspectives on issues of current and ongoing concern, provides international and global perspectives on scientific issues, and encourages productive communication between historians and practicing scientists.

More information about this series at
http://www.springer.com/series/14581

Joanna Boileau

Chinese Market Gardening in Australia and New Zealand

Gardens of Prosperity

palgrave
macmillan

Joanna Boileau
Independent Scholar
Auckland, Auckland, New Zealand

Palgrave Studies in the History of Science and Technology
ISBN 978-3-319-51870-1 ISBN 978-3-319-51871-8 (eBook)
DOI 10.1007/978-3-319-51871-8

Library of Congress Control Number: 2017940479

Printed on acid-free paper

This Palgrave Macmillan imprint is published by Springer Nature
The registered company is Springer International Publishing AG
The registered company address is: Gewerbestrasse 11, 6330 Cham, Switzerland

Acknowledgements

In 2009 I stood on the slopes of Chinaman's Hill in the Brunswick Valley in northern New South Wales, Australia, looking out over the Pacific Ocean. Here in the 1920s Chinese farmers cultivated bananas in the rich volcanic soil, on terraces now covered with regenerating rainforest. Tangible remains of their presence can be seen in the stone walled terraces, wells and steps they constructed. The large timber dormitory in which the workers once lived has long been demolished, but some archaeological remains have been found on the site, including Chinese ceramics and a knife for cutting (dehanding) bananas. This experience sparked my interest in Chinese horticultural methods and how they were transported to new environments. I went on to research Chinese market gardening in Australia and also in New Zealand, which shares a very similar history of Chinese immigration. During this time my intellectual home was the University of New England in Armidale, New South Wales, where I enjoyed the support of teachers, fellow students and friends, and the stimulation of exchanging ideas with them.

I owe a particular debt of gratitude to Associate Professor Janis Wilton and Dr Andrew Piper of the University of New England for their friendship, mentoring and encouragement of my research and writing, and for our stimulating and challenging conversations. Many other people contributed constructive comments and advice which have helped me shape this book: Dr James Beattie, University of Waikato, New Zealand; Associate Professor Warwick Frost, La Trobe University, Melbourne; Dr Roger Launius, Smithsonian National Air and Space Museum, Washington; Emeritus Professor Helen Leach,

University of Otago, Dunedin; Associate Professor Julia Martínez, University of Wollongong; Professor Friedrich Schneider, Johannes Kepler University of Linz; Adrienne Wing, Auckland; Professor Henry Yu, University of British Columbia, Vancouver; and Evelyn Hu de Hart, Browns University, New York.

During the writing of this book I have also enjoyed the support of colleagues and fellow researchers in Chinese history in Australia and New Zealand. They include: John Adam, Kate Bagnall; Dr James Beattie; Sophie Couchman; Melissa Dunk; Ruth Lam, Lily Lee; Barry McGowan; Paul McGregor; Dr James Ng; Michael Williams; Kevin Rains; Jonathan Richards and Sandi Robb. It is always a pleasure to meet and share ideas with them at conferences, seminars and other gatherings.

My special thanks go to Dr James Beattie for his support, advocacy and encouragement in the writing and publishing of this book. I also acknowledge the staff of Palgrave Macmillan for their helpful editorial comments and their support and advice during the publication process, in particular: Molly Beck, Peter Carey, Oliver Dyer and Jim Fleming. My thanks also go to Geoff Dawson and Rachel Salmond for their thoughtful comments on the manuscript.

Many other people assisted in providing research material and access to library and museum collections. My sincere thanks go to Mary Archbold, Anne Dunstan and Peter Pendlebury of the Donald History and Natural History Group, Donald, Victoria; Associate Professor Ian Jack, University of Sydney; Jenni Kemarre Martiniello, Director, Kemarre Arts; Andrea Phippen, Ryde City Libraries, Sydney; and Bozena Wojcik, City of Whittlesea Council. In New Zealand, I would particularly like to thank David Clarke, Director, Lakes District Museum, Arrowtown, and Eva Garbutt, North Otago Museum, Oamaru. The staff of the following libraries and archives were all very helpful in assisting with my research queries. In New Zealand: Auckland City Library and Auckland University Library in Auckland; the Alexander Turnbull Library and Archives New Zealand in Wellington; and the Hocken Library, Dunedin. In Australia: the City of Sydney Archives, State Library of New South Wales and Mitchell Library in Sydney; the National Library of Australia and National Archives in Canberra; the Queensland State Library in Brisbane; and the State Library of Victoria in Melbourne.

Last but not least I would like to thank my husband Kevin McCready for his unfailing support. He not only provided a ready sounding board, but also contributed his expertise in translating from Chinese to English.

A Note on Romanisation of Chinese

Many different systems for the Romanisation of Chinese characters have been adopted over the years. In this work I have used Mandarin Pinyin for the spelling of place names and personal names of well-known historical figures. Mandarin Pinyin was introduced in China in 1958, superseding earlier Romanisation systems such as Wade-Giles, developed in the nineteenth century by Thomas Wade and later H. A. Giles. However, the overwhelming majority of Chinese immigrants to Australia and New Zealand in the nineteenth century spoke various dialects of Cantonese. There are several systems for the Romanisation of Cantonese, and usage has developed informally in Australia and New Zealand, often combining elements from different systems. This makes it difficult to maintain a consistent approach, particularly where personal names are concerned. In most cases I have followed the Romanisation used by the authors of works in which personal names are cited, which generally reflects the common usage of individuals and families. Similarly where authors use the original Wade-Giles spelling when citing the works of Chinese scholars and poets from early Chinese dynasties (prior to the Republican era which began in 1912), I have retained the Wade-Giles spelling. I have also included the common Cantonese Romanisations for the major districts of origin of Chinese immigrants, in brackets, at the first mention.

A Note on Māori Spelling

Māori is an official language of Aotearoa New Zealand. The Māori Language Commission has made efforts to counter the dominance of the majority English language, for example by recommending that

Māori words not be italicised when used within English text. Other conventions are not adding an 's' to pluralise Māori words and marking long vowels with a macron, as a guide to pronunciation. I have followed these conventions in this work.

Contents

List of Figures

List of Tables

Maps

Fig. 1 Map of Victoria and New South Wales showing locations of Chinese market gardens mentioned in the text, and major river systems. (Adapted from Barry McGowan, 'Chinese market gardens Southern and Western NSW', Australian Humanities Review, Issue 36, 2005, Fig. 1.)

Fig. 2 Map of Queensland showing locations of Chinese market gardens mentioned in the text.

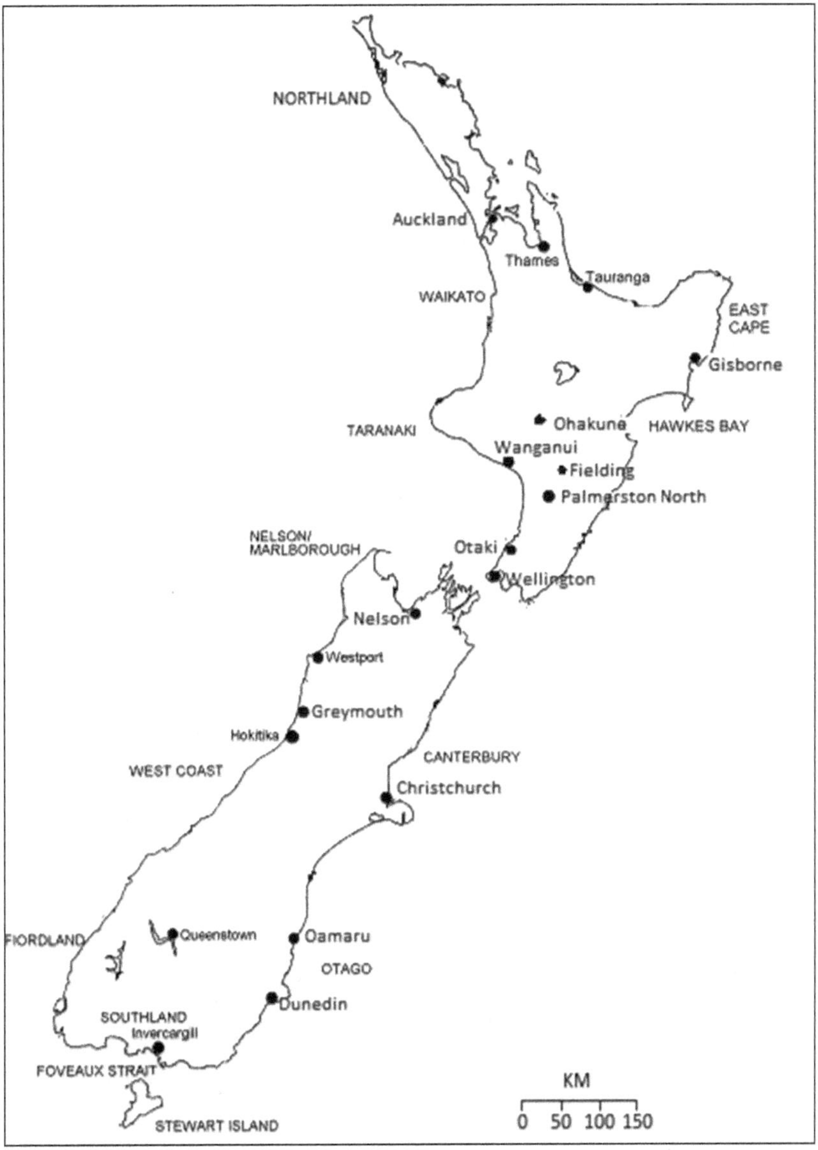

Fig. 3 Map of New Zealand showing locations of market gardens mentioned in the text.

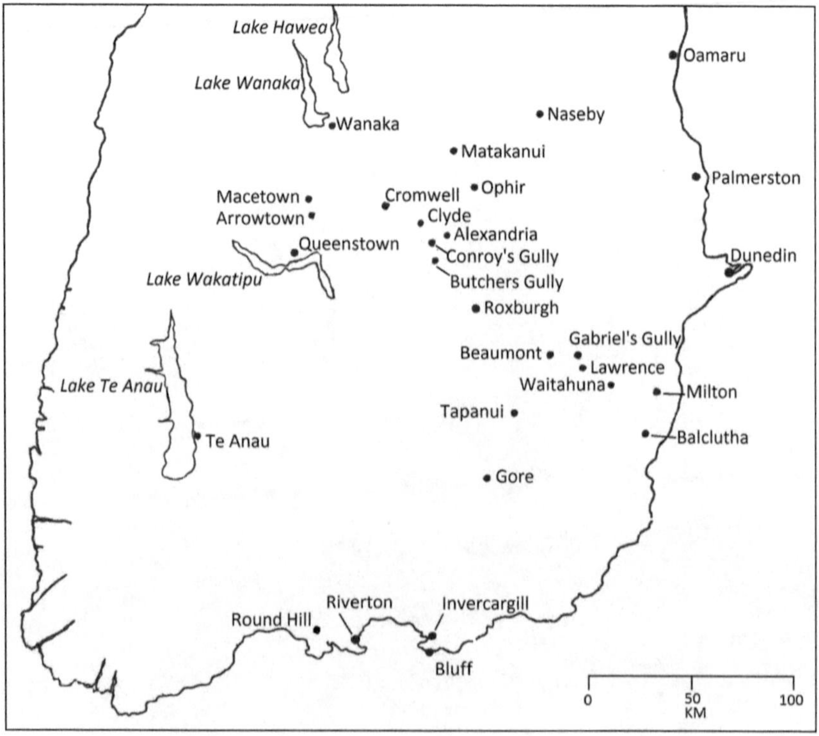

Fig. 4 Map of southern New Zealand (Otago, Fiordland and Southland) showing locations of Chinese market gardens mentioned in the text.

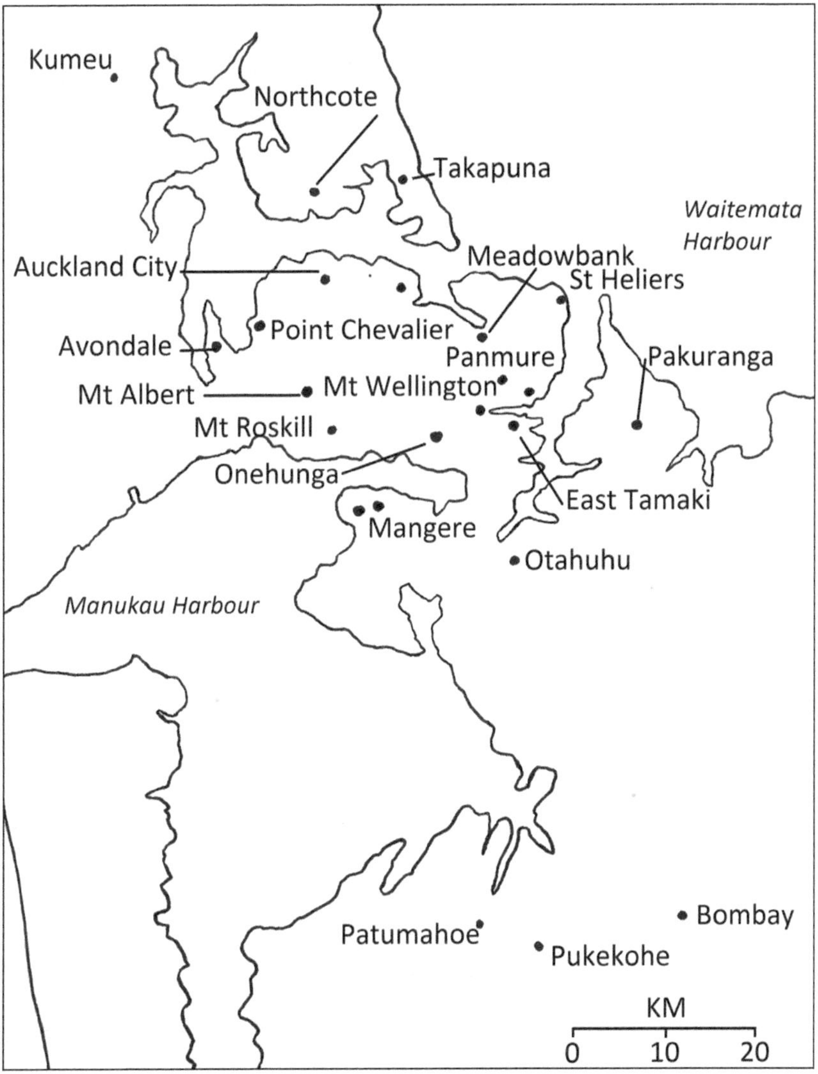

Fig. 5 Map of Auckland region showing locations of market gardens mentioned in the text.

CHAPTER 1

Introduction

> The Chinese are most successful gardeners, or rather farmers, for in their gardens in this colony we really get a specimen of Chinese agriculture. Although we make a distinction between farming and gardening, and conduct them diligently, there is no such distinction made in China, every foot of which is cultivated with as great carefulness as we find displayed here on the land on which they grow vegetables. The Chinese are universally acknowledged to be the best agriculturists in the world. *Sydney Morning Herald*, 2 March 1865, p. 2.

Vegetable gardening is perhaps the most widespread form of gardening in the world, practised by all but a few traditional hunter gatherer societies. As anthropologist Helen Leach has observed, it is an activity as ancient as sowing grains or raising livestock.[1] Leach sees a continuum between small-scale horticulture, where plants are treated individually, and broad-scale agriculture, where they are planted and harvested *en masse*.[2] She argues that the long-running debate on the origins of food production and the domestication of plants and animals has been conducted from the perspective of European-style agriculture, neglecting the role played by gardening in non-Western cultures.[3] More recently, garden historians Michel Conan and W. John Kess reinforce the idea that there is a continuum between agriculture and horticulture. In their view researchers interested in the history of agriculture should be open to the varieties of relationships between humans and plants that have developed through time. They stress

J. Boileau, *Chinese Market Gardening in Australia and New Zealand*,
Palgrave Studies in the History of Science and Technology,
DOI 10.1007/978-3-319-51871-8_1

the important role which domesticated plants have played in the evolution of human societies.[4]

Vegetable gardening has a long tradition in China spanning thousands of years. When Chinese immigrants from southern China travelled to Australia and New Zealand from the mid nineteenth century in search of gold, they brought their horticultural skills with them. From the 1860s until at least the 1960s, they filled an important niche in the horticultural industry, enriched diets and made a significant contribution to public health. The focus here is on commercial gardening, where they cultivated crops on their own account. This is generally termed market gardening in Europe, Australia and New Zealand, or truck farming in North America. Market gardening is a subset of commercial horticulture, which is in turn a subset of the broader agricultural industry. A number of writers have called it 'the Cinderella industry'.[5] Unlike more specialised branches of horticulture such as orchards and vineyards, a wide variety of crops are grown in market gardens, including vegetables, herbs for food and medicinal purposes, and sometimes fruit, flowers and ornamental plants.[6]

Some Chinese gardeners in Australia and New Zealand were employed by Europeans, for example on Australian pastoral stations. However, this study is concerned with the majority who worked for themselves. From the mid nineteenth century, Chinese immigrants established market gardens on the goldfields, serving both Chinese and non-Chinese customers, and market gardens serving Chinese camps, for example in Hay, Albury and Narrandera in southern New South Wales. From the late nineteenth century they established market gardens in towns and cities across Australia and New Zealand.

China's Guangdong province has had a long history of emigration going back to the twelfth century AD. The opening up of the region to foreign trade in the nineteenth century and the spread of news of the gold rushes in California in 1849 and later in Australia and New Zealand promised new economic opportunities overseas. There were also strong push factors. Many scholars have documented the major socio-economic upheavals in Guangdong during the nineteenth century, caused by social unrest and banditry, foreign political and commercial intrusions, and corrupt and oppressive central and provincial governments.[7] June Mei, for example, highlights the problems of overpopulation and the disintegration of the traditional agrarian economy in Guangdong from the mid nineteenth century, as peasants were increasingly unable to maintain their farms and irrigation systems and were forced off the land. Natural disasters such as drought and floods added to their woes. Some turned to non-agricultural jobs such as hawkers, carriers or boatmen, others were left with little choice but to seek work overseas to support their families.[8] New Zealand historian James Ng points out that most Chinese immigrants to New Zealand were not destitute coolies but generally came from rural peasant families who could afford the fare on a ship or had the collateral for a loan. But he notes that many were in fact poor and illiterate, quoting the Presbyterian Minister to the Chinese, Reverend Alexander Don's description: 'Nine tenths of them are small farmers, very small and very poor'.[9]

Chinese people who migrated to Australasia and other Pacific Rim nations from the mid nineteenth century came from a restricted area of southern China, the Pearl River delta region of Guangdong province. The majority came from counties around Guangzhou: Seyip (the 'four counties' comprising Toi Shan (Taishan), Sun Wui, Hoi Ping and Yun Ping), south-west of Guangzhou; the three counties of Poon Yu (Panyu), Nanhai and Shunde (Shuntak), collectively known as Samyip, and Jung Seng (Zengcheng) to the north-east. There were also migrants from neighbouring counties such as Chung Shan (Zhongshan), Dung Goon and Fah Yuen. The largest numbers came from Panyu and Taishan, smaller numbers from Zengcheng and Zhongshan.[10] There were also a number of Hakka from the Sanon and Kueishan districts.[11] Chinese called Australia, and by extension New Zealand, New Gold Mountain (*Sun Gum Shan* in Cantonese) to distinguish it from North America or Gold Mountain (*Gum Shan*) (Fig. 1.1).

For Cantonese, their dream of making their fortunes overseas endured long after gold ran out in California, Australia and New Zealand. Despite multiple hurdles of immigration restrictions, racial prejudice, barriers of culture and language and limited work opportunities, they built new lives and established small businesses, working in market gardens, stores and laundries. Building on the work of Beattie, Reeves and others, this study explores the history of Chinese market gardening in Australia and New Zealand from a comparative perspective, showing how it has been shaped by environmental, economic, technological and social factors. It documents how Chinese market gardeners from subtropical southern China adapted their horticultural techniques, technologies and choice of crops to novel environments and the demands of European consumers. It reveals that they made a significant contribution to the economies of Australia and New Zealand, developing flexible strategies to cope with complex and

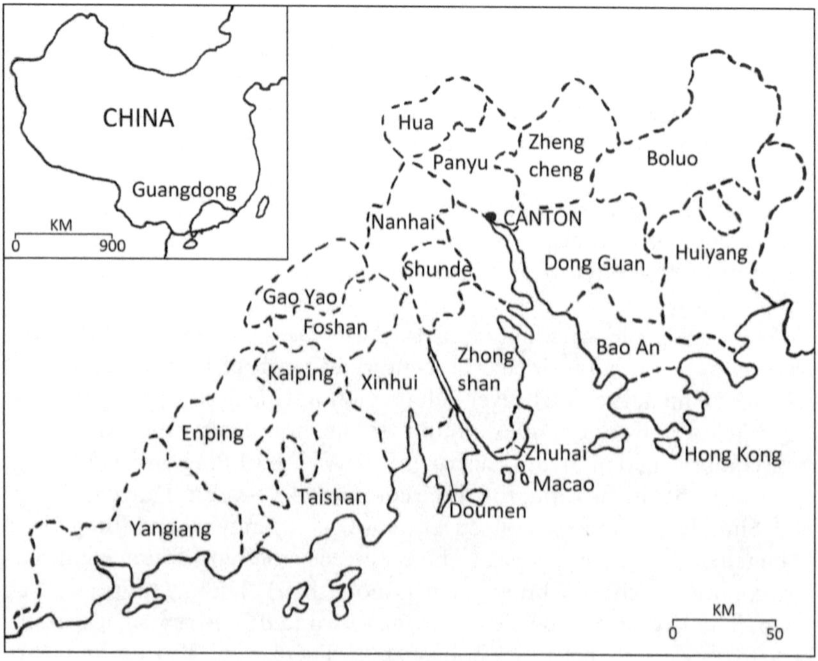

Fig. 1.1 Map of Guangdong showing counties of the Pearl River delta region.
Source: Adapted from Michael Williams, 'Brief sojourn in your native land: Sydney's huaqiao and their links with south China during the first half of the twentieth century', MLetters thesis, University of New England, Armidale, 1998, p. 13.

changing business and social environments and the vagaries of climate. Examining the history of Chinese market gardening in Australia and New Zealand enlarges our understanding of environmental and technological ideas, exchanges and adaptations by adding the views of another group to those of Europeans, Māori and Aboriginal people.

The great majority of Chinese gardeners were initially gold miners, drawn to the goldfields of Victoria and New South Wales in Australia in the 1850s and Central Otago in the South Island of New Zealand in the 1860s. Chinese immigrants came from a culture with centuries-old traditions of water management, and harnessing water was essential to both gold mining and market gardening. This book traces the origins of Chinese market gardens on the goldfields in the 1850s and 1860s and follows their development and survival to the present day in some areas. The broad time span allows us to analyse technological changes over time, and to chart the changing economic and social environments within which market gardens operated. Within this time frame, the primary focus is on the decades between the 1880s and the 1930s, when Chinese market gardening was at its peak.

As many writers on the overseas Chinese have noted, ties of common dialect and place of origin were all-important to overseas Chinese communities, drawing people together and promoting group solidarity.[12] The lineage networks binding family and clan in China extended across oceans and continents. County groupings, native place and clan associations in Australia and New Zealand provided social and financial support to overseas Chinese, including assisting new arrivals to find accommodation and employment, and once they were established, to send letters and remittances home and organise return visits to China. Through these networks travelled money, ideas, news, skills and technology.

Chinese market gardeners in Australasia successfully grew crops in a wide variety of environments, from tropical Northern Australia and semi-arid western New South Wales to the colder climates of Tasmania and Central Otago. In Australia, my main focus in this study is on the eastern states and the Northern Territory because they had the largest populations of Chinese, and because the European population, and the Chinese market gardens which served them, were predominantly located in the relatively well watered and fertile south-eastern corner of the continent and in Tasmania. I discuss Chinese market gardens in a range of environments in New Zealand, from the colder climates of Central Otago, Oamaru and Dunedin in the South Island, to the warmer climates of Otaki, the Bay of Plenty and Auckland in the North Island.

This book positions Chinese market gardeners and the agricultural technologies they brought with them within the complex networks of migration they charted across the Pacific Ocean. It examines the key factors which influenced this circulation of people, ideas, information and technology. These are: the people themselves, their skills, experiences and motivations; the physical, economic and social environments they encountered in Australasia; and developments in technology and industry. Changes in market gardening through time are discussed within each of these themes. Taking up the challenge Edward Beatty makes in his study of technology transfer in nineteenth-century Mexico, my aim is to map the relevant components of the local environments in Australia and New Zealand, the points of contact, conflict and accommodation with the imported technology of Chinese market gardening, and the constraints and opportunities these environments presented to Chinese market gardeners.[13]

The following two chapters of this book trace the transition which Chinese immigrants made from one world to another, from subtropical southern China to a diverse range of physical environments in Australia and New Zealand. Chapter 2 examines the human resources which Chinese and British immigrants brought to their adopted countries: their hopes and aspirations, horticultural skills and experience. It discusses Chinese and British horticultural traditions, and how these traditions interacted with each other and with indigenous subsistence traditions. Chapter 3 contrasts the physical environment of Guangdong, the origin of the majority of Chinese migrants to Australasia from the mid nineteenth century, with the diverse environmental and climatic conditions that they encountered in Australia and New Zealand, and outlines the adaptations they made to them.

Chapters 4 and 5 explore the economic dimensions of Chinese market gardening. Chapter 4 examines the changing occupational structure of Chinese communities in Australia and New Zealand, tracing their movements in search of ways of making a living and processes of urbanisation. Chapter 5 discusses the critical factors involved in running a small business on the land: capital, labour, land, markets and prices, and marketing strategies. Chapter 6 examines the industrial and technological environments within which Chinese market gardeners operated and processes of technological change. It highlights the role of complementary industries and technologies, including suppliers of seeds and fertilisers, agricultural equipment and irrigation systems and developments in transport and communications, in supporting the growth of Chinese market gardening.

The final chapter explores the social environment which Chinese market gardeners encountered in Australia and New Zealand, their responses to government policies of exclusion and changing social attitudes and their diverse social interactions, both within the Chinese community and with the wider society.

Technologies of Mobility

When Chinese immigrants from southern China travelled to Australia and New Zealand from the mid nineteenth century in search of gold, they did not just bring material possessions with them. Their horticultural skills, knowledge and technologies travelled with them across oceans and between cultures. Thus this book is about the mobilisation of knowledge across time and space. It enlarges our understanding of processes of technological change and human mobility, highlighting the mobility of migrants as an essential element in the mobility and adaptation of technologies. It also highlights the mobility of the technologies themselves, taking technology in its broadest sense to include business structures, capital sharing and marketing strategies as well as horticultural skills, techniques and equipment.[14]

This book tells the story of an agricultural technology developed over many centuries in China and transported to two British colonies which share a common heritage of European market gardening. New approaches to what has been termed technology 'transfer' focus on networks and exchanges; therefore a more accurate term would be 'circulation' or 'exchange' of technology. An example is the work of Liliane Hilaire-Perez and Catherine Verna on the dissemination of technical knowledge in medieval and early modern Europe. Building on Nathan Rosenberg's work in this field, Hilaire-Perez and Verna argue that the transmission of technologies is a response to specific needs and expectations, and belongs to 'a world of diversity, contingency and heterogeneity'.[15] They emphasise the complex web of human mobility, shifting the focus to networks and migration of groups. They also argue that geography and choice of scale is crucial – achieving a balance between macro and micro histories.[16]

Earlier studies of technology transfer and diffusion of innovation provide useful background to this study, particularly those which take a historical and contextual perspective. Such approaches are based on the idea that technological change is shaped not only by technological antecedents and economic and technological constraints and incentives, but by an interrelated network of technical, economic, social, environmental and political factors that mediate the relationship between technology and society. Thus technology is seen as an integral part of society and social forces.[17]

Scholars such as Darwin Stapleton and Edward Beatty emphasise that technology 'transfer' is not a single, unidirectional event. It is a process which involves complex pathways and interactions, multiple transfers over time, and continual adaptation and modification to fit into the economy and society of the 'host' country.[18] Arnold Pacey interprets technological change as an ongoing dialectic in which the recipients of new knowledge or techniques interrogate them on the basis of their own experience and knowledge of local conditions, and adapt them to meet their specific needs.[19]

Most studies of technology transfer in the eighteenth and nineteenth centuries have focused on transfers from an industrialised colonial power such as England to its less industrialised colonies including America, Australia and New Zealand, or in the twentieth century, on transfers from technologically developed to less developed countries. Technology transfer theory has been critiqued as an artefact of 1960s modernisation theory, in which the 'developed' West unilaterally transferred technology to so-called 'undeveloped' countries. Rather than something black boxed and unidirectional, it is now seen as something much more fluid and dynamic.[20]

A major shift in academic approaches to technology transfer in agriculture occurred from the 1970s, with critiques of the successes and failures of the 'green revolution' of the twentieth century and its geopolitical implications.[21] There was a move away from a one-way transfer of Western scientific knowledge and expertise from so-called 'developed' nations to developing nations towards greater recognition of indigenous knowledge and technology, and a bottom-up approach, involving local farmers in the design and evaluation of field trials.[22] Influenced by post-colonialism, farmers were increasingly recognised as experimenters and innovators, selecting appropriate technologies or crop varieties according to their own priorities and making continuous adaptations to changing local conditions.

This accords with current approaches to technological change in the social sciences, which see these processes as dynamic, a series of inventions and reinventions. An example is Judith Carney's study of the transfer of rice cultivation technology from Africa to the southern United States by African slaves. Carney analyses the spatial and land management parameters of traditional West African rice cultivation, then charts the introduction of this technology on plantations run by slave labour in South Carolina. She demonstrates a convergence of different knowledge systems, for example of plant species, cultivation strategies and water control, and their recombination into new forms. She concludes: 'As Africans and Europeans faced each other in new territory under dramatically altered and unequal power relations, the outcome was diffusion, technological innovation and novel forms of labour organisation'.[23]

All these approaches have influenced the shape of this book. I endeavour to bring together the stories of individual Chinese market gardeners at the local level and situate them within the broader story of complex global movements of people – across the Pacific between China and Australasia; between Australia and New Zealand; and between England and Australasia. As Canadian scholar Henry Yu describes, Cantonese migrants created complex pathways and networks in their journeys, and recharted their routes in response to the exclusionist policies of the states bordering the Pacific which were designed to curtail or end such migrations. The pace and scale of their movements varied in response to these policies, with stops and starts.[24] A central theme of this book is therefore mobility – not only the mobility of people but the mobility of their modes of living and working, ideas, skills, knowledge, and technology.

Another key theme is the agency of Chinese people themselves as they followed their aspirations in their journeys around the globe and sought to achieve their goals through their own labour and the support of kin and compatriots. Chinese people in Australia and New Zealand and other countries around the Pacific Rim such as Canada and the United States were minority groups in societies where they were marginalised. Their voices too often remain unheard in the mainstream histories of nation states.

There are other insights which can be gained from these studies in the history of technology. Firstly, a given technology or agricultural system is not transferred unchanged, but undergoes adaptation to local conditions – environmental, social, cultural and economic. Secondly, technological change is dynamic, involving complex interactions between different knowledge systems as they traverse multiple pathways criss-crossing the globe. Finally, these studies underscore the need for a multidisciplinary approach to the study of Chinese market gardening.

Multidisciplinary Approaches

A multidisciplinary approach is central to studies of both technological change and the Chinese diaspora. In addition, the sources available for the study of Chinese market gardening in Australia and New Zealand are very diverse, heightening the need for a multidisciplinary approach. This book therefore draws on recent multidisciplinary developments in material culture studies, archaeology, environmental history and garden history.

Material culture is a valuable source of evidence, all too often neglected by historians in their emphasis on written documents. Historians can gain a fuller picture of the past by examining a range of artefacts from different sources, in conjunction with written and pictorial records.[25] In this study I draw on some of the artefacts relating to Chinese market gardening held in museum collections in Australia and New Zealand, bringing together evidence from documentary sources, oral histories and photographs to provide them with a context in time and place. As Janis Wilton notes in Australia, Chinese cultural heritage items are scattered through metropolitan and small regional museums, and often have little information as to their provenance.[26] Items relating specifically to Chinese market gardening are even more dispersed and smaller in number. They are similarly dispersed among museums and private collections in New Zealand.

One of the few collections of artefacts relating to Chinese market gardening held in a single institution, with a well-documented provenance, is the collection which belonged to the market gardener Georgie Ah Ling, housed in the agricultural museum in Donald, Victoria. The artefacts range from agricultural equipment to Chinese language books, Chinese medicine vials and an abacus. This collection is a unique resource which sheds light on the daily life of a Chinese market gardener. The fact that it has been preserved in the Donald Agricultural Museum is a testament to the respect in which Ah Ling was held by the Donald community and their initiative in preserving his belongings following his death in 1987.

I also draw on recent multidisciplinary approaches in historical archaeology which, influenced by the social sciences, take a systemic view of culture. These approaches emphasise the adaptive role of cultural norms and see culture as a differentiated social system mediating relationships between humans and their environment.[27] Again influenced by developments in the social sciences, from the 1970s there was renewed interest in concepts of ethnic identity and processes of acculturation and assimilation, interpreting ethnicity as a political and economic tool to advance a group's welfare.[28]

More recent concepts of ethnic identity stress its fluidity and contingency.[29] Jane Lydon's study of the Chinese community in the Rocks in Sydney, for example, highlights the complexity of cross-cultural encounters, which cannot adequately be explained in terms of ethnicity as allegiance to a bounded group. Lydon argues for more sophisticated approaches towards cross-cultural exchange which consider the dynamic role of material culture in creating, maintaining and transforming culture and which explore its strategic and symbolic meanings.[30] Similarly, Grace Karskens observes that material objects have special significance for migrants in a new land, linking them to who they were and who they are becoming as they adapt to a new society and environment.[31] These approaches have influenced my use of material culture and archaeological evidence, particularly in the way in which I place this evidence in the context of dynamic processes of cross-cultural exchange. This evidence enriches our understanding of the lives of individual Chinese market gardeners, including their agricultural practices, their work organisation, their interactions with Europeans and processes of technological change. It sheds light on the more universal experiences of Chinese immigrants and the cross-cultural connections, exchanges and transformations which are so much part of the immigration process.

Environmental historians of British settler societies have largely ignored Chinese people as agents of environmental change. In the last few decades environmental historians on both sides of the Tasman have regarded Australia and New Zealand as ecological and social laboratories for studying the relations between ecology and history.[32] They have focused on the impact of human activity on the environment, in particular the impact of the introduction of exotic plants and animals and pests on indigenous flora and fauna and the history of extinctions.[33] They have also turned their attention to the moral and social dimensions of European colonisation, exploring the ways in which landscapes have been used and abused, and contrasting indigenous and European attitudes to the land. For example, they have highlighted the ideals, largely unquestioned, inherent in the nineteenth-century British colonial project. These included: the pioneering desire to 'improve' or 'develop' foreign lands; industrial and agrarian ideals of progress; the perceived abundance and inexhaustibility of the natural resources of these new lands; and the economic and political imperatives of exploiting them, to meet the needs of the expanding British Empire.[34] A key theme has been the fundamental mismatch between the expectations of European settlers and the ecological limitations of the new environments they occupied.[35] At the same time, recent scholarship has challenged the simplistic narrative of environmental destruction and human greed during the colonial period, presenting a more nuanced history.[36] This book sheds new light on the role of Chinese immigrants in the transformation of the environments of their adoptive countries and reveals some commonalities in the attitudes of Chinese and Europeans towards the environment, for example their shared desire to 'develop' the land and make it productive.

The history of overseas Chinese people in Australia and New Zealand challenges the overwhelming attention given to Europeans as agents of technological change and environmental transformation. James Beattie's recent work has focused on neglected aspects of global, British imperial and Chinese environmental histories. He has explored the environmental impacts of Cantonese gold miners in New Zealand who adapted water control technology from their homeland in Guangdong and from California and Australia. In New Zealand, along with European miners, they contributed to soil erosion, removed vegetation, reduced timber supplies, and depleted scant water resources.[37]

As Beattie points out, environmental historians of British settler societies have largely ignored Chinese people as agents of environmental change; instead they have focused on how European labour, capital and technology supported by colonial states and their legal, bureaucratic and military machinery facilitated large-scale environmental transformation. This has applied right from the seminal work of Crosby on the Columbian exchange. His study of the global exchange of New World cultigens such as maize, sweet potato, and beans and Old World food staples such as rice, wheat and barley focuses on the impact which the European conquest and colonisation of the Americas had on indigenous populations.[38] In Australia, there have been few studies of the environmental impact of Chinese settlement.[39] Beattie's work in New Zealand indicates the need to extend studies of overseas Chinese to other settler colonies. It also shows that we need to place Chinese environmental history in global perspective. Micoh Muscolino suggests that the interconnectedness of the earth's ecosystems makes it impossible to limit environmental histories to the borders of a single state or society, and historians should stay attuned to transnational and global ecological processes that go beyond human-constructed political boundaries.[40] In addition, as Christine Meisner Rosen points out, environmental transformations cannot be studied in isolation from the business systems and technologies which facilitated them.[41] It is therefore important to explore the environmental dimensions of Chinese business activities, including commercial market gardening.

This book also contributes to research in garden history, another multidisciplinary field which has been rapidly evolving over the last few decades. Garden history is an eclectic mix, ranging across cultures and time periods. It draws on art history, architecture, archaeology, botanical and horticultural science, history and other social sciences in its methodology and approaches. It has evolved from a disparate collection of studies with a narrow national focus towards more comparative approaches and interest in broader themes of gardens as products of human endeavour which meet universal human needs. These needs include contemplation and spiritual renewal, aesthetic appreciation, the display of social status as well as the utilitarian need for fruit, vegetables and herbs. Garden historians are increasingly exploring the symbolic meanings of gardens and their role in social and cultural change. Influenced by anthropology, they are increasingly aware of the need to avoid Eurocentric interpretations.[42]

In particular this book contributes to an area which has been neglected until recently in garden history – gardening technology. As Michael Lee and Kenneth Helphand define it: 'at once purposeful, material, intellectual, active and social, technology is the purposeful organisation of human effort to alter and shape environments. Gardens are one of the most intensively realised of these environments.'[43] They point to a significant gap within garden and landscape studies in investigating and understanding the intimate connections between gardens as designed landscapes and the technologies used to create them. As this study sets out to do, garden historians are beginning to investigate broader questions about technological change: innovation and tradition; the role of individual choices regarding available technologies; and the role of transnational exchange in the adoption of technology.[44]

The key elements of these multidisciplinary approaches that I apply in this study are: using material culture evidence in addition to traditional documentary sources; depicting Chinese market gardeners as active agents in history; and placing the development of market gardening in Australia and New Zealand within the broader contexts of the Chinese diaspora and longer-term processes of social and environmental change.

The Chinese Diaspora and Australasia

Examining the circulation of skills and technology between China and Australia and New Zealand through the lens of market gardening opens up a new perspective on the Chinese diaspora. Most studies of this diaspora have been framed in terms of three broad themes: diverse patterns of adaptation in a wide variety of political and geographic settings; the importance of ongoing links with native place; and the intricate networks connecting overseas Chinese with their home communities.[45] As Madeleine Hsu shows in her study of transnationalism and migration between the United States and South China, the economic and social horizons of Cantonese people expanded from the mid nineteenth century as they became integrated into the global economy. In Australia and New Zealand, as elsewhere, Chinese migrants from the Pearl River delta took advantage of the steadily enlarging web of international trade, improvements in transport and communications technology and the surge in job opportunities created in expanding colonial economies around the world.[46]

In his four-volume anthology, *The Chinese Overseas*, Hong Liu notes that a multidimensional and multidisciplinary perspective is essential for exploring these themes and processes, and can contribute to deeper understanding of the Chinese diaspora and international migration generally, including issues such as identity, ethnicity, multiculturalism and cosmopolitanism. This book contributes to these themes, taking a multidisciplinary approach and focusing on the ideas, skills, technologies and modes of living and working which accompanied the movements of Cantonese migrants across the Pacific between China and Australasia.

This study also builds on recent scholarship on the Chinese diaspora in documenting the way key institutions shaped diasporic Chinese communities, notably the family and clan networks which were the foundation of traditional Chinese society and the voluntary associations and business and trading networks which provided economic and social support for new immigrants.[47] With its focus on mobility, it also highlights the adaptiveness of overseas Chinese communities, the multiple identities they assume, and the flexible strategies they used to negotiate their social relationships in their adopted countries. With its transnational perspective, this study contributes to broader understanding of Chinese migration, settlement and cross-cultural exchange in the Pacific region.

Much has been written about the demographic and social aspects of Chinese immigration to Australia and New Zealand, with a focus on discriminatory European responses towards Asian immigrants and the administrative and legislative restrictions imposed upon them.[48] These studies tend to focus on the gold rush period rather than the range of occupations which Chinese entered after the gold rushes. However, Australian and New Zealand Chinese historiography has broadened in recent years, as historians have extended their inquiries beyond the stereotypes of Chinese as sojourners and gold seekers, market gardeners or storekeepers, revealing more complex social and economic relationships, both within the Chinese community and between Chinese and European communities. They have also moved from depicting Chinese people as passive victims of European prejudice and a multitude of legal and administrative restrictions, to depicting them as active agents in the history of Australasia.[49] For example, Shirley Fitzgerald, Jane Lydon and Grace Karskens chart complex processes of cultural interaction and change in their studies of the Chinese community in Sydney.[50] Other studies include Charles Sedgwick's social history of Chinese people in New Zealand and Julia Bradshaw's history of the Chinese community on the West Coast of New Zealand's South Island, *Golden Prospects*.[51]

These studies open up the possibility of exploring relationships between Chinese and indigenous people as well as between Chinese and Europeans. They provide an inspiration for this study, which considers market gardens as an important locus of cross-cultural exchange. I seek to understand and interpret the experiences of Chinese market gardeners in Australia and New Zealand, by identifying the opportunities and constraints they faced in making a living in new and often hostile physical and social environments. I aim to reveal the complexities of their social relationships, both within the Chinese community and with the wider community, including Europeans and indigenous peoples.

Chinese market gardeners played a major role in supplying fresh vegetables and fruit to the wider community for well over 50 years, yet their contribution has often been neglected in the historiography of Chinese people in Australasia. There is a small but growing body of scholarship addressing Chinese agricultural activities, for example the studies of Zvoncika Stanin and Keir Reeves on the Loddon River, on the Mount Alexander diggings in Victoria, and Neville Ritchie in Central Otago.[52] In New South Wales Barry McGowan has documented Chinese market gardening associated with the goldfields near Braidwood, Araluen, Jembaicumbene and Mongarlowe in the south-east of the State from the 1870s.[53] Through his work here and elsewhere in New South Wales he has demonstrated that significant remains of Chinese agricultural activity exist in rural Australia, and that incorporating cultural landscapes into historical analysis can be an important adjunct to documentary sources.[54]

There is little physical evidence of Chinese market gardens in the urban landscape, for their location is often a distant memory, and many have been obliterated by urban sprawl. However, there have been several historical studies of Chinese market gardening in urban areas of Australia, for example in Sydney, Melbourne and Perth.[55] Awareness of the heritage values of the few surviving market gardens in urban areas has increased in recent years, as shown by the recent debate over the fate of the

Chinese market gardens in Botany in Sydney, threatened by the proposed expansion of the adjoining Eastern Suburbs cemetery.[56]

A key inspiration for this study is Warwick Frost's 2002 paper on Chinese farming in Australia between 1850 and 1920, which considers agricultural technology and methods, labour organisation and interactions with Europeans.[57] Frost challenges the notion that market gardening arrived fully formed from China during the gold rushes and changed little through the rest of the nineteenth century, demonstrating that farming practices evolved over time. He argues that Chinese farming was complex and dynamic, ranging from large-scale commercial agriculture to smaller-scale market gardening. In his view the success of Chinese farmers was due not only to the simple transfer of agricultural skills from China, but to their entrepreneurship and ability to adapt their techniques to the range of environments they encountered in Australia.

These studies highlight the need for more detailed research into the technology, horticultural methods, labour organisation and social interactions involved in market gardening, and the adaptations Chinese gardeners made to the varied environmental conditions of Australia and New Zealand. This book aims to help fill this gap. Bringing together existing local and regional studies of Chinese market gardening in diverse locations in Australia and New Zealand, it explores the adaptations which Chinese migrants made to the physical, economic and social environments they encountered in Australasia. Taking Frost's paper as a starting point, it explores how Chinese market gardening developed over time, and the extent to which Chinese market gardeners adopted advances in European technology.

Notes

1. Helen Leach, *1,000 Years of Gardening in New Zealand*, Auckland, 1984, p. ix.
2. Ibid., p. 4.
3. Helen Leach, 'The terminology of agricultural origins and food production systems – a horticultural perspective', *Antiquity*, Vol. 71, No. 271, March 1997, pp. 135–148.
4. Michel Conan and W. John Kress (eds.), *Botanical Progress, Horticultural Innovations and Cultural Change*, Washington D.C., 2007, pp. 9–10.
5. Malcolm Thick, *The Neat House Gardens: Early Market Gardening Around London*, Totnes, 1998, p. 11; Nigel Murphy, 'A history of the horticultural industry in New Zealand', unpublished ms., 2010, p. 20.

6. R. Sakiyama, 'What is Horticulture?', *Chronica Horticulturae*, International Library for Horticultural Science, Vol. 31, No. 2, 1991, p. 17.
7. June Mei, 'Socioeconomic Origins of Emigration: Guangdong to California, 1850–1882', in James Gerber and Lei Guang (eds.), *Agriculture and Rural Connections in the Pacific*, Aldershot, 2006, pp. 195–234; Michael Williams, *Chinese Settlement in New South Wales*, Sydney, 1999, p. 9; and C. Y. Choi, *Chinese Migration and Settlement in Australia*, Sydney, 1975, p. 3.
8. Mei, 'Socioeconomic Origins', pp. 204–208.
9. Alexander Don, *Outlook*, 4 March 1905, p. 29, quoted in James Ng, *Windows on a Chinese Past*, Vol. 1, Dunedin, 1993, p. 87.
10. Ng, *Windows on a Chinese Past*, Vol. 1, Dunedin North, 1993, p. 123; Neville Ritchie, 'Traces of the past: archaeological insights into the New Zealand Chinese experience in southern New Zealand', in Manying Ip (ed.), *Unfolding History, Evolving Identity: The Chinese in New Zealand*, Auckland, 2003, pp. 31–32.
11. James Ng, 'The sojourner experience: the Cantonese gold seekers in New Zealand', in Ip (ed.), *Unfolding History, Evolving Identity*, p. 7.
12. Choi, *Chinese Migration and Settlement*, pp. 5, 11–13; Michael Williams, *Chinese Settlement in New South Wales*, Sydney, 1999, p. 11.
13. Edward Beatty, 'Approaches to technology transfer in history and the case of nineteenth century Mexico', *Comparative Technology Transfer and Society*, Vol. 1, No. 2, August 2003, p. 188.
14. Henry Yu, pers. comm. My thanks go to Professor Henry Yu for broadening my thinking on the mobility of peoples and technology.
15. Liliane Hilaire-Perez and Catherine Verna, 'Dissemination of technological knowledge in the Middle Ages and early modern era: new approaches and methodological issues', *Technology and Culture*, Vol. 47, No. 3, July 2006, p. 537.
16. Ibid., pp. 538–539, 559.
17. Beatty, 'Approaches to technology transfer', pp. 170, 183. See also Darwin Stapleton, *The Transfer of Early Industrial Technologies to America*, Kansas, 1987.
18. Darwin Stapleton, *The Transfer of Early Industrial Technologies to America*, Kansas, 1987, p. 27; Beatty, 'Approaches to technology transfer', pp. 172, 193.
19. Arnold Pacey, *Technology in World Civilisation*, Cambridge Massachusetts, 1993, p. viii.
20. Hilaire-Perez and Verna, 'Dissemination of technological knowledge in the Middle Ages', pp. 536–537; Nathan Rosenberg, *Exploring the Black Box: Technology, Economics and History*, Cambridge, 1994, p. 5; Nathan

Rosenberg, *Inside the Black Box: Technology and Economics*, Cambridge, 1982, pp. 9, 272.

21. While access to new high yielding, disease-resistant varieties of staple cereals and inputs of fertilisers and pesticides increased agricultural production in some areas, the results were unevenly spread and acted to increase inequalities, further impoverishing the lowest income groups. See, for example, Gordon Conway, *One Billion Hungry: Can We Feed the World?*, Ithaca, London, 2012; Keith Griffin, *The Political Economy of Agrarian Change: An Essay on the Green Revolution,* London, 1979; and *Keith Hoggart, Agricultural Change, Environment and Economy: Essays in Honour of W. B. Morgan,* London, 1992.
22. For example, Robert Chambers, Arnold Pacey, Lori Ann Thrupp (eds.), *Farmer First: Farmer Innovation and Agricultural Research*, London, 1989; Ian Scoones and John Thompson (eds.), *Beyond Farmer First: Rural People's Knowledge, Agricultural Research and Extension Practice*, London, 1994; Paul Richards, *Indigenous Agricultural Revolution: Ecology and Food Production in West Africa*, London, 1985. In particular see the following contributions in *Farmer First*: Robert Rhoades, 'The role of farmers in the creation of agricultural technology', pp. 3–8; D. M. Maurya, 'The innovative approach of Indian farmers', pp. 9–14; Roland Bunch, 'Encouraging farmers experiments', pp. 55–60.
23. These traditional methods involved a range of cropping systems along a gradient from tidal to upland, and sophisticated manipulation of fresh and salt water regimes. Judith Carney, 'Landscapes of technology transfer: rice cultivation and African continuities', *Technology and Culture*, Vol. 37, No. 1, January, 1996, pp. 12–20; 35.
24. Henry Yu, 'The Rhythm of Cantonese trans-Pacific migration' in Donna Gabaccia and Dirk Hoeder (eds.), *Connecting Seas and Connected Ocean Rims: Indian, Atlantic and Pacific Oceans and China Seas Migrations from the 1830s to the 1930s*, Leiden, Boston, 2011, pp. 393–414.
25. Janice Tauer Wass, 'Teaching history with material culture', http://www.lib.niu.edu/1998/iht529802.m html, accessed 3 July 2012; Richard Grassby, 'Material culture and cultural history', *Journal of Interdiscplinary History*, Vol. 35 No. 4, Spring 2005, pp. 597, 602. See also Karen Harvey (ed.), *History and Material Culture: A Student's Guide to Approaching Alternative Sources*, London, New York, 2009, p. 5.
26. Janis Wilton, 'The Chinese History and Heritage of Regional New South Wales', in Henry Chan, Ann Curthoys and Nora Chiang (eds.), *The Overseas Chinese in Australasia: History, Settlement and Interactions*, Canberra, 2001, p. 96.

27. See Louis Binford, 'Archaeology as anthropology', *American Antiquity* Vol. 28, No 2, 1962, pp. 217–225; Ian Hodder, *Symbols in Action: Ethnoarchaeological Studies of Material Culture,* Cambridge, 1982, p. 7.
28. See Randall McGuire, 'The study of ethnicity in historical archaeology', *Journal of Anthropological Archaeology,* Vol. 1, 1982, pp. 159–178.
29. Jane Lydon, *Many Inventions: The Chinese in the Rocks 1890–1930,* Clayton, Victoria, 1999, p. 15.
30. Ibid., pp. 17, 191.
31. Grace Karskens, 'Engaging artefacts: urban archaeology, museums and the origins of Sydney', *Humanities Research,* Vol. 9, No. 1, 2002, pp. 41–42; and *The Cumberland Street/Gloucester Street Site, the Rocks: An Historical Discourse*, Sydney, 1994.
32. Libby Robin and Tom Griffiths, 'Environmental History in Australasia', 2004, pp. 1–2. http://ceh.environmentalhistory-au-nz.org/wp content/uploads/Environmental_History_in_Australasia_2004, accessed 8 February 2013. See also Eric Pawson and Tom Brooking, 'Introduction', in *Environmental Histories of New Zealand*, Auckland, 2002, p. 5.
33. An important recent example is Tim Flannery, *The Future Eaters: An Ecological History of Australasian Lands and People,* Sydney, 1994.
34. Kevin Frawley, 'Evolving visions: environmental management and nature conservation in Australia', in Stephen Dovers (ed.), *Australian Environmental History: Essays and Cases*, Melbourne, 1994, pp. 55–56, 60; J. M. Powell, *Mirrors of the New World: Images and Image Makers in the Settlement Process*, Folkestone/Hamden Connecticut, 1977, pp. 62, 70–78; Robin and Griffiths, 'Environmental History in Australasia', pp. 4–6.
35. See, for example, in Australia, Frawley, 'Evolving visions', p. 55; and in New Zealand, Pawson and Brooking, *Environmental Histories of New Zealand*, pp. 5–7.
36. Writers such as Tim Bonyhady and J. M. Powell highlight the minority voices of politicians, scientists, academics and individual landowners expressing their appreciation of the natural environments of Australia and New Zealand and advocating measures for their conservation. See Tim Bonyhady, *The Colonial Earth*, Melbourne, 2000, pp. 2–5, 11; J. M. Powell, *Environmental Management in Australia 1788–1914: Guardians, Improvers and Profit*, Melbourne, 1993, pp. 100–101, 113–116. See also Michael Roche, 'The state as conservationist, 1920–60: "wise use" of forests, lands and water', in Pawson and Brooking, *Environmental Histories of New Zealand*, pp. 183–99; and Paul Star and Lynne Lochhead, 'Children of the burnt bush: New Zealanders and the indigenous remnant, 1880–1930', in Pawson and Brooking, *Environmental Histories of New Zealand*, pp. 119–135.

37. James Beattie, 'Hungry dragons: expanding the horizons of Chinese environmental history–Cantonese gold-miners in colonial New Zealand 1860s–1920s', *International Review of Environmental History*, Vol. 1, 2015, n.p.
38. Alfred Crosby, *The Columbian Exchange: Biological and Cultural Consequences of 1492*, Westport, Connecticut, 1972.
39. See Fei Sheng, 'Environmental experiences of Chinese people in the mid-nineteenth century Australian gold rush', *Global Environment, A Journal of History and Natural and Social Sciences*, 7–8, 2012, pp. 99–127 and Keir Reeves, 'Tracking the dragon down under: Chinese cultural connections in gold rush Australia and Aotearoa New Zealand', *Graduate Journal of Asia-Pacific Studies*, Vol. 3, No 1, 2005, pp. 49–66.
40. Micah Muscolino, 'Global dimensions of modern China's environmental history', *World History Connected* 6 (2009): 31 pars, http://www.history cooperative.org/journals/whc/6.1muscolino.htm, accessed 19 January 2012.
41. Christine Meisner Rosen, 'The business–environment connection', *Environmental History*, Vol. 10, No. 1, 2005, pp. 77–79.
42. Michel Conan (ed.), *Perspectives on Garden Histories*, Washington D.C., 1999, pp. 14–15; Michel Conan and Jeffrey Quilter (eds.), *Gardens and Cultural Change: A Pan-American Perspective*, Washington D.C., 2007, pp. 3–4. See also Conan (ed.), *Gardens and Imagination: Cultural History and Agency*, Washington D.C., 2008.
43. Michael Lee and Kenneth Helpland (eds.), *Technology and the Garden*, Washington D.C., 2014, p. 1.
44. See, for example, the following studies in Lee and Helpland (eds.), *Technology and the Garden:* Mark Laird, '"Much better contrived and built then any other in England": stoves and other structures for the cultivation of exotic plants at Hampton Court Palace, 1689–1702', pp. 79–107 and Jan Woudstra, 'Garden hydraulics in pre-Sistine Rome: theory and practice', pp. 111–128.
45. Hong Liu, 'Towards a multidimensional exploration of the Chinese overseas', in Hong Liu (ed.), *The Chinese Overseas*, Vol. 1, Abingdon, 2006, pp. 2–3.
46. Madeleine Hsu, *Dreaming of Gold, Dreaming of Home: Transnationalism and Migration between the United States and South China, 1882–1943*, Stanford, 2000, p. 2.
47. See the following in Hong Liu (ed.), *The Chinese Overseas*, Vol. 2: Edgar Wickerberg, 'Overseas Chinese adaptive organisations, past and present', pp. 87–103; and Maurice Freedman, 'Immigrants and Associations: Chinese in nineteenth century Singapore', pp. 104–127.
48. Key examples are: Kathryn Cronin, *Colonial Casualties: Chinese in Early Victoria*, Melbourne, 1982; Ann Curthoys, '"Men of all nations, except

Chinamen": Europeans and Chinese on the gold-fields of New South Wales', in I. McCalman, A. Cook and A. Reeves (eds.), *Gold: Forgotten Histories and Objects in Australia*, Cambridge, 2001; Ng Bickleen Fong, *The Chinese in New Zealand*, Hong Kong, 1959; David Horsfall, *March to Big Gold Mountain*, Melbourne, 1985; Andrew Markus, *Fear and Hatred: Purifying Australia and California 1850–1901*, Sydney, 1979.

49. Cathie May, *Topsawyers: The Chinese in Cairns 1870 to 1920*, Townsville, 1984; Jan Ryan, *Ancestors: Chinese in Colonial Australia*, South Fremantle, 1995; Janis Wilton, *Golden Threads: The Chinese in Regional New South Wales 1850–1950*, Sydney, 2004.
50. Shirley Fitzgerald, *Red Tape Gold Scissors: The Story of Sydney's Chinese*, Sydney, 1996; Lydon, *Many Inventions*; Karskens, *The Cumberland Street/Gloucester Street Site*, and *The Rocks: Life in Early Sydney*, Melbourne, 1997. See also John Fitzgerald, *Big White Lie: Chinese Australians in White Australia*, Sydney, 2007.
51. Charles Sedgwick, 'The politics of survival: a social history of the Chinese in New Zealand', PhD Thesis, University of Canterbury, Christchurch, 1982; Julia Bradshaw, *Golden Prospects: Chinese on the West Coast of New Zealand*, Greymouth, 2009.
52. Zvonkica Stanin, 'From Li Chun to Yong Kit: A Market Garden on the Loddon, 1851–1912', *Journal of Australian Colonial History*, Vol. 6, 2004; Keir Reeves, 'Historical neglect of an enduring Chinese community', *Traffic*, Issue 3, July 2003, pp. 53–77; and Neville Ritchie, 'Archaeology and history of the Chinese in southern New Zealand during the nineteenth century: a study of acculturation, adaptation, and change', PhD Thesis, University of Otago, Dunedin, 1986.
53. Barry McGowan, 'Chinese market gardens in Southern and Western NSW', *Australian Humanities Review*, Issue 36, 2005, n.p.
54. McGowan, 'Chinese market gardens'; Reeves, 'Tracking the dragon', p. 51.
55. Colleen Morris, 'Chinese market gardens in Sydney', *Australian Garden History*, Vol. 12, No. 5, March/April 2001, pp. 4–7; Eardley and Eileen Gifford, 'The Chinese market gardeners of St George district', *St George Historical Society Bulletin*, December 1970, pp. 75–81; Karla Whitmore, 'Willoughby's Chinese market gardens', unpublished ms., January 2004; Sandra Pullman, 'Along Melbourne's rivers and creeks', *Australian Garden History*, Vol. 12, No. 5, March/April 2001, pp. 9–10; and Anne Atkinson, 'Chinese market gardening in the Perth metropolitan region, 1900–1920', *Western Geographer*, Vol. 8, 1984, pp. 38–52.
56. The Botany gardens were listed on the State Heritage Register in 1999 and by the National Trust of Australia (NSW) in 2011. Shirley Fitzgerald, 'Desktop review of heritage significance of market gardens', June 2012, http://laperousemarketgardens.wordpress.com/history/, accessed 4

February 2013; see also Karl Zhao, 'Chinese market gardens', *Heritage New South Wales*, Vol. 6, No. 3, October 1999, p. 15. See also Andrea Humphries, Michael Wright, Adam Fowler, Paul Tinslay, Denis Gojak and Mark Tooker, 'Rockdale Market Gardens Conservation Management Plan', Sydney, 2001; and Joanne Monk, 'The diggers in the trenches: a history of market gardens in Victoria 1835–1939', *Australian Garden History*, Vol. 4, No. 1, July/August 1992, pp. 3–5.

57. Warwick Frost, 'Migrants and Technological Transfer: Chinese farming in Australia, 1850–1920', *Australian Economic History Review*, Vol. 42, No. 2, 2002, pp. 113–131.

CHAPTER 2

The Human Resource

Scholars in the field of the history of technology highlight the crucial role which movements of people play in the circulation of ideas, skills and technology. For example, Liliane Hilaire-Perez and Catherine Verna observe that recent research focuses on complex webs of human mobility – individual artisans, entrepreneurs or engineers acted within networks of kin, compatriots and friends.[1] Through the ongoing links which they maintained with their homelands, these migrants opened up multiple routes along which information, ideas, tools, technology and other migrants flowed.

During the Chinese diaspora across Southeast Asia and the Pacific in the mid nineteenth century, thousands of people left their homeland to escape civil war, famine and social upheaval, and seek a better life overseas. They brought with them skills, knowledge and experience and a set of beliefs, values and customs which helped them survive and adapt in foreign lands. These ideas, beliefs and skills circulated via the intricate web of contacts which they maintained with their homeland and their multiple journeys between China and their adopted countries. Their paths crossed with those of the thousands of European migrants, the majority from the British Isles, who settled in Australia and New Zealand.

This chapter begins by comparing two immigrant groups from very different cultural backgrounds – Chinese and British – who settled in Australia and New Zealand. It highlights their socioeconomic backgrounds,

J. Boileau, *Chinese Market Gardening in Australia and New Zealand*,
Palgrave Studies in the History of Science and Technology,
DOI 10.1007/978-3-319-51871-8_2

skills and experience as well as their hopes and aspirations. The following two sections contrast Chinese and British horticultural traditions, while the subsequent section discusses the indigenous gardening traditions which European and later Chinese settlers encountered in Australia and New Zealand and the interactions between them. The final section illustrates the continuity of Chinese horticultural traditions in these new lands.

Journeys to the Antipodes

Chinese immigrants came to Australia and New Zealand in small numbers from the early days of European settlement.[2] Between 1847 and 1853 some 3,000 Chinese indentured labourers were imported to the colony of New South Wales in Australia, to fill a major shortage of labour on pastoral properties. They worked as cooks, shepherds, shearers and labourers, clearing land and fencing.[3] It was not until the gold rushes of the 1850s and 1860s that large numbers of Chinese men made the long sea journey to Australia and New Zealand in search of wealth. The majority came from the southern province of Guangdong, and in particular from a few counties in the Pearl River delta around Canton, the modern city of Guangzhou. Most who arrived in Australia between the 1850s and 1870s travelled to the goldfields of Victoria and New South Wales. They arrived in large numbers from 1854, and like the miners of many other nationalities who flocked to the goldfields, joined a highly mobile population, moving from field to field as new discoveries were made.[4]

The first Chinese miners arrived on the goldfields of southern New Zealand in 1866, following other miners from Australia. They came from the Victorian goldfields at the invitation of the Otago Provincial Council, sent via leading Chinese merchants in Melbourne. Gold had been discovered in 1861 in Central Otago, but by 1864 returns were declining, and there was an exodus of miners to new fields in Marlborough and on the West Coast. Otago business people were concerned about the impact of this dramatic decrease in population on the local economy and after much debate decided to recruit experienced Chinese miners from Australia. As news of the gold rush in New Zealand spread, more men travelled there directly from China. Like those who travelled to Australia, the majority came from counties around Guangzhou. These men had close links with their compatriots on the Australian goldfields, and many passed through Australia on their way to New Zealand.[5]

While they spoke different dialects and had varying levels of prosperity, Chinese immigrants to Australia and New Zealand shared a common farming and peasant background. Most arrived indebted to family or labour agents for their passage, and had little capital.[6] Their skills and labour were therefore all-important, and they made use of their traditional skills in horticulture and water management in gold mining and market gardening. Market gardening and mining – at least of the more easily won alluvial gold which was the focus of early Chinese mining activity – both initially required relatively little capital. As noted above, they were independent migrants, driven to seek their livelihood overseas by deteriorating economic conditions and political instability at home. They were sojourners; their goal was to support their family in China and eventually save enough money to return home wealthy men. As Henry Yu observes, the image of *Gum Shan*, the Gold Mountain, signified more than just the precious specks of gold they could win from alluvial gravels, but a lifelong quest to better themselves – 'a life cycle of upward mobility built out of labour and mutual investments in each others' aspirations of wealth'.[7]

Nineteenth-century Chinese immigrants encountered British colonial societies which were still in the process of establishing their hegemony over the indigenous owners of the lands they had settled. These societies were peopled by predominantly British settlers, part of a large-scale diaspora from Europe to the New World which had begun in the seventeenth century and populated Australia, New Zealand, the Americas and southern Africa.[8] British migrants came from all over the British Isles and from all classes, from labourers to gentry. James Belich, writing about the peopling of New Zealand in the nineteenth century, describes the fluidity and fragmentation of society in Victorian Britain as people moved from country to city and from the city to the colonies, blending regional and ethnic groups and blurring class lines.[9]

While economic motivations were paramount, they do not fully explain this mass exodus. Geographer Joe Powell highlights the role of illusion in the settlement process, bound up with the hopes, fears, uncertainties and challenges that are so much part of the human condition. Early European settlers to the Antipodean colonies had aspirations to better themselves and many were attracted by visions of a rural arcadia where there was natural abundance and fertility, and where they could become independent small land holders. British and colonial governments, entrepreneurs and immigration agents promoted the vision of the independent yeoman farmer in the colonies, where, in the words of Powell, it became 'a central symbol in a form of popular and politically useful agrarian idealism'.[10] Similarly Belich emphasises the powerful influence of these myths and prophecies in the peopling of New Zealand, identifying 'a convergence between private and public interests in the colonising crusade'.[11] New Zealand was said to share a similar temperate climate with Britain; colonisers, politicians and publicists waxed lyrical about its natural abundance and the benefits of rural life, and promoted it as a 'Better Britain', a climatic paradise.[12]

Despite these agrarian ideals, government endeavours to create a new class of small farmers were a failure in Australia, and to some extent in New Zealand. The story of Australian agriculture so often presented is a saga of men and women on the land, small farmers battling the power and political influence of wealthy squatters (land holders who occupied vast areas of Crown land on which they produced wool, wheat, mutton and beef for the British market). Large-scale commercial agriculture and pastoralism for export dominated, taking up most of the productive land.[13] Land selection legislation introduced in Victoria and New South Wales in 1860 and 1861, and later in other colonies, left many squatters still in control of their vast holdings and failed to benefit small farmers. The soldier settlement schemes after World War I were also unsuccessful – they were controlled by state governments which failed to apply consistent standards in the selection of settlers, the quality of land allotted or the farming advice and support given to returned servicemen.[14]

In New Zealand there was a similar movement in the late nineteenth century to break up the large sheep runs established in the 1850s and 1860s, to allow small farmers to gain a foothold on the land. This policy gained the Liberal Party great popular support, particularly after advances in refrigeration technology in the 1880s opened up overseas markets for meat and dairy produce.[15] Refrigeration radically changed the nature of farming; now farmers could raise sheep for meat and wool, which made their small farms considerably more profitable. New Zealand's economy became increasingly based on the export of meat and dairy produce to Britain.[16]

Given the dominance of large-scale pastoralism and agriculture for export, market gardening formed a small part of the agricultural sectors in Australia and New Zealand. Most European farmers grew fruit and vegetables as part of mixed farming operations.[17] As Nigel Murphy notes, small-scale, intensive market gardening as a sole source of income was considered a low status occupation, in the same class as agricultural labouring, and this prejudice was carried from Britain to the colonies.[18] This attitude towards market gardening facilitated the entry of Chinese immigrants into market gardening, an economic niche where they faced relatively little competition from Europeans.

The Chinese and British immigrants who came to the Antipodes from the mid nineteenth century came from different cultural backgrounds and had different aspirations. They brought with them distinctive horticultural traditions, developed over centuries of experimentation. These traditions are the focus of the following sections.

Chinese Horticultural Traditions

One of the major sources on Chinese agriculture is Francesca Bray's work, part of Joseph Needham's massive multi-volume study, *Science and Civilisation in China*. Bray draws on the many written texts produced in China over the centuries, including agricultural treatises going back as far as the second century AD.[19] Her work has been criticised by later writers such as Xu and Peel, who argue that she relies too heavily on written sources at the expense of practical application.[20] However, Bray's work provides a useful background to first-hand accounts of Chinese agricultural practices, such as those recorded by the New Zealand missionary Alexander Don and the American agricultural scientist Franklin Hiram King. Both men travelled in southern China in the late nineteenth and early twentieth centuries and watched farmers at work.

Bray divides Chinese agriculture into two broad systems, based on the two major climatic zones in the country: dry land grain cultivation in the continental climate of the northern plains, and wet rice cultivation in the subtropical south.[21] The focus of this chapter is on the horticultural systems that developed in southern China by the mid nineteenth century and which were brought to Australia and New Zealand by Chinese immigrants.

Vegetable cultivation has been an integral part of Chinese agriculture from its beginnings. Archaeological research indicates that two broad centres of domestication of plants and animals arose, based on millet and wheat in the north and rice in the south. There were a number of distinct regional clusters of Neolithic cultures by 5000 BC, and over the next two millennia they formed spheres of interaction, expanding and merging.[22] There is strong evidence for gardening, for example from the Ta-p'en-k'eng culture in Taiwan (c. 4450–4350 BC), and early farmers continued to hunt, fish and gather wild foods throughout the period.[23] From the written sources Bray concludes that the characteristic pattern of Chinese cropping systems was established as early as 5000 BC, and had surprising continuity through time. Over the centuries the majority of arable land in China continued to be devoted to grain production (millet and wheat in the north, rice in the south), while roots, tubers, legumes and other vegetable crops took up the remainder. For example, in the early 1900s over 70 percent of the total area under crops in China was planted in grains.[24] In contrast to the European tradition of mixed farming, animal husbandry played a minor role in the rural economy of the Han heartland of China, and little land was used for pasturing livestock.[25]

The importance of grain and vegetable production in the Chinese economy is reinforced by the fact that the Chinese diet has always been predominantly vegetarian, supplemented by meat and fish. The Chinese word for meal or food is *fan,* a general term for grains, including rice, millet, wheat and sorghum. This staple component of every meal is enlivened by a wide variety of fresh vegetables, sauces, vinegar, ginger, chilli and other spices. Vegetables are probably the second most important food group, while additional protein comes from bean curd, or small quantities of fish or meat.[26]

Although most arable land in China was devoted to cereal cultivation, vegetable cultivation played a key role in agricultural innovation. Xu and Peel note that intensive agricultural systems were initially developed in vegetable gardens, and later introduced into field cropping.[27] The predominance of free, small-scale farmers in China, many of whom owned their own land, also encouraged agricultural innovation and experimentation with new crops. Individual households were the basic unit of production, allowing both tenant and land owning farmers to make their own decisions about what crops to grow and how and where to sell their produce. Thus institutional barriers to innovation were minimised.[28]

By the early nineteenth century an integrated farming system was established in the densely populated southern provinces of China such as Guangdong. Individual farmers grew rice and vegetables, raised chickens and pigs, planted fruit trees, including mulberry trees to raise silkworms, and farmed fish. Nothing was wasted – the mulberry trees fed the silkworms, the silk worm waste fed the fish and the manure from humans, pigs and chickens fertilised the fields. Irrigated rice fields were highly productive. Farmers produced two crops of rice per year, and in between harvests grew a wide variety of other crops such as green vegetables, tobacco, peas, rape, sweet potatoes, sugar cane, peanuts and ginger.[29] Guangdong province was renowned for its lychees, ginger and oranges which were exported throughout China.[30]

This intensive agricultural system was the product of centuries of increasing population pressure on existing arable land. As population increased, food needs increased, and a larger labour force was needed to increase production. There was no scope to increase the size of their plots, so farmers' only course was to increase the yield from the acreages they had.[31] Robert Marks shows how climatic change, population growth, commercialisation and state intervention in southern China interacted to cause environmental and economic change, leading to an unsustainable agro-economic system and resulting in severe environmental degradation and loss of biodiversity. The limits of cultivable land in Guangdong were reached by 1850. At this critical point people were living on ever smaller plots of land, agricultural techniques were as developed as they could be, and yields were stagnating. This ecological crisis led to increasing disputes over access to land, water, and trees for fuel and timber and contributed to feuds and social uprisings such as the Taiping rebellion in the 1850s.[32] It also led to mass emigration from southern China to countries around the Pacific Rim.

Marks argues that the integrated system of raising fish and silkworms was ultimately unsustainable, as by the mid sixteenth century it required inputs from an outside system – the developing regional market economy. Unable to produce enough rice to feed themselves, farmers relied on rice and other staples imported from hundreds of miles away.[33] In 1898 Alexander Don visited Guangdong and travelled to Upper Panyu county where he planned to establish a mission. He visited the home villages of many of the Chinese miners he had met in New Zealand on his annual tours of the isolated gullies of Central Otago and saw first-hand the daily struggle for survival. He commented on the intensive use of every available acre of land for food production, the scarcity of animals and the high population densities:

> Here are no wastelands. The rice fields are in stubble, some of them being ploughed for the spring crops. Sweet potatoes and water caltropes are being gathered in, mostly by women and girls. Small plantations of tall sugar cane appear here and there; and many peach orchards . . . There are no fences. A strip of turf some six or eight inches wide divides the fields. A strange landscape where all is fields and tombs – where there are no highways or

> hedges, only byways and banks. Why every inch of ground is laid under tribute is evident when the density of population is considered. Within eight miles square are forty villages with fully a quarter of a million inhabitants... Animals are scarce. The struggle for human subsistence is too severe to allow many such.[34]

The key horticultural techniques which had developed in southern China by the mid nineteenth century are discussed below. They provide a baseline for identifying these features in the practices of Chinese market gardeners in Australia and New Zealand, and assessing the extent to which they were modified in response to the new physical, economic and social environments they encountered.

Field Systems and Multicropping

Crop rotation, intercropping (where several crops are planted at the same time alongside each other) and inlaid cropping (where two or more crops with different but overlapping growing periods are grown in the same field) make maximum use of space and solar energy. These systems allow farmers to produce crops continuously, without having to leave land fallow and therefore unproductive.

As protection against drought or flood, Chinese farmers raised the soil of their plots into ridges and furrows. Crops were planted on the ridges in long, level rows, and the furrows drained off excess water in times of heavy rain, or could be filled with water during droughts, protecting the roots of the growing plants from drying out. Franklin King outlines the advantages of the ridged fields he saw on his travels in southern China in 1909, in making the most efficient use of natural rainfall. The wet soil in the bottom of the furrows promotes the lateral percolation of water and nutrients to the roots of the plants, and the furrows act as long reservoirs to prevent soil erosion. While the ridges and furrows increase the surface area of the plot, any additional loss of soil moisture is balanced by the moisture retained in the furrows.[35] King also describes how ridges and furrows were changed over between crops, so that the soil was worked over deeply and thoroughly.[36] Bray notes that when sown in long rows, plants mature rapidly, do not interfere with each other's growth, and receive maximum sunlight. They are also easily accessible for planting, weeding, hoeing and harvesting.[37]

Manuring

In intensive agricultural systems where the land was not left fallow, manuring was essential to maintain soil fertility over centuries of continuous cropping. King declares that the first lesson that the West can learn from the East is how the farmers of China, Japan and Korea solved the problem of conserving natural resources and maintaining soil fertility. He decries the short-sightedness of European societies which let valuable fertiliser go to waste in rivers and oceans.[38] In his account of his travels in Guangdong he describes boat loads of human manure being shipped by canal from Canton. In the fields armies of men waded barefoot in the water in the furrows between the beds, applying diluted manure to crops of leeks, using long-handled dippers holding a gallon (4.5 litres).[39]

In China livestock played a much lesser role in producing manure than in Europe. Chinese farmers never wasted the manure produced by the few pigs, chickens or oxen they kept, even saving the water used to wash the stone floors of pig pens to use on their gardens.[40] They relied heavily on human manure and urine to supplement animal manure, a practice which repelled some European visitors to China. However, modern chemical analyses of this ancient practice show that human and animal manure is rich in essential plant nutrients such as nitrogen, potassium and phosphorous.[41]

Collecting, preparing and applying fertilisers required considerable care and labour. Traditionally human waste was stored in large terracotta urns near houses, or sometimes privies were combined with pig pens so that human and animal manure was efficiently collected in one location.[42] The manure was then left to ferment aerobically in large pits for several days to a few weeks before being used. Recent research has shown that the heat generated during the composting process, which reaches 55–60 °C, is high enough to kill many of the harmful micro-organisms present in the manure so that the application of human and animal excreta to the fields is in fact less of a health hazard than has often been supposed.[43] A variety of other organic materials were added to manure to make compost: chaff, straw and husks; composted city rubbish; sediments dredged from ponds and canals; wood ash; and green manures such as legumes, weeds and aquatic plants.[44] Farmers recognised that the direct application of manure is harmful to plants, so they diluted the fermented manure with water before it was used. Liquid manure was carefully applied to the plants using a long-handled dipper; it was also added to irrigation water.[45]

Thorough preparation of soil and enriching it with compost was a hallmark of Chinese horticulture. Frequent hand hoeing aerated the soil and spread organic matter through it, and by reducing it to a fine tilth, facilitated the growth of seedlings. King watched a farmer near Kuching carefully hoeing the soil, noting how he finely pulverised it, without injuring any of the multitude of worms his hoe disturbed.[46]

Individual Care of Plants

A hallmark of gardening, in contrast to broad-scale farming, is the care and attention paid to individual plants. Unlike farmers who sow and harvest *en masse*, gardeners treat their plants individually throughout their growth cycle and propagate them from seeds, rhizomes, bulbs or offshoots.[47] In the Chinese context, Xu and Peel term this 'meticulous cultivation'.[48] When plants are handled individually, variations between plants can be more easily spotted and those with higher yields or greater resistance to disease or pests can be selected, encouraging gardeners to develop new varieties. In comparison, mass sowing and harvesting of cereal seeds encourages uniformity.[49]

In China, gardening began with the careful selection and storage of the seeds or tubers of the best specimens for planting in the following season. A common technique to test the viability of stored seeds was to soak them in water and discard those that float.[50] Pregermination and transplanting techniques are associated with wet rice cultivation across Southeast Asia, but they were applied to a variety of other crops, particularly small-seeded vegetables.[51]

Frequent weeding and hoeing around growing plants aerates the soil and reduces competition for water, sunlight and nutrients. Hand tools such as hoes, mattocks and spades were indispensable for individually caring for plants, as well as opening up new land and digging drainage channels. Many Western travellers to China in the nineteenth and early twentieth centuries saw the preponderance of hand tools and the limited use of draught animals as evidence of the 'primitive' nature of oriental agriculture. But as Bray points out, this form of cultivation was adapted to small individual vegetable plots rather than field agriculture, where ploughs had been used for many centuries in China.[52]

The individual care given to plants extended to controlling insect pests. In 1919 King observed a farmer near Kuching searching his plot of potatoes for signs of insect damage:

> There were no potato beetles and we saw no signs of injury but the gardener was scanning the patch with the eye of a robin. He spied the slightest first drooping of leaves in a stem; went after the difficulty and brought and placed in our hand a cut-worm, a young tuber the size of a marble and a stem cut half off, which he was willing to sacrifice because of our evident interest.[53]

Irrigation

A constant theme in Chinese history has been controlling the country's rivers, which are characterised by marked variations in water levels, bringing regular floods and heavy loads of silt from the high snow fields of the Tibetan plateau. The earliest hydraulic engineering works, constructed in the Zhou Dynasty (1066–1256 BC), were dams impounding runoff from hills and dykes controlling the flow of rivers in the lower Yangtze and Yellow River basins. Networks of irrigation canals diverting flows from rivers and cross-country navigation canals were developed over the succeeding centuries, and foreign observers were dazzled by their scale and complexity. They included perennial canals, drawing water from rivers throughout the year, and others designed to capture water during seasonal floods. China's achievements in hydraulic engineering, particularly the use of artificial waterways for transporting heavy goods, were not equalled until the industrial revolution in Europe in the eighteenth century.[54]

Most major engineering works were constructed in the drier north and east of the country, but in the south intricate systems of canals and waterways supplied the water necessary for wet rice cultivation. On his first visit to Guangdong in 1880 Alexander Don described the great natural fertility of the Pearl River delta and the complex network of irrigation systems criss-crossing the plains south of Guangzhou:

> The hundreds of natural streams of the Canton delta are supplemented by thousands of artificial ones running in every direction and so is formed a perfect network of canals and rivers so close that fully one fifth of the entire surface may be reckoned as occupied by water . . . Of course rice is the staple product but there are many other products and we passed through four well-defined belts of vegetation – the first and fourth rice, the second mulberry

> and the third fan palm. When we come to see the results of the crops in this place it will be seen that there are good grounds for reckoning the Canton delta among the most fruitful spots on earth's surface.[55]

Some 20 years later, King described how water was supplied via channels or ditches to small reservoirs in the corners of terraced fields or gardens, where it was stored for hand watering, and the surplus was led down from terrace to terrace.[56]

Machinery for raising water from one level to another is essential to any system of irrigated agriculture. The simplest device, known across the ancient world, was the counterbalanced bailing bucket or well sweep, generally known by the Arabic name *shaduf*. Consisting of a long pole with a bucket suspended on one end, and a counterweight on the other, it lightened the labour of dipping, carrying and emptying buckets. The *shaduf* was known in China from the fourth century BC.[57] An elaboration of this device consisted of the addition of a gearing system and the elongation of the bucket spout into a flume (water chute), often made of a dug out palm trunk or bamboo, which automatically emptied into the irrigation channel on the upward motion. It could be operated by human or animal power. Animal power was also used to operate the well windlass, or whim, which used the rotary motion provided by a pulley or drum at the mouth of a well. These simple machines were common in the Han Dynasty (206 BC–220 AD).[58]

Probably the most characteristic Chinese water raising machine was the square pallet chain pump, which originated around the first century AD and diffused worldwide after the sixteenth century. It consisted of an endless chain carrying a succession of pallets which drew water along as they passed upwards and discharged into an irrigation canal or field at the top. The chain was turned by human, animal or water power. Human power was probably the earliest as radial treadles could easily be placed on the upper sprocket wheel. Chain pumps were described in Chinese agricultural treatises from the twelfth century AD, and their design changed little over the centuries. To use animal power, the animal was harnessed to a cogged drive wheel which engaged at right angles with a gear wheel on the sprocket wheel axle.[59] Similar devices were used to dredge canals and water channels.[60] Franklin King photographed three men operating a chain pump on his travels in China and Japan in 1909–1910 (Fig. 2.1).

Fig. 2.1 Chain pump in use in Southern China, 1909–1910, photographed by Franklin Hiram King. Published in King, *Farmers of Forty Centuries*, Fig. 46, p. 78.

British Horticultural Traditions

In Britain a mixed farming economy based on cultivating cereals and vegetables and raising livestock was established well before Roman times. Crops included wheat, barley, rye, oats, the Celtic bean (a type of broad bean) and wild herbs gathered for medicinal and culinary purposes.[61] When the Romans conquered Britain in 43 AD they brought with them many new vegetable crops: cabbage, kale, lettuce, salad greens, parsnip, turnip, radish, leeks, onions, garlic and cucumber.[62]

The horticultural techniques brought to England by the Romans had many parallels to those developed independently in China. They included transplanting, taking cuttings and offshoots, grafting fruit trees, deep digging, the use of raised beds, manuring, and even raising cucumbers in mica glazed frames to protect them from cold weather.[63] Over the following centuries gardening was carried out in monasteries, royal pleasure gardens and country manors, in conjunction with broad-scale farming.[64]

In contrast to China where livestock production played a relatively minor role, the development of agriculture in Britain is essentially the story of broad-scale grain and livestock production. Historians of the so-called agricultural 'revolution' in England have emphasised expansion of output and the contribution of agriculture to economic progress.[65] In his study of market gardening around London from the sixteenth century, Malcolm Thick highlights the neglect of market gardening in general histories of agriculture. He points out that the very factors identified as the essential features of the English agricultural revolution, namely the development of capitalist, market-oriented agriculture by the mid nineteenth century, were already present in market gardening from the sixteenth century.[66]

The English kitchen garden, an integrated system involving raising fruit and vegetables and animal husbandry, had many similarities to the horticulture of southern China. This tradition developed from the sixteenth century, and had its origins in France and the Low Countries. The typical French kitchen garden was a complex including a pond for fish and ducks, a well, two manure heaps (one for fresh, one for old), an orchard and two main garden plots, one for vegetables and pot herbs, the other for ornamentals. There may also have been additional plots for pulses such as peas and beans, cucumbers, melons and other crops such as hemp or saffron.[67] This gardening system was the product of influences from abroad, enriched by the introduction of new crops from Europe and the New World, and improved techniques for prolonging the cropping season and making more intensive use of arable land.[68]

From the 1670s, English gardeners sowed seeds in rows or drills, on raised beds with channels in between, a technique used in China for many centuries. The trenches surrounding each bed doubled as irrigation channels and paths.[69] However, by the nineteenth century the Chinese and English horticultural traditions diverged. The English shifted away from beds and channels to large flat areas sown in rows.[70]

Water is as crucial to the vegetable garden as good soil and shelter. In the damp climate of Britain where rainfall was relatively evenly spread throughout the year, extensive irrigation systems such as those developed by the Arabs, Romans and Chinese were unnecessary. While the networks of canals and irrigation channels typical of southern China were not duplicated in Britain, Campbell records the strategically placed ponds, dipping pools, wells and water tanks in nineteenth-century English kitchen gardens, similar to the reservoirs in Chinese gardens described by King.[71]

The use of hotbeds to raise crops such as melons, cucumbers, nasturtium seeds, tomatoes, capsicums, early salad greens, pumpkins and cauliflowers was well established in nineteenth-century market gardening. From the Elizabethan period seeds were planted on a bed of finely mixed, well decayed manure covered with a layer of fine soil. The use of fermenting manure to generate heat was known to many ancient cultures including the Egyptians, Romans, Chinese and Arabs.[72] This technique greatly extended the growing season of crops and hastened the growth of seedlings. To retain the heat a cover was placed over the seedlings until they germinated and when the seedlings were large enough they were transplanted. By the end of the seventeenth century hotbeds were becoming more sophisticated. Cloches in the form of glass jars or glazed tent-like structures, rectangular cases constructed of glazed panes and cold frames covered with glass or oiled paper were widely used.[73] From the 1650s glass became cheap enough to be viable for use on market gardens.[74]

Transplanting techniques, similar to those used in China, were well developed by the Elizabethan period.[75] The art of saving seed was well understood and commercial nurseries were in existence from the 1590s.[76] Commercial seed growers offered a much wider variety of seeds, propagated for specific characteristics and with greater consistency in quality through their trade with foreign suppliers.[77]

In eighteenth-century England intensive horticulture was practised within the walled kitchen gardens on the estates of the wealthy. A few of these gardens still survive, as reminders of this golden age of gardening.[78] Estate gardens made use of abundant labour to cart and spread manure, hoe, weed, raise and transplant seedlings, prune and water. There was a growing demand for exotic fruit and vegetables, and professional gardeners and nurserymen were sought after and accorded high status.[79] The wealthy desired a reliable supply of fresh fruit and vegetables year round and this stimulated experimentation and innovation in gardening techniques, for example extending the growing season of vegetables through the use of hotbeds and greenhouses and propagating earlier maturing varieties. Many new varieties of early potatoes, early peas, gooseberries and other fruit were developed in the eighteenth century.[80]

Commercial market gardening was well established in England by the nineteenth century. Malcolm Thick documents the spread of market gardening around London from the early 1600s, as the rapidly growing population placed huge demands on the surrounding countryside to produce food.[81] For over 200 years from around 1600, the Neat House Gardens, an area of around 100 acres beside the Thames, in what is today the suburb of Pimlico, supplied Londoners with fresh produce.[82] At their peak between 1688 and 1721, the Neat House gardens were renowned for their advanced horticultural techniques. These included careful cultivation, thorough digging and hoeing of the ground, heavy manuring, multiple cropping, crop rotation and producing vegetables out of season under glass.[83] London market gardeners were renowned for their intensive use of small plots of land. The variety and rotation of the crops they grew protected them from the problems associated with monoculture, in particular the spread of disease and insect pests.[84]

By the late nineteenth century simultaneous cropping was widely adopted where space was at a premium. Common crop combinations were radishes, lettuces or endives sown on the banks in between the trenches dug for celery, broad beans and early potatoes, tall and dwarf peas, and rhubarb and sea kale.[85] Methods of grafting, dwarfing and espaliering – training fruit trees and vines against a supporting wall – were well known.[86] The art of grafting enables the gardener to create trees with desired characteristics such as drought or frost tolerance, early or late fruiting, or bearing fruit of specific size, shape and flavour.[87] These techniques were brought to England by Huguenots, Protestant refugees from France, Holland and Belgium, who fled in their thousands to England in the sixteenth and seventeenth centuries to escape Catholic oppression. Their artisanal skills contributed to the growth of many industries, including silk weaving, gold and silver smithing, clock making, paper making and glass blowing.[88] Many were skilled gardeners and they set up commercial market gardens all over southern England and particularly around the expanding city of London.

The Huguenots also brought refined techniques of 'forcing' or stimulating the growth of seedlings using hotbeds, which was well known in seventeenth-century France and also in China. The adoption of hotbeds and glass frames, enabling crops to be produced without regard to season, was the major innovation of intensive market gardening. Gardeners could supply a succession of roots, herbs, green vegetables and fruit throughout the year.[89] The increasing demands for novelty from rich customers stimulated innovation and ingenuity, as gardeners could reap great profits from supplying exotic produce such as asparagus, artichokes, cherries, apricots, raspberries and strawberries, out of season.[90] At the same time gardeners took the opportunity to reduce the prices of common vegetables, increasing the demand for them from the working classes, who consumed relatively little fresh fruit and vegetables, traditionally eating them in stewed form as potage.[91]

As their name implies, market gardens in Britain were commercial operations, and they made use of capitalist forms of enterprise such as rented land, large outlays on hired labour, and as time went on, investment in glasshouses and soil additives. Access to markets was the main determinant of location, and responsiveness to the changing demands of consumers drove innovation and maximised profits. By adding manure and fresh soil, draining and irrigating, gardeners could transform their soils to meet the needs of production.[92]

Chinese and British horticultural traditions developed under very different circumstances and had distinctive histories. However, the discussion above shows that they shared many common elements, in particular the attention given to individual plants, deep cultivation of the soil, techniques of raising seedlings in hotbeds, transplanting, multicropping and crop rotation, and the extensive use of manure. Thus, when Chinese horticultural traditions were transported to Australia and New Zealand, many of the methods employed by Chinese market gardeners were familiar to Europeans. It is probable that this familiarity promoted European acceptance and recognition of Chinese horticultural skills. There are some further parallels in shared Chinese and British attitudes to the land, in particular their abhorrence of 'wasteland', glossed in Chinese as *huang*.[93] This point is discussed further in a later chapter.

While Chinese and British horticultural traditions shared basic techniques, their application in Australia and New Zealand took place under different circumstances and followed different trajectories. As a result of the tensions between the yeoman ideal and the reality of the dominance of large-scale pastoralism, commercial market gardening was relegated to a minor element of the agricultural economies of Australia and New Zealand. Although Chinese market gardeners never fully supplanted Europeans in market gardening, they played a significant role in the industry over many decades and in some areas of New Zealand such as

Oamaru and the Auckland urban area, they cultivated larger acreages overall than European gardeners.[94]

Indigenous Horticultural Traditions

British and Chinese horticultural traditions encountered distinctive indigenous subsistence traditions in Australia and New Zealand, and interacted with these traditions in distinctive ways. In Australia, Aboriginal people remained predominantly hunters and gatherers while the majority of the rest of the world became cultivators. They had no tradition of true horticulture, although they replanted the tops of edible yams and other tubers after harvesting and deliberately scattered the seeds of edible fruit trees in refuse heaps near camp sites. They have been characterised as the original affluent society, successfully adapted to survival in the world's driest inhabited continent. It is probable that these unique environmental challenges made any investment in agriculture uneconomic.[95] Indigenous Australian subsistence traditions had relatively little impact on imported British and Chinese horticultural systems, although, as discussed in the following chapter, some indigenous plant species were incorporated into British and Chinese horticulture.

In New Zealand, British and Chinese horticulturalists encountered an established Māori horticultural tradition, which originated in tropical Polynesia. Māori people had a long tradition of gardening, based on the tropical crops they brought with them from the Pacific islands. Polynesian gardeners specialised in tropical root crops such as yam (*Dioscoria batatas*), taro (*Colocasia esculenta*), sweet potato (*Ipomoea batatas*), ginger (*Zingiber officinale*) and Polynesian arrowroot (*Tacca leontopetaloides*) as well as tree crops such as bananas (*Musa* sp.), breadfruit (*Artocarpus altilis*), coconut (*Cocos nucifera*), ti tree (*Cordyline* spp.) and paper mulberry (*Broussonetia papyrifera*).[96] However, the Polynesian voyagers who settled in New Zealand some 900 years ago had to adapt their agricultural techniques to a much larger and cooler landmass than the tropical islands they left behind. A number of cultigens in the basic Polynesian repertoire, including breadfruit and banana, did not survive. But six key crops were successfully grown in New Zealand: sweet potato (kumara), yam, taro, ti tree, paper mulberry and the bottle gourd (*Lagenaria siceraria*).[97]

Māori gardeners quickly adopted European crops such as maize, potatoes, cabbages, turnips and leeks, introduced by early explorers, whalers and sealers and later by missionaries.[98] These crops provided them with a variety of new vegetable foods, particularly in the warmer North Island, and they grew them not only for their own consumption but also for trade with Europeans. By 1840, when Auckland was founded as the capital of New Zealand, entrepreneurial Māori had established a thriving trade in garden produce, growing beetroot, radish, artichokes, cucumbers, celery and endive to supply early European settlers and visiting ships. They also grew maize, potatoes, watermelons and peaches for their own consumption and for Europeans. By 1850 Māori tribes in the Thames and Waikato districts supplied the rapidly growing town of Auckland with most of its vegetables.[99] Later they supplied the goldfields of New South Wales and Victoria with potatoes and wheat.[100] The scale of their enterprise can be judged by the fact that in 1853 alone the Hauraki tribes sent at least 1,300 canoe loads of fresh vegetables, estimated to be worth over £8,000, to Auckland.[101] The production of these huge amounts of food, in addition to the quantities required for the competitive feasts held between neighbouring tribes, was carried out using traditional labour intensive cultivation techniques. This commercial boom was short-lived. The application of manure ran counter to Māori beliefs and without it the fertility of the soil declined and yields dropped so that by the early 1860s large-scale Māori gardening virtually ceased. The New Zealand Wars of the 1860s displaced Māori from their traditional lands and European settlers spread out over the countryside, establishing home gardens and orchards. Europeans established commercial market gardens close to population centres, and were followed by Chinese market gardeners later in the nineteenth century.[102] As this book will demonstrate, the social and economic relationships that developed between Chinese market gardeners and indigenous people were significant factors in the success of their businesses, particularly in New Zealand.

Chinese Horticulture in the Antipodes

Overlaying indigenous and British horticultural traditions, the traditional Chinese horticultural techniques transported to Australia and New Zealand were remarkably enduring. The cultivation of plants in rows on raised beds with long furrows in between was a feature of Chinese market gardens in Australia and New Zealand noted by contemporary European observers, for example in Sydney in 1867 and Brisbane in 1882.[103] Ridges and furrows are distinctive archaeological features that survive to the present day as indicators of Chinese market gardening activity in more remote rural areas, for example on Mona and Manar stations near Braidwood in south-eastern New South Wales.[104] They can also be seen in early aerial photographs of urban areas, taken before Chinese market gardens were swallowed up by housing. Aerial photographs of the Brisbane suburbs of Ashgrove and Enoggera, taken in the 1940s, show characteristic crop marks on either side of the banks of Ithaca and Enoggera creeks, vestiges of the Chinese market gardens which once flourished there. Chinese market gardeners settled here and in the neighbouring suburbs of Bardon and Kelvin Grove from the 1880s and cultivated vegetables until the 1950s.[105]

Chinese manuring methods aroused both approval and opprobrium in European observers. In 1867 Sydney horticulturalist John Gelding visited a Chinese market garden at Rushcutters Bay where he observed liquid manure stored in a long array of Chinese jars.[106] As early as 1862, Chinese market gardeners were growing crops around the Victorian gold mining towns of Castlemaine, Ballarat and Sandhurst. While his report was couched in the language typical of the day and depicted Chinese market gardening as an 'exotic' practice, an observant journalist recorded useful details of the way in which Chinese gardeners in Victoria handled liquid manure:

> It is in market-gardening, apart from digging, that they have chiefly made themselves useful in the colony . . . He [the Chinese gardener] is great in the use of liquid manure, and round his garden gate at all times long rows of peculiar round long-necked pipkins are to be seen – miniatures of those oil jars in which the forty thieves who tormented Ali Baba were so summarily disposed of by quick-witted Morgiana. These pipkins the Chinese gardener circulates throughout his neighbourhood when empty, and recovers when full, and, with the help of the river water, to thoroughly disseminate this manure about the roots of his plants, he performs marvels in the art of kitchen gardening.[107]

Some archaeological evidence of Chinese manuring methods has survived on the goldfields of Central Otago. A 'short drop' privy associated with Ah Lum's store in Arrowtown was still extant in the early 1980s.[108] It is thus quite probable that he collected the waste from the privy for use on his garden.

Individual care of plants was another hallmark of Chinese gardening in Australia and New Zealand, and was noted in oral histories and newspaper reports. George Lee Kim, who grew up on his grandfather's market garden in Bendigo in the 1920s and 1930s, describes his grandfather's painstaking methods:

> Always on the go when you look at it, rows and rows of vegetables, there was always something to do, as they got bigger and he distributed them into different beds, like cabbages and caulis... Once you had planted them you had to nurse them, weed them, hoe them, keep them going with applications of very strong horse manure, the produce them days was very strong and good.[109]

Esma Smith, the daughter of a Chinese market gardener, grew up in Forbes, New South Wales in the 1940s. She recalled that her father did not use pesticides, but walked up and down the rows of vegetables painstakingly collecting beetles in a small funnel-shaped wooden container. He would then pour hot water through it.[110] In 1888 a New Zealand agricultural expert Mr T. Kirk reported on the devastating effects on food crops of the cabbage moth *Plutella cruciferarum*, which was introduced from England into the Wellington district around 1880 and rapidly spread, devastating European and Chinese market gardens. He described how Chinese gardeners were able to bring the pest under control through their 'untiring industry' and the application of lime dust, which to be effective must be frequently and lightly applied. He attributed their success to the care with which they treated individual plants, hand picking insect pests from them:

> They spare neither time nor pains. They work from daylight to dark, their gardens are scrupulously clean and neat, no accumulations to harbour vermin; they carefully watch their crops, and apply 'hand picking' to an extent that would cause an Englishman to lose patience.[111]

Careful preparation of the soil prior to planting and regular hoeing and weeding was another key feature of Chinese market gardening in Australia and New Zealand. One of the earliest and most detailed descriptions of Chinese market gardening methods, as they were applied in Australia, was published in the *Melbourne Age* in 1865. The report describes Chinese market gardening in the Castlemaine district on the Victorian goldfields, and outlines the methods of preparing the soil and the typical layout of Chinese gardens, which facilitated easy access to the growing plants:

> The gardens are models of neatness . . . They are thoroughly clean of weeds . . . The ground is prepared by being trenched to a depth of from two and a half to three feet. The garden is divided by a number of narrow walks into long beds or plots which are of such a width that the gardener, sitting, or rather crouching by the side of it, as the Chinese do, can easily stretch half way across the plot, so that he can weed it without having to step on the plot. His beds are raised to a height of about eight inches above the level of the walks. It has to be observed that the soil is always beautifully pulverised, and it may be presumed that it would be sufficient to drive a Chinese gardener frantic if any one were to tread and leave a footprint on his plots.[112]

China's long history of water control stood Chinese immigrants to Australia and New Zealand in good stead, in their occupations of mining and market gardening. The Californian pump, named from its use on the Californian goldfields, derived directly from the traditional Chinese chain pump. It was a continuous belt, usually made of canvas, with buckets attached, and was commonly used on the goldfields in Australia and New Zealand.[113] In 1877 a newspaper reporter described Chinese miners operating a Californian pump in Central Otago. When viewed from a distance they appeared to be dancing:

> Passing along Gabriel's Gully I observed about twenty-five Chinamen at work, in parties of three and four. They were engaged in dancing operations. What this means we fail to understand excepting they were pumping water out of their claims by foot – that is, driving California pumps with a wheel a la treadmill, a system very common with Chinese.[114]

The traditional layout of gardens with water reservoirs and a network of irrigation channels endured in Australia and New Zealand, and their remains can still be seen as archaeological features in some places (for example, on the site of Ah Toy's garden on the Palmer River in far north Queensland, discussed in the following chapter). Their remains are also visible in Whittlesea in Victoria, where Chinese market gardeners settled along the Plenty River and Bruce's Creek from the 1870s. In an aerial photograph of a Chinese market garden site in Whittlesea, taken in 2007, the water reservoirs and parallel rows of irrigation channels through the gardens are clearly visible (Fig. 2.2).

Fig. 2.2 Aerial photograph of Chinese market garden site, Whittlesea, Victoria, January 2007. Courtesy City of Whittlesea.

One of the most detailed accounts of Chinese market gardening methods in Australia in the nineteenth century is a description of the making of a garden on 4.5 acres of land on the banks of the Barwon River in Geelong, Victoria in 1870. It illustrates many of the methods discussed in this chapter.

> His first operation was to dig a pit or trench 12 feet by 20 and 5 feet deep. He used some of this soil to level a small portion, and made the rest into a bank. The bank of soil he got turned once a fortnight. I was puzzled to think what he intended to do with it, never thinking that John was aware that dry clay was a good deodoriser. He could then talk no English, and at last one of his men – he employed three – explained, 'He carry dung'.... During the first three months all were engaged in plotting out and levelling; raising the ground into ridges, with a narrow trench of 10 inches width. The work was entirely done with fork-shaped hoes, and

> heavy long-headed common hoes. He bought, or procured all the manure, urine, or human excrement which could be got easily from nightmen, and filled the tank; sent his men to the gas works and tanneries for refuse, and mixed all together, and sunk some barrels perforated in the sides beside the tank. He next cut a trench from the river, and erected a Callfornian pump, worked by the feet on treadles, and by a simple system of watercourses could irrigate every inch of his ground when required. In hot dry weather he irrigated every second day. The ridges were made slightly concave – that is, the sides were a little higher than the middle. When he wished to force the young plants he used the liquid manure, and when they had well struck root he covered the ridges with the solid manure, mixed with the dry clay, and let the water run over the ridges. The water or rain carried the manure to the root of the plant, and the insoluble portion protected the soil and roots from the heat of the sun. No ammonia was lost, as it had became fixed by the dried clay. The two years after he began were seasons of drought. While you could not see a cabbage, or scarcely a blade of grass, for miles, this garden was one map of verdure.... The crops were wonderful – cabbages, lettuces, radishes, red beet, and Chinese peas and cabbages growing as close as they could stand, and not a yellow or unhealthy plant to be seen.[115]

Conclusion

This chapter has explored how two immigrant gardening traditions with very different origins and histories were transported to Australia and New Zealand, where they came into contact with each other and with indigenous subsistence traditions. Chinese immigrants maintained their traditional Chinese gardening practices in these new environments. In contrast, the intensive horticultural traditions which were imported from Britain were overshadowed by broad-scale agriculture for export. Most British migrants who took up agriculture had aspirations to become yeoman farmers. Broad-scale agriculture and pastoralism came to dominate the economies of Australia and New Zealand, with market gardening relegated to a relatively minor role.

While there was considerable continuity in traditional Chinese agricultural technologies and practices in Australia and New Zealand, this study will go on to show that Chinese market gardeners adopted European technological developments which were appropriate to the scale and work organisation of their enterprises. It will explore how Chinese gardening traditions interacted with European and earlier indigenous gardening traditions in Australia and New Zealand, and were blended and modified. The following chapter discusses the diverse environments which the Chinese immigrants who took up market gardening in Australia and New Zealand encountered, and the adaptations which they made to their horticultural techniques and choice of crops.

Notes

1. Liliane Hilaire-Perez and Catherine Verna, 'Dissemination of technological knowledge in the Middle Ages and early modern era: new approaches and methodological issues', *Technology and Culture*, Vol. 47, No. 3, July 2006, p. 559.
2. One of the earliest recorded arrivals in Australia was Mak Sai Ying (John Shying) who arrived in Sydney in 1818 (Michael Williams, *Chinese Settlement in New South Wales*, Sydney, 1999, p. 6). The earliest recorded arrival in New Zealand was Wong Ah Poo Hoc Ting (Appo Hocton), a seaman who jumped ship in 1842 and settled in Nelson (Nigel Murphy, *A Chronology of Events Relating to the History of the Chinese in New Zealand*, Wellington, 2002, p. 2).
3. Maxine Darnell, 'The Chinese Labour Trade to New South Wales 1783–1853', PhD Thesis, University of New England, Armidale, 1997.
4. Up to 50,000 Chinese arrived on the Australian gold fields in the 1850s, the majority in Victoria. Warwick Frost, 'Migrants and Technological Transfer: Chinese farming in Australia, 1850–1920', *Australian Economic History Review*, Vol. 42, No. 2, 2002, p. 116. See also C.Y. Choi, *Chinese Migration and Settlement in Australia*, Sydney, 1975, pp. 11–13.
5. Keir Reeves highlights the fluidity and transfer of ideas throughout Australasia during the gold rush period (Keir Reeves, 'Tracking the dragon down under: Chinese cultural connections in gold rush Australia and Aotearoa New Zealand', *Graduate Journal of Asia-Pacific Studies*, Vol. 3, No 1, 2005, pp. 49–66).
6. James Ng, *Windows on a Chinese Past*, Vol. 1, Dunedin, 1999, pp. 11–12; Choi, *Chinese Migration and Settlement*, p. 32; Williams, *Chinese Settlement*, pp. 10–11.
7. Henry Yu, 'The Rhythm of Cantonese trans-Pacific migration' in Donna Gabaccia and Dirk Hoeder (eds.), *Connecting Seas and Connected Ocean Rims: Indian, Atlantic and Pacific Oceans and China Seas Migrations from the 1830s to the 1930s*, Leiden, Boston, 2011, p. 399.
8. During the nineteenth century some 50 million people emigrated from Europe to the New World (James Belich, *Making Peoples: A History of the New Zealanders from Polynesian Settlement to the End of the Nineteenth*

Century, Auckland, 1996, p. 278). Around 10 million of them came from Britain (Joe Powell, *Mirrors of the New World: Images and Image-makers in the Settlement Process*, Folkestone/Hamden Connecticut, 1977, p. 45). By comparison some 2.35 million people left China during the same period (Zhu Guohong, 'A historical demography of Chinese migration', in Hong Liu (ed.), *The Chinese Overseas*, Vol. 1, London, New York, Routledge, 2006, p. 156).

9. Belich, *Making Peoples*, pp. 101, 278–279.
10. Powell, *Mirrors of the New World*, p. 48.
11. Belich, *Making Peoples*, p. 283.
12. Belich, *Making Peoples*, pp. 300–301, 305.
13. See, for example, Manning Clark, *A Short History of Australia*, Camberwell, 2006, pp. 162–168; and Stuart Macintyre, *A Concise History of Australia*, Port Melbourne, 2009, pp. 97–99.
14. Robin Bromby, *Unlocking the Land: The Saga of Farming in Australia*, Port Melbourne, 1989, pp. 31–32, 102–103; and Andrea Gaynor, *Harvest of the Suburbs: An Environmental History of Growing Food in Australian Cities*, Perth, 2006, p. 56.
15. Michael King, *The Penguin History of New Zealand*, Auckland, 2003, pp. 260–261.
16. King, *Penguin History of New Zealand*, pp. 237–238; and Nigel Murphy, 'A history of the horticultural industry in New Zealand', unpublished ms., 2010, p. 18.
17. For example, in Auckland by the 1950s, European market gardens were operated in conjunction with viticulture, orcharding or dairying and tended to concentrate on a smaller range of crops (Donald Hunt, 'Market Gardening in Metropolitan Auckland', MA Honours Thesis, University of Auckland, Auckland, 1956, pp. 3, 24).
18. Murphy, 'History of the horticultural industry', p. 18.
19. The earliest work that has survived in its entirety was written by Chia Su Hsieh around 535 AD (Francesca Bray and Joseph Needham, *Science and Civilisation in China Vol. 6, Part II: Agriculture*, Cambridge, 1984, p. 55).
20. Guohua Xu and L. J. Peel (eds.), *The Agriculture of China*, Oxford, 1991, p. 166.
21. Bray and Needham, *Science and Civilisation*, 6:II, pp. 19–20.
22. Kwang-chih Chang, *The Archaeology of Ancient China*, 4th edn, Harvard, London, 1986, pp. 234–237; and Michael Chazan, *World Prehistory and Archaeology: Pathways Through Time*, Boston, 2011, pp. 241–242.
23. Tianlong Jiao, *The Neolithic of South East China: Cultural Transformation and Regional Interaction on the Coast*, Youngstown, 2007, p. 256; Ekaterina Pechenkina, Robert Benfer and Xiaolin Ma, 'Diet and health in the Neolithic of the Wei and middle Yellow River basins, Northern China',

in Mark Nathan Cohen and Gillian M. Crane-Kramer (eds.), *Ancient Health: Skeletal Indicators of Agricultural and Economic Intensification*, Gainsville, 2007, p. 259; and Chang, *Archaeology of Ancient China*, p. 231.

24. Bray and Needham, *Science and Civilisation*, 6:II, pp. 423–424.
25. Ibid., pp. 3–4.
26. Bray and Needham, *Science and Civilisation,* 6:II, pp. 4–5; Xu and Peel, *Agriculture of China,* p. 93; and Jacqueline Newman, *Food Culture in China*, Westport, 2004, pp. 35, 87–88.
27. Xu and Peel, *Agriculture of China*, p. 70.
28. Evelyn Sakakida Rawski, *Agricultural Change and the Peasant Economy of South China*, Cambridge, Massachusetts, 1972, p. 3; and Xu and Peel, *Agriculture of China*, p. 71.
29. Xu and Peel, *Agriculture of China*, pp. 69–70; Bray and Needham, *Science and Civilisation*, 6:II, pp. 111–112; and Franklin Hiram King, *Farmers of Forty Centuries,* London, 1949, p. 87. Maize, peanuts and sweet potato all originated in the Americas; they were introduced to southern China after the middle Ming Dynasty (fourteenth century AD). Another variety of sweet potato was introduced from the Philippines towards the end of the sixteenth century (Xu and Peel, *Agriculture of China*, p. 69).
30. Bray and Needham, *Science and Civilisation*, 6:II, p. 542.
31. Xu and Peel, *Agriculture of China*, pp. 66, 70–71.
32. Robert Marks, *Tigers, Rice, Silt, and Silk*, Cambridge, 1998, p. 308.
33. Ibid., p. 120.
34. Alexander Don, *Under Six Flags*, Dunedin, 1898, pp. 91–99.
35. King, *Farmers of Forty Centuries*, p. 105. See also Bray and Needham, *Science and Civilisation*, 6:II, pp. 105–106.
36. King, *Farmers of Forty Centuries*, pp. 90–91.
37. Bray and Needham, *Science and Civilisation*, 6:II, p. 254.
38. King, *Farmers of Forty Centuries*, pp. 15, 171.
39. Ibid., p. 74.
40. Ibid., p. 73.
41. Bray and Needham, *Science and Civilisation*, 6:II, pp. 289–290; King, *Farmers of Forty Centuries*, p. 171.
42. Bray and Needham, *Science and Civilisation*, 6:II, p. 292; King, *Farmers of Forty Centuries*, p. 175.
43. Bray and Needham, *Science and Civilisation,* 6:II, p. 290; and Thomas McCalla and Donald Plucknett, 'Collecting, transporting and processing organic fertilisers' in Donald Plucknett and Halsey Beemer (eds.), *Vegetable Farming Systems in China*, Boulder, Colorado, 1981, pp. 23–24.
44. Anderson, *Food of China*, New Haven, London, 1988, p. 102; Bray and Needham, *Science and Civilisation,* 6:II, p. 290; McCalla and Plucknett, 'Collecting, transporting and processing', p. 19.

45. King, *Farmers of Forty Centuries*, pp. 67, 175; McCalla and Plucknett, 'Collecting, transporting and processing', p. 24.
46. King's Kuching was possibly modern Xinxing, in Guangdong (King, *Farmers of Forty Centuries*, p. 181).
47. Helen Leach, *1,000 Years of Gardening in New Zealand*, Auckland, 1984, p. 3.
48. Xu and Peel, *Agriculture of China*, p. 66; Leach, *1,000 Years of Gardening*, p. 3.
49. Jaques Barrau, 'Histoire et préhistoire horticoles de l'Océanie tropicale', *Journal de la Société des Océanistes*, Vol. 2, 1965, p. 56, quoted in Leach, *1,000 Years of Gardening*, p. 3.
50. Bray and Needham, *Science and Civilisation*, 6:II, p. 245.
51. Ibid., pp. 247–248.
52. Ibid., p. 197.
53. King, *Farmers of Forty Centuries*, p. 182.
54. Chinese engineering treatises and dynastic histories testify to the centralised planning and financing required for these engineering works (Joseph Needham in collaboration with Wang Ling and Lu Gwei Djen, *Science and Civilisation in China, Vol. 4, Part III, Civil Engineering and Nautics*, Cambridge, 1971, pp. 374–376, 378).
55. Alexander Don, 'Account of travels in Guangzhou: Kwong Chayu to Sanning', *New Zealand Presbyterian*, 1 February 1881, p. 149.
56. King, *Farmers of Forty Centuries*, p. 6.
57. Needham in collaboration with Ling and Djen, *Science and Civilisation* 4: III, p. 332.
58. Ibid., pp. 334–335.
59. Ibid., pp. 339, 362.
60. Ibid., pp. 78–79.
61. Resources for History, 'Farming in Celtic Britain', http://resourcesforhistory.com/Celtic_Farming_in_Britain.htm, accessed 25 March 2011.
62. Leach, *1,000 Years of Gardening*, p. 73.
63. Susan Campbell, *A History of Kitchen Gardening*, London, 2005, pp. 70, 83–84, 125, 145; and Leach, *1,000 Years of Gardening*, p. 74.
64. Campbell, *History of Kitchen Gardening*, pp. 84–87; and Leach, *1,000 Years of Gardening*, pp. 76–85.
65. Robert Allen, 'Agriculture during the industrial revolution', in Roderick Floud and Donald McCloskey (eds.), *The Economic History of Britain Since 1700*, Vol. 1, 1700–1860, Cambridge, New York, 1994, pp. 100–110; Mark Crafts, 'The industrial revolution', in Floud and McCloskey (eds.), *The Economic History of Britain Since 1700*, Vol. 1, pp. 57–58; Norman McCarol, *British History 1815–1906*, Oxford, 1991, p. 81; and Patrick O'Brien and Roland Quinault, 'Introduction' in Patrick O'Brien and

Roland Quinault (eds.), *The Industrial Revolution and British Society*, Cambridge, 1993, pp. 19, 22–26.
66. Malcolm Thick, *The Neat House Gardens*, Totness, 1998, pp. 12–13.
67. Leach, *1,000 Years of Gardening*, p. 81.
68. During the 1500s and 1600s the gooseberry, globe artichoke and asparagus were introduced into England from Europe, and potatoes, capsicums, maize, tomatoes, kidney beans and yams from the New World (Campbell, *History of Kitchen Gardening*, pp. 114–115; and Leach, *1,000 Years of Gardening*, pp. 78–79, 83, 87).
69. Campbell, *History of Kitchen Gardening*, p. 35.
70. Ibid., pp. 87–88, 117.
71. Ibid., p. 33.
72. The Arabs brought this technique to Spain in the eighth century AD, but it did not spread to the rest of Europe for around another three centuries (Campbell, *History of Kitchen Gardening*, pp. 119, 123–126; and Thick, *Neat House Gardens*, pp. 103–104).
73. In horticulture, a row cover or cloche is any material used as a protective covering to shield plants, usually vegetables, from cold, wind, and insect damage (Wikipedia, 'Row cover', http://en.wikipedia.org/wiki/Row_cover, accessed 26 March 2011).
74. Campbell, *History of Kitchen Gardening*, p. 129; Leach, *1,000 Years of Gardening*, p. 84; and Thick, *Neat House Gardens*, p. 103.
75. Leach, *1,000 Years of Gardening*, pp. 82–83.
76. Trade lists included fruit trees and ornamental trees and shrubs; vegetable seeds do not appear until 1677 (Ibid., p. 85).
77. Campbell, *History of Kitchen Gardening*, pp. 237–238.
78. Ibid., p. 11.
79. Leach, *1,000 Years of Gardening*, p. 89.
80. Ibid., pp. 89–91, 96. See also Joan Thirsk, *Alternative Agriculture: A History from the Black Death to the Present Day*, Oxford, 1997, p. 36.
81. Thick, *Neat House Gardens*, pp. 41–58.
82. Ibid., p. 93.
83. Ibid., pp. 47, 63–65.
84. Ibid., p. 64.
85. Campbell, *History of Kitchen Gardening*, p. 111.
86. Ibid., pp. 70–71; and Thick, *Neat House Gardens*, p. 42.
87. Campbell, *History of Kitchen Gardening*, p. 70.
88. Hilaire-Perez and Verna, 'Dissemination of technological knowledge in the Middle Ages', p. 563.
89. Thick, *Neat House Gardens*, p. 12.
90. Thick, *Neat House Gardens*, pp. 49–51.
91. Thirsk, *Alternative Agriculture*, p. 173.

92. Thick, *Neat House Gardens*, pp. 12–13.
93. James Beattie, pers. comm., July 2014.
94. Figures collected in 1925, 1945 and 1964 in the Auckland urban area show that in each of these years, the total acreage cultivated by Chinese market gardeners was larger than that cultivated by Europeans (Mohommod Taher, 'Asians in New Zealand: A Geographical Review and Interpretation', PhD Thesis, University of Auckland, Auckland, 1965, Table 3.5, p. 184).
95. Josephine Flood, *Archaeology of the Dreamtime: The Story of Prehistoric Australia and its People*, Pymble NSW, 1995, pp. 259–260, 282; John Mulvaney and Johan Kamminga, *Prehistory of Australia*, Washington, London, 1999, p. 87; and Macintyre, *Concise History of Australia*, p. 15.
96. Louise Furey, *Maori Gardening: An Archaeological Perspective*, Wellington, 2006, p. 10; and Leach, *1,000 Years of Gardening*, p. 17.
97. Leach, *1,000 Years of Gardening*, pp. 53–56.
98. Ibid., p. 101.
99. In 1840 New Zealand's first governor, William Hobson, named Auckland as the site of the nation's first capital (Margaret McLure, 'Auckland region', Te Ara – the Encyclopedia of New Zealand, updated November 2012, http://www.teara.govt.nz/en/auckland-region/, accessed 20 June 2013).
100. Bee Dawson, *A History of Gardening* in New Zealand, Auckland, 2010, p. 97; Furey, *Maori Gardening*, p. 111; and Leach, *1,000 Years of Gardening*, p. 106.
101. Dawson, *History of Gardening*, p. 97.
102. Furey, *Maori Gardening*, pp. 111–112; and Dawson, *History of Gardening*, pp. 97–101.
103. For example, John Gelding, 'John Chinaman and his garden', *Horticultural Magazine and Gardener's Calendar of New South Wales*, Vol. IV, no. 44, August 1867, p. 192; and *Brisbane Courier*, 11 October 1882, p. 2.
104. Barry McGowan, 'Chinese market gardens in Southern and Western NSW', *Australian Humanities Review*, Issue 36, 2005, n.p.
105. Desley Drevins, 'Chinese market gardens and market gardeners on Enoggera Creek and Ithaca Creek', presentation given to Ashgrove Historical Society, 1 September 2007, pp. 1–4.
106. Gelding, 'John Chinaman and his garden', p. 192.
107. *Argus*, 27 August 1862, p. 5.
108. Andrew Piper, pers. comm., April 2011.
109. George Lee Kim interviewed by Paul McGregor, 17 May 1994, Australia–China oral history project, National Library of Australia, ORAL TRC 3522/4/4, Tape 1.
110. Esma Smith interviewed by Rob Willis, 4 September 2009, Forbes, NSW. National Library of Australia, Voices of the Bush oral history project, ORAL TRC 6125/11.

111. *Hawke's Bay Herald,* 13 November 1888, p. 3.
112. *Sydney Morning Herald,* 2 March 1865, p. 2, reprinted from the *Melbourne Age*, 24 February 1865.
113. Barry McGowan, 'The economics and organisation of Chinese mining in colonial Australia', *Australian Economic History Review,* Vol. 45, Issue 2, 2005, p. 124.
114. *Dunstan Times,* 7 December 1877, p. 2.
115. *Bruce Herald,* 24 August 1870, p. 7.

CHAPTER 3

The Physical Environment

Fields, gardens and cultivated landscapes embody a continual interaction between human beings and the rest of nature. In the words of Kathryn Gleason, they are never-finished projects – part of a cycle of hourly, daily, seasonal and annual changes, which governs the rhythm of human activity.[1] In her study of Chinese agriculture, Francesca Bray observes that agriculture is the prime technological system mediating between the natural environment and human society, and its history represents a continual interplay between the forces of nature and those of state and society. Environmental conditions influence the development of agricultural technology and therefore the economic and political relations of the wider society. At the same time social and political constraints influence the patterns of development of agricultural systems.[2]

The Chinese immigrants who took up market gardening in Australia and New Zealand encountered diverse environments, generally very different from their homeland in subtropical southern China. They modified their horticultural techniques and choice of crops accordingly. In common with European farmers and pastoralists, they transformed the landscape, clearing forests and draining wetlands. This chapter explores the environmental factors which influenced the course of Chinese market gardening in Australia and New Zealand. It begins with a brief introduction to the physical environments of southern China, Australia and New Zealand. It then traces the development of Chinese market gardening in the more temperate areas of Australia and New

J. Boileau, *Chinese Market Gardening in Australia and New Zealand*,
Palgrave Studies in the History of Science and Technology,
DOI 10.1007/978-3-319-51871-8_3

Zealand, where most of the population and therefore most market gardens were located. This section focuses on the environmental factors determining the location of market gardens, in particular availability of water, soil fertility, aspect and slope. The final section explores how Chinese gardeners adapted their horticultural techniques and choice of crops to some of the more extreme environments in which they settled, ranging from the tropics of northern Australia and semi-arid western New South Wales to the cool temperate climates of Central Otago and Tasmania.

Home Soil: The Physical Environment of Southern China

Bray divides the vast land area of China into two contrasting physical environments: the continental zone of the northern plains and the subtropical zone south of the Yangtze River. Guohua Xu and L. J. Peel present a more detailed picture of China's environment, dividing it into three natural geographic regions: the Eastern Monsoon region, the Xinjiang-Inner Mongolia arid region and the Qinghai-Tibetan Plateau or alpine region. The Eastern Monsoon region, the Han heartland of China, can be subdivided into four subregions on the basis of temperature differences due to latitude. The province of Guangdong spans two of these subregions: subtropical humid central and south China and tropical humid south China.[3]

Bounded in the north by the Nan Ling mountain range, Guangdong has mean annual temperatures between 12 °C and 30 °C and high rainfall ranging between 1,500–2,000 mm per annum, mostly occurring between summer and autumn (mid April to mid October). Winter and spring can be dry. Most of the region is frost-free, so crops can be cultivated year round. Much of the province is hilly or mountainous, limiting the amount of arable land. Agriculture is mainly confined to the irrigable river valleys, including the broad Pearl River delta, while the long coastline provides a wealth of marine resources.[4] The region has been the nation's main point of contact with foreign powers, through the trading enclaves of Guangzhou, Macao and Hong Kong, and also the gateway for millions of departing Chinese emigrants.

As discussed in the previous chapter, the fertile coastal plains of Guangdong, which produce two crops of rice a year, have supported high population densities over many centuries, resulting in significant environmental degradation by the mid nineteenth century. In addition to the staple crops of rice and barley, many other crops were produced in rotation, including sweet potato, peanuts, tobacco, ginger, arrowroot, water chestnut, taro, melons and green vegetables. In the rich Pearl River delta area, criss-crossed by canals, clusters of small villages were surrounded by fields, ponds for raising fish, orchards containing a great variety of fruit trees (including banana, plum, lychee, mulberry, peach, guava and pear) and stands of bamboo and timber trees.[5]

Despite Guangdong's rich environment, there were often times of drought and flood, bringing famine and disease. On his fifth visit to the villages of Upper Panyu (Poon Yu) County north of Guangzhou in 1902–1903, the New Zealand missionary George Hunter McNeur found them in the grip of a severe drought. He vividly describes the people treading their water wheels, trying to raise water to save their crops:

> Every well sweep is swinging up and down, up and down, all day... Many have had to give up, and have left their fields to bake in the sun, and now the ground is covered with great thirsty cracks, while the paddy is tinged with yellow.[6]

Chinese immigrants' accounts of their homeland are rare. In about 1900 a Chinese miner, Yut Kui Kong, was living near Alexandra in Central Otago, eking out a living searching for gold. Aged 59, he had been in New Zealand for 32 years. Recalling his home village, he told Alexander Don that the family field was over 10 *mu* (approximately 2 acres), but part of it was hilly. If there was a dry season or major floods, there was no harvest, and the family had to either beg or starve. While the loss of one rice crop meant months of hunger, the loss of two crops in succession inevitably resulted in many deaths from starvation. Diseases such as cholera, smallpox, typhoid and malaria were never far away.[7]

The Chinese immigrants who travelled from their home villages to the Antipodes left behind them a rich subtropical environment which was bountiful, but could also be cruel. In Australia and New Zealand they encountered new environments and faced new challenges. Just as they had done in their homeland, they had to cope with the unpredictable forces of nature.

New Soils: The Physical Environments of Australia and New Zealand

The geological histories of Australia and New Zealand diverged markedly from the time of the breakup of the ancient Gondwanaland supercontinent. Today they have strikingly different geologies and ecologies. In the words of Libby Robin and Tom Griffiths, Australia is 'a wide, brown land, flat and worn', while New Zealand is 'a cluster of green and black, high and steaming islands'.[8]

Australia is an ancient, geologically stable continent with generally poor soils, low relief and slow-flowing rivers. About 70 percent of the land area is arid or semi-arid; around 50 percent receives less than 300 mm of rain per annum.[9] The continent has a variety of distinct regional climates with transition zones often extending over a considerable distance. These range from a tropical monsoonal climate in the north, with high summer rainfall and mild, dry winters, to the temperate climates of southern New South Wales, Victoria, Tasmania and south-west Western Australia. Central Australia has an arid desert climate, with annual rainfall averaging only 200 to 250 mm, frequently in the form of heavy thunderstorms. North of the Tropic of Capricorn the terrain varies from desert with vast stretches of sand dunes and clay pans, to grassland and tropical rainforest.

The most notable feature of Australia's climate, and a major limiting factor for agriculture and human settlement, is its extremely variable rainfall, with frequent droughts lasting several seasons. This is influenced by the Southern Oscillation and the accompanying El Niño/La Niña effect, driven largely from the surrounding oceans and their overlying atmosphere. The rainfall pattern is strongly seasonal, dominated by the subtropical high pressure belt which moves north and south with the seasons.[10] In his history of Australia, Stuart Macintyre notes that climatologists have detected the El Niño effect in records going back to the early nineteenth century and suggests that it has probably shaped the Australian environment for much longer.[11]

Water has thus been a crucial determinant of subsistence throughout the human history of Australia.[12] Archaeological evidence indicates that long before the arrival of Europeans, Aboriginal populations were concentrated in better watered areas where food resources were most abundant – coastal areas and inland lakes and rivers. In the well-watered eastern zone an average of 2.5 square kilometres was needed to support one person, while in the central desert as much as 260 square kilometres per person was required.[13] European settlement followed a similar pattern, as many historians and human geographers have observed. J. M. Powell and Kevin Frawley, for example, document the expansion of pastoralism in the 1820s and 1830s across a 'fertile crescent' in the southeast corner of the continent stretching over 2,000 kilometres from Brisbane to Melbourne and Adelaide. By 1850, two-thirds of the white population still lived in this fertile zone, within 200 kilometres of the coast, and this proportion continues to hold today. Most of Australia's primary production occurs in the temperate south-east and relies on winter rainfall.

Powell and Frawley describe the settling of Australia as a process of trial and error, testing the limits of the physical environment. They highlight the environmental challenges faced by European settlers – the attempts of pastoralists to expand their holdings into the arid and semi-arid grasslands of mulga and saltbush, which often ended in retreat, and the attempts to establish settlements in the tropical north which were often defeated by extremes of heat, humidity and torrential rains.[14]

In contrast to Australia, New Zealand is a young, geologically active land with high mountains, fjords, glaciers, geysers and rushing rivers. Importantly for agriculture, flat land is limited and soils in these areas are enriched by nutrients scoured from the surrounding rugged terrain.[15] A relatively small, narrow land mass in the South Pacific, surrounded by oceans, New Zealand has a temperate maritime climate. Mean annual temperatures range from 10 °C in the south to 16 °C in the north. The climate is dominated by three main factors: the surrounding oceans; the prevailing westerly winds; and the mountain ranges which trend north to south and form a barrier to the westerlies. Most of the country is near the sea, with mild temperatures, moderate rainfall and high sunshine hours.

However, New Zealand's climate is highly variable over short distances and encompasses a great diversity of regional microclimates; it has been classified into as many as 18 climatic zones. In broad terms the top half of the North Island has a warm temperate climate with warm summers, mild winters and an annual rainfall of 1,000 to 1,500 mm, with a winter maximum. The coastal areas of the central and southern North Island (such as the Bay of Plenty and Hawke's Bay) also fall into this climatic zone. The central plateau of the North Island and the Southern Alps of the South Island have cold temperatures and high rainfall, and local conditions vary greatly according to altitude, distance from the sea and exposure. For much of the southern region, winter is the driest season.

The South Island has distinct climatic zones due to the orographic effect of the Southern Alps. The West Coast has very high rainfall, as much as 10,000 mm per annum on the divide of the Southern Alps and 6,000 mm at Milford Sound on the coast. To the east of the Alps annual rainfall is much lower – 500 to 800 mm on the Canterbury Plains and 300 to 500 mm in Central Otago.[16] An elevated region relatively distant from the coast, Central Otago has a more continental climate than any other area of the country. It has some of New Zealand's highest and lowest temperatures and lowest rainfall. In Alexandra, for example, the average maximum temperature in July, the coldest month, is only 8 °C, one of the lowest in New Zealand, and the maximum temperature in January, the hottest month, can reach over 37 °C.[17]

Since European settlement the dense forests which once covered large parts of New Zealand have been mostly clearfelled, transforming the landscape in the drive to create pasture to produce wool, meat and dairy products for export. Environmental historians have also documented the destruction of the wetlands which once abounded in lowland New Zealand, in the same drive to convert seeming 'wastelands' into productive use and in response to ever-increasing demands for flat land.[18] Chinese market gardeners, who selected sites for their gardens close to watercourses, also played a part in clearing native forest and draining wetlands, changing the hydrological features of the surrounding catchment areas. The environmental impact of Chinese market gardening and its sustainability has not been researched in any detail. For example, there are questions to be answered about the extent to which the intensive agricultural practices they brought with them depleted soil fertility and the extent to which traditional manuring and irrigation practices countered this effect.

For the first Chinese immigrants who arrived in search of gold, the cool temperate climate of New Zealand would have presented a stark contrast to subtropical southern China. Central Otago, the region to which they first came, had a particularly extreme climate, with hot, dry summers and cold winters. Similarly in Australia, Chinese immigrants encountered diverse environments, ranging from tropical to arid and semi-arid. The following sections describe how Chinese market gardeners in Australia and New Zealand adapted their crops and horticultural techniques to these novel environments.

Chinese Gardens in New Lands

This section explores how Chinese market gardeners selected sites for their gardens and modified their horticultural techniques and selection of crops in the more temperate areas of Australia and New Zealand. Historically, market gardens in Australia followed the spread of European settlement and consequently were predominantly located in the relatively well-watered and fertile south-eastern corner of the continent and in

Tasmania. Townships grew up close to watercourses, often at key locations such as river crossings or the confluence of rivers, and the market gardens serving them were generally located beside these watercourses.[19] In New Zealand, market gardens also followed the spread of European settlement and major centres of market gardening developed in areas with favourable local climates, flat land and fertile soils, such as Oamaru in the South Island and Otaki and Hawke's Bay in the North Island. Thus the siting of market gardens on the fringes of towns and cities was at least partly, if not significantly, influenced by environmental factors.

Writing in the *Agricultural Gazette of New South Wales* in 1916, A. J. Finn provided advice on the selection of suitable sites for market gardens.[20] He highlighted environmental factors such as soil quality, water supply and aspect. Chinese gardeners considered similar environmental factors in selecting the location of their gardens. They followed the traditional environmental principles used in the siting of villages and fields in southern China, taking into account access to water, soil fertility, drainage, orientation, terrain, shelter from prevailing winds and land and water transport routes.[21] Thus their market gardens were generally located near water courses, freshwater lakes or swamps, on relatively flat areas of fertile alluvial soil. These factors have been highlighted in recent studies of market gardening in Sydney, Perth, Melbourne, rural Victoria, and rural New South Wales.[22] As early as 1865, a report in the Melbourne newspaper the *Age* noted that the primary consideration in the location of market gardens was an adequate supply of water:

> The Chinese display much skill and intelligence in the choice of their garden ground, both as regards the character of the soil, and the nature of the situation. The first consideration with them is the supply of water, and they choose a site... on the bank of a creek, in which they may obtain water in the driest seasons. Failing this, they must have wells dug in their gardens, from which they can get a constant supply of pure water.[23]

Water was an essential commodity for both alluvial gold mining and market gardening. In his study of Chinese market gardens in southern and western New South Wales, Barry McGowan notes the close links between market gardening and mining. The dams and water races constructed for sluicing alluvial gravels and separating the precious grains of gold were also used by market gardeners to channel irrigation water to their gardens. For example, at Majors Creek near Braidwood in southern New South Wales a Chinese market garden was located some distance away from the main creek, which was devoted entirely to mining. A tail race reticulated water from the dam constructed for mining purposes, diverting it down a gully to run along the upper length of the garden.[24] Similar examples can be seen in western New South Wales. At Gilgunnia McGowan recorded a water race which once conveyed water from an old gold battery dam to a Chinese market garden. At Mount Drysdale Station water was reticulated via a pipe from a mining dam to a holding tank sunk in the middle of a Chinese market garden.[25] Water was also vital for Chinese miners and gardeners in relatively arid Central Otago in New Zealand. They drew water from the same dams and water races, and competed with Europeans for access to this critical resource.

In Australia, Chinese market gardeners demonstrated the value of irrigation and intensive cultivation in areas where few Europeans attempted to grow vegetables and provided an argument for proponents of major irrigation schemes 'to make the desert bloom'. Advocates of closer settlement argued that European farmers should adopt the labour-intensive methods of Chinese gardeners and try to equal their successes. In 1899 John West, a proponent of irrigation along the Murray River, bemoaned the fact that: 'the average farmer . . . is in the unfortunate position . . . with the belief . . . [that] irrigation will do no good in the hands of any other than a Chinaman'.[26] In his lavishly illustrated *Australia Unlimited*, published in 1918, the journalist Edwin Brady writes enthusiastically of the untapped potential of Australia for agricultural development and records the successes of Chinese farmers in Nowra, New South Wales, and in Cairns, Atherton, Hughenden and Texas in Queensland, citing their high yields as evidence Europeans could achieve the same results.[27]

Even critics of the proselytisers for irrigation and patriotic nation builders, such as the geographer Thomas Griffith Taylor, acknowledged the inventiveness and technological skills of Chinese people.[28] Taylor had no fears of an 'Asian invasion' and his views on race aroused as much controversy as his views on irrigation. It is difficult to separate irrigation and settlement policy from the complexities of the social climate, with its intersection of class, gender and race. Europeans recognised the horticultural skills of Chinese market gardeners and their achievements in producing crops in arid areas, but their willingness to adopt Chinese methods was constrained by their racial attitudes.

In New Zealand, with its generally temperate climate and reliable rainfall, there was less emphasis on large-scale irrigation schemes. Irrigation was a priority only in the drier regions of the South Island, where a number of major government-funded irrigation projects were undertaken in Central Otago in the 1920s and on the Canterbury Plains in the 1930s and 1940s.[29] From a search of digitised New Zealand newspapers dating from the 1840s to the 1940s, it appears that less attention was paid to the irrigation techniques of Chinese market gardeners and they had less direct influence on agricultural policy making than in Australia. A few newspapers noted the successes of Australian irrigation schemes and the successes of Chinese market gardeners. In 1920, for example, the *Northern Advocate* in Whangarei described the major irrigation scheme constructed in the Murrumbidgee Irrigation Area in New South Wales as a useful model for the irrigation schemes proposed in Central Otago and noted that Chinese market gardeners on the Yanco Experimental Farms claimed to produce tomatoes which yielded a gross return of £400 per acre.[30]

Soils and aspect were also important factors in the locations of market gardens. In New Zealand, Chinese and European market gardeners established gardens in sheltered coastal areas with fertile alluvial or volcanic soils, for example in Auckland, Otaki, the Hutt Valley and the Bay of Plenty. Otaki, 70 kilometres north of Wellington, became an important centre of market gardening in the early 1900s. As Wellington expanded, land values and land rents rose with urban sprawl and market gardeners in the city and the Hutt Valley were forced to move on.[31] Otaki was relatively close to the Wellington market, and the location was ideal – a relatively sheltered alluvial plain with deep, fertile soils running several miles inland and a mild, sunny climate. The Otaki River provided a reliable water supply and easy access inland.[32] In November 1909 the *Bay of Plenty Times* reported that there were about 150 Chinese in Otaki, most working on market gardens, and 200 were expected by Christmas.[33]

In Auckland, Chinese and Europeans established market gardens on the fertile volcanic soils derived from the more than 50 volcanoes in the Auckland isthmus. Now dormant, these volcanic cones dominate the Auckland landscape. The gardens were typically located on gentle north or east facing slopes with good drainage and receiving maximum solar energy.[34] In the late nineteenth and early twentieth centuries market gardens were located in the inner city and in the suburbs of Avondale, Mangere, Takapuna, Mount Wellington and Panmure.[35] In Mangere Chinese gardeners followed in the footsteps of indigenous gardeners.

With rich volcanic soils, a mild subtropical climate and sheltered location on the Manukau harbour, Mangere was a prime gardening location where Māori had cultivated crops of taro and kumara for centuries before the arrival of Europeans.[36] The volcanic stones ejected in successive eruptions were used to demarcate garden plots and warm the soils, extending the growing season.[37]

Nelson in the north and Oamaru on the central east coast of the South Island were important horticultural centres from early European settlement and remain so today, due to their favourable climates. Oamaru, with its well-drained, fertile, clay loam soils and mild winter temperatures, allowed gardeners to produce carrots, winter lettuce and early crops of potatoes and Brussels sprouts which realised premium prices.[38] According to the *Dunstan Times* the first Chinese settlers established a market garden in Oamaru in 1870.[39] The numbers of Chinese market gardeners remained small until the 1920s, fluctuating between 8 and 20. Between the mid 1920s and late 1930s, seven new market gardens were established in Oamaru. During World War II market gardening was declared an essential service. The numbers of Chinese market gardeners increased to around 13 by 1948 and they increased the acreages they had under crops. From the 1950s to the 1990s Chinese market gardeners in the district gained national prominence for the quality of their produce, particularly their Brussels sprouts.[40] As Ruth Lam and Lily Lee demonstrate in their history of Chinese market gardening in New Zealand, Chinese growers have successfully adapted to the myriad changes in the horticultural industry and continue to play a part in it today.

Market gardening was always a risky business. Chinese market gardeners were subject to the vagaries of the weather, which included floods, drought and frosts. In cooler areas of Australia, such as Tasmania, and much of New Zealand south of Auckland, frosts were a constant danger to garden crops and orchards. Early or late frosts could destroy crops, robbing gardeners of the profits they might gain from supplying seasonal produce to the markets ahead of, or later than, other growers.[41]

In Auckland there was less risk of frost, but it was still a gamble growing frost-sensitive crops such as lettuces and cucumber in winter. Adapting ancient Chinese horticultural techniques, Chinese market gardeners used growing frames covered with paper or hessian to protect delicate seedlings from frost. Percy Luen, whose father Jack Luen established a market garden in the Auckland suburb of Mangere in 1950 after his garden in Gisborne was flooded, recalled his father's methods of protecting early cucumbers:

> Early cucumber planting was done by compost . . . the seeds were sown and covered with a 30 centimetre by 30 centimetre wooden frame and then covered with glass. If a frost was imminent then the nightly chore would be to cover the glass with a sheet of paper, such as a page from the phone directory. Though time consuming, it was a way to have short green cucumbers first on the market floor.[42]

In another adaptation to their new environments, Chinese gardeners began growing European crops suited to temperate climates. In New Zealand, they also added indigenous cultigens to their repertoire, blending indigenous and Chinese gardening traditions. In warmer areas such as Auckland, the Bay of Plenty and Taranaki, they grew kumara, the same species as the sweet potato grown in southern China.[43] This was a sound business decision as kumara had been successfully acclimatised to New Zealand conditions by generations of Māori gardeners and found a ready market among Māori and Europeans.[44] Unlike green vegetables, which had to be sold immediately after harvesting, kumara could be stored and sold when out of season for higher prices. Joe and Fay Gock, for example, grew kumera in Mangere for over 50 years, after their neighbour Hiko Raniera Wilson gave them some spare kumara plants in 1952. Her generosity began a long lasting friendship.[45]

Chinese market gardeners also grew puha (*Sonchus kirkii* or New Zealand sow thistle), generally for their own consumption rather than for sale.[46] This green vegetable, a staple of the Māori diet, often grew wild in market gardens among the other vegetables, for example in Auckland and Gisborne.[47] Fig. 3.1 is a photograph of three Chinese market gardeners in their garden in the Auckland suburb of Western Springs, taken in the early 1900s. Puha can be seen growing in the foreground.[48]

From Monsoon to Desert

Wherever they settled, from tropical northern Australia to the cooler climates of southern New Zealand and Tasmania, Chinese market gardeners grew vegetables for their own consumption and for sale to Europeans. This section explores how they adapted their horticultural techniques and

Fig. 3.1 Chinese market gardeners in their garden on Great North Road, Western Springs, Auckland, early 1900s. Photographer Ernest Charles Binns. Courtesy Sir George Grey Special Collections, Auckland Libraries, 80-BIN 185.

choice of crops to some of the more extreme environments they encountered.

Gardens in the Tropics: Queensland and the Northern Territory

One of the few Chinese market garden sites in Australia to be investigated by archaeologists is Ah Toy's garden on the Palmer River in Cape York, North Queensland.[49] Combined with documentary and oral history evidence, the archaeological evidence from the site provides much information about the daily lives of Chinese gardeners in a remote tropical environment. The climate of Cape York is one of extremes – during the summer wet season the rivers and creeks are raging torrents, making travel difficult, while in the winter dry season they dwindle to a trickle. The terrain of the Palmer River basin is rugged and the soil is poor, supporting only sparse vegetation.[50]

This site is an excellent example of Chinese market gardening on the goldfields. An Gee and Ah Ung, the original Chinese leaseholders, combined mining with market gardening. They were among the thousands of Chinese who flocked to the Palmer River after gold was discovered there in 1872. By 1877 Chinese people far outnumbered Europeans on the Palmer River; there were about 18,000 compared to 1,500 Europeans. In this remote area where communications were difficult and natural food supplies were limited, fresh fruit and vegetables were vital commodities. Ian Jack and Katie Holmes note that in 1883 there were 68 market gardeners on the Palmer River paying rates, all Chinese. Their numbers declined as the fortunes of the goldfields declined, to 57 in 1885 and 36 in 1887.[51] By 1900, when Ah Toy took over the lease of the garden, officially only 393 Chinese remained on the Palmer River, and only 13 gardens. For more than 30 years Ah Toy supplied vegetables to local miners, the population of Maytown and a Chinese storekeeper in Cooktown. He also grew pannikin grass to supply horse fodder to the packer who regularly passed the garden on his route between Maytown and Laura.[52]

Ah Toy's gardens were located on either side of an unnamed creek running into Lone Star Creek. They were irrigated by water channelled from a sturdy log dam across the creek, fed by a natural spring. The dam collapsed in the late 1970s, but sections of the water race carrying the water to the gardens remained in 1984. It was an earth-dug ditch, with hollow logs supported on a wooden trough carrying the water across side gullies. At several places additional sluices led south into the creek, giving greater control over the water flow. The vegetable garden was on the north side of the creek, with a complex system of parallel irrigation channels running south-east from the water race. The garden on the south bank of the creek was planted mainly with fruit trees (custard apples, oranges and mandarins) together with taro, yams (also known as Chinese yam) and rough leaf pineapples.[53]

The evidence uncovered at this isolated site illustrates the necessity for adaptation and improvisation common to both Chinese and Europeans living in the Australian outback. The design of the irrigation system shows ingenuity in coping with climatic extremes and the great variations in water volume between the wet season and the dry. Percy Willmett describes Ah Toy's irrigation system in his reminiscences of his journey to the Palmer goldfields as a school boy, published in 1932.

> Tommy Ah Toy has a nice garden in the valley: grows pumpkins, corn, oranges, mandarins etc, and perhaps the most interesting feature here is the Chinaman's irrigation scheme. From a spring a quarter of a mile up the valley he runs a drain around the side of the hill and finally crosses a creek some 15 feet deep by running the spring water across it in a hollow log, finally to irrigate the whole farm.[54]

As Ah Toy's garden illustrates, Chinese gardeners in tropical Queensland grew crops suited to the climate. Here they could grow the tropical crops familiar to them from southern China, such as yam, taro, sweet potato and ginger. It is thought that Chinese immigrants introduced yam, taro and sweet potato to far north Queensland during the gold rushes.[55] However, the histories of the origins and dispersal of these crops are complex and it is likely that there were multiple introductions. Chinese were probably only one of a number of different groups to introduce domesticated varieties to Australia over the last 200 years, including Europeans and South Sea Islanders. Fishermen from Papua New Guinea, Indonesia, Malaysia and China visiting the north coast of Australia could also have introduced these crops well before European settlement.[56] To further complicate the picture, there are wild species of sweet potato (commonly called bush potato) and taro native to Australia, which were cultivated by Aboriginal people long before the arrival of Europeans.[57] Unlike taro, the yam is a cultivar with no known wild ancestor and it is hypothesised that it was originally introduced to Australia from Papua New Guinea and island South East Asia sometime during the last 11,000 years.[58]

Sweet potatoes could be a lucrative crop. In the Mackay district Chinese market gardeners were the main producers of sweet potatoes for the sugar cane plantations, which they supplied on contract at a fixed price. Sweet potatoes were the staple diet of the kanakas (South Sea Islanders 'imported' as indentured labourers from the islands of Melanesia to work on the plantations), guaranteeing a constant demand.[59] According to an 1885 report in the *Queenslander*, the market price in the early 1880s ranged from £3.10s to £5 per ton, making it a profitable enterprise. Using their skills in intensive horticulture Chinese gardeners obtained two crops of sweet potatoes a year, realising an annual yield of around 14 tons per acre. However, by the mid-1880s the numbers of South Sea Island workers were beginning to decline. Sweet potato prices fell and larger acreages were devoted to growing European potatoes. Even at the 1885 prices of around £2 or £2.10s per ton for sweet potatoes, it was estimated that gardeners could realise a yearly gross return of about £30 per acre.[60]

Yams and taro were less commercially viable crops and were generally grown by Chinese gardeners for their own consumption. In 1878 Sir Samuel Wilson read a paper on the Chinese yam to the Council of the Zoological and Acclimatisation Society in Melbourne. He concluded that it was ideally suited to Chinese horticultural techniques – it was deep rooted, making it labour intensive and costly to harvest, and most appropriate for small-scale gardening where the soil was thoroughly trenched and well manured.[61] In 1932 a Townsville newspaper reported that the cultivation of yams in the north was mainly carried out by Chinese people, for their own consumption.[62] According to a newspaper report on the Melbourne Centennial Exhibition in 1888, taro was grown only occasionally in Queensland, mainly as an ornamental plant.[63]

Ginger has played an important role in Chinese cuisine and traditional medicine for centuries, in fresh, dried and preserved form. It was introduced to Australia independently from Britain by the first European settlers and from China by the Chinese immigrants who began arriving in the early nineteenth century.[64] Ginger thrived in the tropics and subtropics and Chinese settlers planted ginger rhizomes for their own consumption wherever they established gardens. There are many reports of Chinese growing ginger in Queensland and the Northern Territory, and some evidence that they grew ginger in New South Wales; for example an article in the *New South Wales Agricultural Gazette* in 1922 reported that Chinese were growing ginger using seed stock imported from China.[65]

Chinese merchants also imported large quantities of preserved and dried ginger. Leonie Ryder's history of ginger in Australia suggests that there was little communication between European and Chinese growers and the European ginger industry developed independently.[66]

In addition to selecting appropriate crops for the tropical environment, in planning their gardens Chinese gardeners balanced the need for young seedlings to be sheltered from the full heat of the sun with the need for growing plants to receive maximum sunlight. Together with manuring and frequent watering, this ensured the rapid growth of crops, quick turnaround and maximum economic return. A report in the *Queenslander* in 1878 described the methods of Chinese gardeners.

> In making their seed beds, great judgment is employed. The open ground is chosen in preference to shady places near fences. The ground is turned over several times, and no matter how rich it may be manure seems to be dug in each time; and after every digging the bed is soaked full of water . . . At this time of the year, while the sun's heat is so powerful, forked stakes or pieces of hardwood are driven at the sides of the seed beds; battens or saplings are put on to the stakes, and dried twigs are laid over the battens, thus forming a bush house protection, with its invigorating shade over the seed bed.[67]

Even in southern Australia, summers could be hot. Similar methods were applied elsewhere in the country, for example in Geelong in Victoria and Milparinka in western New South Wales (as discussed below). In Geelong, Chinese gardeners erected temporary paling roofs over their garden beds to protect young plants from the sun during their early growth.[68]

While Chinese farmers and market gardeners played a key role in land clearing and 'developing' the Australian tropics, like Europeans, they had a major impact on the environment. This was particularly true where Chinese farmers engaged in large-scale cultivation of monocrops, for example bananas and sugar cane in Queensland. Powell observes that the groups of Chinese who cleared tropical rainforest to cultivate bananas in north Queensland practised a form of shifting cultivation, deliberately exhausting the soil before moving on to a new block of virgin land.[69] It was easier to abandon a plantation after five or six years rather than devote time and labour to removing old plants, restoring soil fertility and dealing with insect pests. Powell cites the environmental damage caused by the Hop Wah syndicate's pioneering sugar growing venture. Planting and overcropping on steeper slopes would have accelerated soil erosion and runoff, contributing to the silting up of watercourses, damaging stream banks and altering hydrological characteristics.[70]

In his history of Australia's multi-racial north, Henry Reynolds highlights the mobility and versatility of Chinese immigrants as they sought economic opportunities across the region, working at a wide range of occupations, including market gardening, large-scale farming, mining and store keeping.[71] Thus they were not only geographically mobile but economically mobile, always seeking new opportunities to increase their wealth. In the Northern Territory, market gardening was the most common form of Chinese economic activity in the late nineteenth century apart from mining. Chinese people greatly outnumbered Europeans – in 1888 the Chinese population numbered approximately 7,000 and Europeans 1–2,000.[72] In 1901 the European population numbered only 864. Among the approximately 3,000 Chinese in the Territory there were nearly 1,700 miners, 400 market gardeners, 160 cooks and 120 shopkeepers and their assistants.[73]

In the monsoonal 'Top End', Chinese immigrants encountered a climate similar in many respects to that of their homeland. A rare account by a Chinese immigrant to the Northern Territory, written by Shi Kwong Ho, describes how a Chinese gardener chose a site for his garden and the crops he grew.

> He saw a river there and appreciated its view. He trampled on the bank of the river and found an area down the stream suitable for gardening. He was fully determined to stay in this yet-to-be-developed Eden. He did his best to use the water, the soil and everything he could find locally. Not long afterwards the banana trees, paw paw trees, cedars and other fruit trees were grown up into the air. His beans, melons and other vegetables were spread out along the riverside and up to the mountain's base.[74]

As they did in Queensland, Chinese market gardeners in the Northern Territory grew a diverse range of tropical crops, many familiar to them from southern China. Their produce included Chinese beans, yams, taro, sweet potatoes, sugar cane, peanuts, ginger, bananas, mangoes, papayas and pineapples. They also successfully grew the temperate crops familiar to Europeans, such as cabbages, cucumbers, turnips, tomatoes, carrots and beetroot. In 1878 the editor of the *South Australian Register* observed that without Chinese farmers there would be scarcely any vegetables to feed the residents of the

Northern Territory. He noted the large quantities of sweet potatoes they grew, along with a great variety of other crops:

> Chinese beans, pumpkins, cucumbers, melons and the like thrive very well, while at the Chinese garden in Palmerston where they have command of water for irrigation the English cabbage, lettuce, radishes, onions, carrots are produced with very little trouble . . . the pineapple, plantain, pawpaw and custard apple are now becoming tolerably plentiful in both town and country and at Palmerston the jackfruit, mango, tamarind, orange, lime, coconut and sugarcane are progressing favourably.[75]

The prizes regularly won by Chinese gardeners in the Agricultural and Horticultural and Industrial Association (AH&I) shows in Darwin illustrate the variety of produce they grew. The report of the AH&I show in 1903 noted the variety of crops suitable for a tropical climate grown by Chinese gardeners, such as taro, yam, sweet potato, mango, soursop, coconuts and sugar cane.[76] At the 1906 AH&I show Chinese gardeners won prizes for the following produce: Gee Sing, English cabbages and bananas; Wah Sing, cucumbers, turnips, yams and paddy rice; Ah Sam, carrots and tomatoes; Mow Sing, sweet potatoes and sugar cane; Yee Sing Yin, peanuts, watermelon and pumpkin; and Pang Quee, mangoes.[77]

In a report on the agricultural industry in the Northern Territory in 1897, the staunch Territorian F. R. Finniss argued that the neglect of cultivating the soil was the chief cause of the Territory's lack of economic development. He attributed this neglect to the inability or unwillingness of Europeans to work in tropical climates and their lack of knowledge of tropical agriculture. He contrasted the dismal failures of European planters with the success of Chinese gardeners and acknowledged that they had been keeping the European community supplied with fresh fruit and vegetables for many years. He also noted the large-scale ventures of some Chinese farmers growing rice and maize.

> Some are successfully cultivating rice at Millner's Lagoon, and a year or two ago a rice farm near Union Town was most successful, producing a ton of grain per acre, I have been told, of No. 1 rice, which brought a better price up country than the imported article.[78]

The extremes of the tropical climate in the Northern Territory affected both mining and gardening. Alluvial mining was mainly limited to the wet season and the early weeks of the dry when water was plentiful.[79] During the wet market gardeners had to cope with large volumes of water surging through their irrigation channels, often causing flooding and crop losses. In the dry they harnessed scarce water resources using a variety of irrigation techniques including dams and reservoirs. In February 1879, towards the end of the wet, a correspondent to the *Northern Territory Times and Gazette,* writing from Yam Creek, described how floods had swept all crops clean away in country districts and bemoaned the loss of the expected harvest of sweet potatoes. He expressed sympathy for the Chinese gardeners in the district and warned that people would soon be short of food.[80]

Gardens in the Desert: Western New South Wales and South Australia

In the late nineteenth century, Chinese immigrants followed the expanding pastoral frontier in western New South Wales. They were also drawn by the mineral boom in copper, gold, silver and lead which lasted around 40 years. Chinese men engaged in mining and market gardening on the fringes of country towns. Several thousand were employed as contract workers on pastoral stations, clearing vast tracts of land and working as gardeners, fencers and shearers. Although the numbers of independent Chinese market gardeners were small compared to those engaged in land clearing, they were an essential part of rural life, supplying fruit and vegetables to their own community, pastoral stations and isolated mining settlements.[81]

The expansion of European settlement into western New South Wales followed the Darling River. By 1859 paddle steamers were operating as far north as the port of Wilcannia. However, the far west of the State remained remote and unattractive to pastoralists – distances to markets were great, transport costs were high, and the region was marginal for agriculture or pastoralism. The terrain is largely desert, with less than 200 mm of rain a year. Rainfall is extremely variable, and when it does occur it is often in the form of heavy storms, which can cause localised flooding. Temperatures range from over 40 °C in summer to minus 6 °C in winter.[82]

A few Chinese market gardeners managed to make a living in this inhospitable region, their fortunes tied to those of the settlements they served. By 1882 a group of about eight Chinese had established two market gardens near the small settlement of Milparinka, in the far north-west corner of New South Wales. The first site, known as Chinaman's Garden Well, was located on Evelyn Creek about 2.5 kilometres upstream from Milparinka. The second garden was 3 kilometres away at Chinaman's Well, on a tributary of Evelyn Creek.[83]

Milparinka was established in 1880 as a service town for the nearby Mount Browne goldfield. It existed on the very frontier of European settlement until World War I, and then went into rapid decline. Today it is a ghost town. The Chinese community were central to the life of Milparinka, despite general opposition to their presence as gold diggers. In 1882 the Mining Warden reported that the health of inhabitants was 'in some degree . . . attributable to the good supply of vegetables raised by the Chinese gardeners, of whom there are eight employed on two gardens'.[84]

Throughout its history Milparinka was plagued by chronic water shortages and competition over this vital resource caused friction between European and Chinese residents. In his study of this marginal community, Geoffrey Svenson records how the small group of gardeners gained a monopoly over the supply of scarce garden produce and also over a significant proportion of Milparinka's water resources.[85] In March 1896 the local newspaper published an editorial deploring the state of the water supply:

> The settlement which might, and would be with a proper water supply, an oasis in the desert, is still a barren waste. There are no cottage gardens to educate and purify the minds of the young and act as a solace to the old. The vegetable and fruit supply, which might be so abundant, is in the hands of a few Chinese, who charge 1/- for a small bunch of vegetables, and from 9d to 1/6 per pound for grapes, and in like proportion for melons and tomatoes and such like when they are to be had.[86]

It appears that Chinese gardeners in this remote region turned the challenges of life in an arid desert environment to their advantage, obtaining good prices for the scarce and highly sought after fruit and vegetables they grew. Drawing on their horticultural and water management skills, they grew grapes, potatoes, apples, peaches, pears, melons, tomatoes, cabbage, carrots and turnips. They sold their produce through the only store in Milparinka and direct to the residents of the town. To fertilise their garden they scoured the town for pig and fowl manure.[87]

As there were no permanent watercourses residents relied on wells for their water supply. At Chinaman's Garden Well the gardeners raised water using a windlass and mining buckets, and hand watered their crops. Among the surface artefacts which Svenson recorded on the site were a windlass barrel, tar ladle, part of the handle of a watering can and horse tackle, indicating that the gardeners used horses, perhaps to drive the windlass.[88] He also recorded medicine bottles and vials, opium smoking equipment and Chinese earthenware and porcelain tableware.[89] Overall the assemblage was similar to those recorded by Neville Ritchie on Chinese mining sites in Central Otago.[90]

Svenson draws parallels between the horticultural techniques employed by the Chinese gardeners in Milparinka and the traditional methods observed by Franklin King in his travels in China and Japan in the early twentieth century. King reports that cucumbers and other vine crops were commonly grown above other vegetables, and suggests that apart from economising in the use of land, vine crops served to shelter more susceptible crops.[91] Svenson argues that such a system would have a clear application at Milparinka, where the Chinese gardeners probably grew melon vines and tomato plants to shelter other crops from excessive exposure to the sun in summer and sub-zero overnight temperatures in winter.[92]

Contemporary newspaper reports record that there were a number of storage tanks at the Chinese gardens at Milparinka. Svenson suggests that these had a similar function to the glazed terracotta urns described by King in southern China which were used to store liquid manure or irrigation water.[93] A notable feature of the garden at Chinaman's Garden Well was a cellar excavated into the ground, probably used to store grapes and other produce and protect them from the extreme desert temperatures. In Chihli Province (modern Hebei Province) in northern China, King notes the use of 'dry earth cellars which can be maintained at very uniform temperature, the separate fruits being wrapped in paper'. This enabled farmers to preserve perishable fruit such as grapes and pears and keep them on the market almost year round.[94]

The last Chinese in Milparinka died during World War I after successfully maintaining a market niche for almost 35 years. The demise of the Chinese gardeners compounded the demise of Milparinka itself. According to Svenson, the collapse of the isolated town was due to a number of factors, including the economic recession of the 1890s, the accompanying bank failures and the mass departure of the young men who represented the business future of the town to the war.[95]

Elsewhere in the arid west of the State many other Chinese market gardeners succeeded in making a living. In his study of southern and western New South Wales Barry McGowan highlights the inventiveness and adaptability of Chinese market gardeners in harnessing scarce water resources and modifying and adapting the technology and implements they had available to them. At Nymagee Station, for example, a wing dam (a barrier only extending part way into a water course) and water race were constructed to divert water to a market garden from a water soakage and overflow. The remains of internal water channels and storage reservoirs were still evident on the site in 2005. At two market garden sites in Cobar storm water flows were captured from several directions via railway culverts and eroded gullies on the roadsides. The storm water was channelled via sluice gates into large dams on the gardens and later pumped into storage tanks placed at strategic points around the gardens.[96] These techniques parallel those observed by King in southern China.[97]

In the 1880s, 30–40 Chinese lived in a camp at Hillston on the banks of the Lachlan River. Adjoining the camp was a garden known as Chong Lee's. Local resident Tom Parr, who worked there during his school holidays for 5 shillings a week, recalled that the Chinese gardeners raised water from the river using small buckets holding about 2 litres fastened to an endless chain driven by a horse going round and round. This description is very similar to King's 1911 account of agriculture in China, except a horse was used instead of a human powered treadmill.[98] The gardeners in Hillston stored most of their water in holding ponds of about 1,350 litres capacity and flood irrigated their vegetables and trees by opening and closing off the irrigation channels as required. Plants were also watered individually using two large watering cans carried on a shoulder yoke.[99] The Chinese gardens at Bourke were irrigated in a similar way, the water being raised from the Darling River and channelled to open 200 gallon (about 920 litre) holding tanks.[100]

Growing fresh vegetables in arid central Australia was an equal challenge. From the 1890s Ned Chong ran a market garden at Hookey's waterhole near Oodnadatta in the north of South Australia, initially in partnership with a countryman, Cherry Ah Chee. Chong cultivated a wide variety of fruit and vegetables and used a horse-driven windlass to irrigate his crops. A visitor to Chong's garden in 1913, known only by the initials E.S.A., was amazed by its fertility and the variety of crops that flourished, including green peas, tomatoes, lettuce, French beans, celery, beetroot, cabbages and watermelons. But this account also highlights Chong's precarious existence:

> We were met at the entrance to the garden by Ned Chong . . . His philosophy was plainly visible in the wonderful Edenic garden which unfolded itself like a marvellous carpet before my astonished eyes. We were conducted through a maze of vegetable beds to view the remarkable water system. This consisted of a revolving wheel, a huge wooden affair, driven by a horse, connected with a windlass, whereby vast quantities of water were obtained from the stream nearby. Huge trenches completed the irrigation scheme. We had come at a low-ebb time, when very little water was available, and when Ned Chong was in deep anxiety concerning whether he could keep the garden going much longer. This seemed cruel luck, considering the almost overpowering odds which he had fought in that wilderness of stone and sand.[101]

It is not surprising that Chong branched out into other enterprises to supplement his income from market gardening. He raised pigs and later established a butcher's shop and bakery. In the 1940s he ran a store and boarding house in Alice Springs for several years. Ned Chong died in Oodnadatta in 1949, at the age of around 100.[102]

Gardens in Cold Climates: New Zealand and Tasmania

The goldfields of Otago and Southland are the southernmost goldfields in the world. The Chinese miners who travelled there in the nineteenth century had to adapt to a climate very different to their homeland. In 1897 Alexander Don spoke with a Chinese miner from Upper Panyu County in Guangdong, who was fossicking at Waipori in Central Otago. The miner compared the barrenness of the steep Otago ridges with the 'fatness' of his native district, where three crops annually were taken off the same ground.[103] Discussing the insights that archaeological evidence can provide into the lives of Chinese miners in southern New Zealand, Neville Ritchie highlights their adaptability and resourcefulness. They built shelters from whatever local materials were available in the remote, rugged gorges of Central Otago, walling up rock overhangs or building huts of river cobbles, rock rubble, adobe or mud bricks. They also used flattened kerosene tins, rice sacks and whatever other materials they could scavenge, as recorded at Lawrence. In the far south at Round Hill which has a more temperate coastal climate and where timber is plentiful, they built timber huts with shingle roofs. Chinese miners quickly adopted the practical European miners' work clothes and boots to combat the cold.[104]

In a similar way the miners who took up market gardening adapted their horticultural practices and choice of crops to the hot summers and cold winters of Central Otago. They were unable to grow crops year round; in winter rain, snow and sleet meant that they had to cease regular gardening activity.[105] Many Chinese miners made a small garden to supply themselves and then perhaps seized the opportunity to supply their Chinese and European neighbours. James Ng has documented gardens in many of the scattered mining camps in Central Otago and Southland, including Wetherstones, Macetown, Conroy's Gully, Butchers Gully and Round Hill. There were also gardens in townships such as Waitahuna, Tapanui, Lawrence, Roxburgh, Cromwell and Queenstown. Large market gardens were recorded at Beaumont, Clyde, Ophir, Matakanui, Arrowtown and Palmerston.[106]

In the rugged terrain of Central Otago Chinese gardens were established on terraces cut into the hillside or formed by widening naturally occurring river terraces.[107] A photograph of the Chinese camp in Arrowtown (Fig. 3.2) shows gardens on flat land above the Arrow River. The method of terracing slopes for agriculture is common throughout the world, particularly in densely populated areas such as Asia. Visiting Hong Kong in 1911, King marvelled at the efficiency with which Chinese horticulturalists surface fitted their gardens to the landscape, achieving the maximum ground area for cultivation.[108] Terraced gardens constructed by immigrant Chinese communities have been recorded elsewhere in the world, notably in the remote Salmon River mountains of Idaho. Perhaps the most extensive Chinese terraced gardens outside Asia, these commercial scale gardens were constructed by Chinese miners who were attracted to Idaho by the gold rushes in 1869. They were cultivated until the 1920s.[109] There are many parallels between the Chinese history of this region and Central Otago.

Chinese gardeners in Central Otago grew crops which were suited to the local climate and found a ready market among Europeans, including potatoes, corn, cabbage, peas, turnips, gooseberries, strawberries and green vegetables.[110] They also grew familiar Chinese vegetables such as Chinese cabbage, turnip, radish and bean sprouts for their own use.[111] They probably brought the seeds with them from China or purchased them from Chinese merchants in New Zealand. According to Ng, Chinese

Fig. 3.2 View of Arrowtown Chinese camp overlooking the Arrow River. Market garden plots can be seen on the terraces above the river. Courtesy Lakes District Museum, EL1870.

cabbage was the most common Chinese vegetable grown. It is a hardy plant which grows readily throughout New Zealand and acquires its best flavour in the colder south.[112] On his regular 'inland tours' through Central Otago Alexander Don often stayed and ate with Chinese miners. On his 1905–1906 tour he climbed over a 1,000 foot high range to Bannockburn where three men had opened a claim in the river bed of the great gorge, where they faced 'a grim battle with rock and torrent and frost'.[113] He reported: 'At 8.15 pm we all riced together with an abundance of delicious white cabbage grown of seed from Canton'.[114] In 1901 he sat down to a meal of rice, bacon, peas, eggs and 'Shantung cabbage' at Mareburn.[115] Earlier, in 1883, Don wrote that *wong nga paak*, a variety of

Chinese cabbage, was being grown at Round Hill and that 3 pounds of bean sprouts or *nga tsoi* could be bought there for a shilling.[116]

Chinese gardeners in southern New Zealand continued the horticultural traditions of their homeland, raising pigs and planting fruit trees as well as growing vegetables. In Otago they grew stone fruit ideally suited to the cool climate such as peaches, plums and apricots, as well as apples and pears.[117] Fruit could be a lucrative supplement to vegetable growing, and some Chinese market gardeners became successful orchardists. One was Lye Bow who, like thousands of his countrymen, originally came to Central Otago in search of gold. However, he quickly saw the demand for fresh food to feed the thousands of miners who flocked to the goldfields. He began a market garden in Butchers Gully and branched out into what became a very successful orchard business. By the 1890s he was regularly receiving prizes at horticultural shows for his magnificent apples, pears and plums.[118] In May 1896 the *Otago Witness* described Lye Bow as 'a straight out enthusiast for fruit culture, ever ready with praise and energy to assist the development of the industry'.[119] The editor noted how he had succeeded in keeping the codling moth pest away from his orchard for the previous few years, probably drawing on traditional Chinese horticultural knowledge. It had destroyed other orchards in the neighbourhood, but caused relatively little damage in Lye Bow's orchard.

By 1903 it was reported that Lye Bow had 1,000 apple, 2,000 apricot, 200 peach, 200 greengage plum, and several hundred pear trees on his 14-acre orchard. Despite unfavourable weather the previous year, he expected to net £3,000 that season, a substantial sum.[120] Lye Bow grew vegetables between the rows of trees in his orchard, using every foot of soil.[121] However, other Chinese market gardeners who combined vegetable growing and orcharding were less successful, particularly where they planted vegetables too closely between their fruit trees and cultivated the soil too deeply. A report on the fruit industry in Central Otago in 1903 noted that Ah Wie's orchard near Earnscleugh Station had relatively light crops of fruit, due to too close planting of vegetables between the trees and a resulting depletion of soil fertility.[122]

Water was a critical resource for both miners and gardeners in relatively arid Central Otago. In 1911 the Reverend Alexander Don was on one of his regular 'inland tours' visiting Chinese miners in the mountains. In the remote Nokomai valley he recorded a couplet on the wall of a miner's hut:

'When water runs gold is produced and all get profit. The dragon comes 1,000 miles and opens wealth's door'.[123] The Chinese word for water race is 'water dragon'.

Historians of the gold rushes in New Zealand suggest that Chinese miners commonly used dams on their creek and river claims.[124] James Ng records Chinese dams at Big Beach on the Shotover River north of Queenstown in 1869 and also a late description of Chinese wing dams on the upper Clutha River in 1899.[125] Wing dams, built of wood or stone or a combination of both, diverted the flow of a creek or river against a bank so that the bank could be broken down and worked, or diverted it away from one side so that the bed could be dry worked.[126] Hydraulic sluicing and elevating techniques which originated on the gold fields in California and Victoria were also used in Otago.[127]

In his study of gold mining in the Alexandra district, John McCraw documents the battles over water rights that were fought out in the warden's courts and the struggles of local authorities and residents to obtain adequate town water supplies when legislation gave priority to water for mining.[128] Like his counterparts on the Australian goldfields, Lye Bow used mining infrastructure to convey water to his gardens. He drew water from the same dam in Butchers Gully used by his countrymen who were searching for gold in the bed of Butchers Creek. During the 1890s Lye Bow and his compatriots fought long drawn out legal battles to retain their water rights, against the European mining companies which were operating large steam and water operated dredges upstream of their own operations.[129] At the same time the Alexandra Borough Council was attempting to construct an adequate town water supply to serve the town's growing population and approached Lye Bow to relinquish some of his water rights. In 1907, exasperated by the harassment he had received from European mining companies, Lye Bow finally sold part of his water rights to the Council.[130]

Like all people on the land in Central Otago, Chinese gardeners had to cope with unpredictable weather and fluctuating prices for their produce. In October 1898 one of the few remaining Chinese market gardens in Central Otago, run by Charlie Lock Chong at Matakanui, was destroyed by an unseasonal snow storm. The report in the *Dunstan Times* commented on the high demand for his produce.

> The prevailing weather last week was rain, hail, snow and frost, accompanied by thunder and lightning. The storm did great damage to the gardens. Our

> local gardener, Ah Lock, told me that the crop of strawberries and plums in his garden will be a complete failure. This is a pity . . . Last year his garden was a favourite resort of the townsfolk and visitors from a distance when his strawberries were ripe. His loss will be considerable, especially with the strawberries, as the demand for the fruit was always greater than the supply.[131]

High winds were a danger to horticulturalists in many parts of New Zealand, requiring the use of fences or hedges to shelter crops. In 1874 the *Tuapeka Times* in Otago reported that a party of Chinese had started a garden by the Wetherstones town boundary, notable for the fact that they used thistle stalks to line the wires of their fences as protection from the wind.[132] In Outram, near Dunedin, Chinese market gardeners surrounded their plots with manuka brush fences. Ted Broad recalled visiting the gardens of Sam Young and Co in Outram with his parents when he was growing up in the 1930s and early 1940s: '. . . the fences around the property were all manuka brush, laced close together to protect the vegies from the wind'.[133] Fences were probably an adaptation to the more extreme climate of southern New Zealand, for in southern China market gardeners traditionally left their fields open to the elements with no enclosing walls or fences. One possible explanation for this is that allowing winds to blow freely through growing crops discouraged insect pests.[134] In the Auckland region, gardeners left their gardens unfenced. Donald Hunt observes that they preferred to have the prevailing cold winds from the south and west than the warm, humid winds from the north and east which facilitate the spread of blight and insect pests.[135]

Strong dry winds, usually from the north-west or south-west, are a feature of the Canterbury climate, particularly in spring and summer when they can create drought conditions. The Chinese market gardeners who settled in the Christchurch district from the 1870s favoured the warm, sheltered north-facing valleys in the Port Hills and the sheltered bays of the Banks Peninsula such as Akaroa.[136] Similarly in Wellington, located on Cook Strait, high winds were a regular occurrence. In June 1892 cold southerly gales destroyed vegetation across the city, particularly affecting market gardens. The Chinese gardeners reported that all vegetables above ground had been destroyed and nothing was left but root crops.[137] To minimise this risk, market gardeners chose sheltered locations where their crops were protected from the prevailing winds, for example in Island Bay, Seatoun on Worser Bay and Miramar on Evans Bay. When a large Chinese market garden in Island Bay, which had been cultivated probably from the late 1880s, was subdivided into residential sections and advertised for sale in 1905, the advertisement noted the advantages of its location.

> In the early days of the Bay, when the Chinaman commenced gardening operations there, practically the whole district was at their disposal, but with their national shrewdness, they decided on this block of land . . . The reasons are obvious. It is sheltered from prevailing winds. It basks in the rays of 'bright Phoebus' from the moment of his morning entry until his evening exit.[138]

There are parallels between the entrepreneurial orchardist Lye Bow in Central Otago and James Chung Gon in Tasmania. Chung Gon was born in Shandong Province in about 1854, and according to family history he arrived in Australia at the age of 18 with only a shilling in his pocket, speaking no English.[139] He received a hostile reception from European larrikins but managed to make his way to kinsmen who gave him European clothes and money to travel to the Bendigo goldfields. Disliking the rough life of the Victorian diggings he travelled to Tasmania with Chinese friends where he worked as a tin miner, market gardener and wood cutter. He finally settled in Launceston where he began a small market garden near the junction of the Tamar and North Esk Rivers. Business prospered and Chung Gon leased 10 acres in South Launceston. He employed ten of his countrymen and had three hand carts supplying fresh produce to shops, hotels and the Launceston General Hospital.[140]

Chung Gon ran market gardens in Launceston for over 50 years. He went on to become a successful businessman, investing in tin mines in north-eastern Tasmania, running a 200-acre orchard near Lilydale which was one of the first commercial orchards in northern Tasmania and opening fruit and vegetable stores for his children.[141] He read constantly to keep abreast of innovations in horticulture; he is said to have pioneered the commercial use of irrigation in Northern Tasmania and to have adapted agricultural machinery for use on his market gardens and orchards.[142] After World War II, in partnership with Messrs Springer and Pinner, he began a new business venture manufacturing agricultural pipes and flower pots from locally sourced clay.

He was still working in his garden, weeding beds of onions, well into his nineties.[143]

By the 1950s Chung Gon found that market gardening was no longer economically viable, due to increased costs of labour and artificial fertilisers and the difficulty of obtaining adequate supplies of stable manure. Land values within the Launceston city boundaries were rising and he finally sold his 18 acres of land for residential development. In 1953 the oldest market gardens in Launceston were subdivided into 100 housing allotments.[144] Throughout his life Chung Gon maintained traditional Chinese agricultural methods: rotating root crops such as carrots and parsnips with green crops such as lettuce, cabbages and peas to maintain soil fertility; planting by the moon so that seedlings would mature earlier; thoroughly preparing the soil before planting; and enriching it with hundreds of cart loads of stable manure a month.[145] But he also adopted new technology and European gardening techniques. He grew tomatoes in hot houses and in the 1950s he experimented with various types of artificial manures to replace the cart-loads of stable manure he once used.[146]

Chun Gon showed great flexibility in adapting his horticultural methods to the cool climate of Tasmania. In addition to the traditional range of vegetables familiar to Europeans he also grew Chinese cabbage and marketed it as a winter alternative to summer lettuce. In his orchards he grew cherries, plums, apples and pears which thrive in a cool climate.

Conclusion

As Lesley Head, Pat Muir and Eva Hampel observe, gardens can be seen as places of environmental engagement, where immigrants and plants continue traditions from their homelands and accommodate to new biophysical and social environments.[147] By the late nineteenth century Chinese market gardeners were widely dispersed across Australia and New Zealand and could be found in the most remote settlements and the most marginal areas for agriculture. Drawing on their traditional skills in horticulture and water management, they successfully tilled the soil in environments ranging from the rugged mountains of Central Otago to the tropics and the deserts of Australia.

Chinese market gardeners created productive gardens in inhospitable terrain which Europeans did not attempt to cultivate, for example in Milparinka where they turned the impediments of life in an arid environment to their advantage, obtaining good prices for the scarce and highly sought after fresh fruit and vegetables they grew. Similarly in the Northern Territory they capitalised on Europeans' lack of confidence in the prospects for agriculture in a tropical environment. In addition to supplying the fruit and vegetables familiar to Europeans, which they did not traditionally grow in China, Chinese market gardeners introduced new crop varieties to Australia and New Zealand, and incorporated indigenous plants into their repertoire, for example kumara and puha in New Zealand.

Chinese farmers and market gardeners played a key role in land clearing, 'developing' the land and making it productive. They were therefore responsible with Europeans for the environmental impacts they wrought on the landscape. The environmental impact of Chinese market gardening and its sustainability has not been researched in any detail. Despite these potential negative effects on the environment, Europeans came to rely on Chinese gardeners for fresh fruit and vegetables, particularly in remote areas far from major population centres. The occupational shift from gold mining to market gardening and the economic niche which Chinese market gardeners were successfully able to occupy over many decades are examined in more detail in the following two chapters.

Notes

1. Kathryn Gleason, 'To bound and to cultivate: an introduction to the archaeology of gardens and fields', in Naomi Miller and Kathryn Gleason (eds.), *The Archaeology of Garden and Field*, Philadelphia, 1994, pp. 2–3.
2. Francesca Bray and Joseph Needham, *Science and Civilisation in China,* Vol. 6: Biology and Biological Technology, Part I: Agriculture by Francesca Bray, Cambridge, 1984, p. xxiv.
3. Guohua Xu and L. J. Peel, *The Agriculture of China*, Oxford, 1991, pp. 1–4.
4. Ibid., pp. 135–136; Bray and Needham, *Science and Civilisation*, 6:II, pp. 19–20.
5. James Ng, *Windows on a Chinese Past*, Vol. 1, Dunedin, 1999, pp. 12, 16–19.
6. George Hunter McNeur, letter, *Outlook,* 28 February 1903, p. 15.
7. Ng, *Windows on a Chinese Past*, Vol. 1, p. 33.
8. Libby Robin and Tom Griffiths, 'Environmental History in Australasia', p. 20, http://ceh.environmentalhistory-au-nz.org/wpcontent/uploads/Environmental_History_in_Australasia _2004.pdf, accessed 8 February 2013.

9. Ibid., p. 20. See also Andrew Sturman and Nigel Tapper, *The Weather and Climate of Australia and New Zealand*, South Melbourne, 2006, pp. 344–345.
10. Sturman and Tapper, *Weather and Climate*, pp. 344–345; Bureau of Meteorology, http://www.bom.gov.au/lam/climate/levelthree/ ausclim/zones.htm, accessed 10 May 2011.
11. Stuart Macintyre, *Concise History of Australia*, Port Melbourne, 2009, p. 8.
12. Michael Cathcart's 2010 study explores how water has shaped the history of Australia, imposing a fundamental limit on how and where Australians live (Michael Cathcart, *The Water Dreamers: The Remarkable History of Our Dry Continent*, Melbourne, 2010, p. 2).
13. Josephine Flood, *Archaeology of the Dreamtime: The Story of Prehistoric Australia and its People*, Pymble, 1995, p. 195; and John Mulvaney and Johan Kamminga, *Prehistory of Australia*, Washington, 1999, pp. 62–63.
14. Powell, *Historical Geography of Modern Australia*, pp. 133–135; and Frawley, 'Evolving visions', pp. 62–63. See also Macintyre, *Concise History of Australia*, pp. 100–103.
15. Robin and Griffiths, 'Environmental History in Australasia', p. 21.
16. Sturman and Tapper, *Weather and Climate*, p. 345; and Brett Mullan, Andrew Tait and Craig Thompson, 'Climate', *Te Ara – the Encyclopaedia of New Zealand*, updated November 2012, http://www.teara.govt.nz/en/ climate, accessed 4 May 2011.
17. Mullan, Tait and Thompson, 'Climate'.
18. Since European settlement there has been an 85 percent decline in wetlands in New Zealand, one of the most dramatic in the world (Geoff Park, 'Swamps which might doubtless be easily drained: swamp drainage and its impact on the indigenous', in Pawson and Brooking (eds.), *Environmental Histories of New Zealand*, pp. 151–158; and Pawson and Brooking, *Environmental Histories of New Zealand*, p. 10).
19. Examples include the Chinese market gardens on the Mulwaree River at Goulburn, the Lachlan River at Hillston and the Castlereagh River at Gilgandra in New South Wales (Barry McGowan, 'Chinese market gardens in southern and western New South Wales', *Australian Humanities Review*, Issue 36, July 2005, n.p.).
20. A.J. Finn, 'Suitable sites for market gardening', *Agricultural Gazette of NSW*, Vol. XXVII Part 9, 2 September 1916, pp. 633–634.
21. Ronald Knapp (ed.), *Chinese Landscapes: The Village as Place*, Honolulu, 1992. p. 6.
22. See Zvonkica Stanin, 'From Li Chun to Yong Kit: A Market Garden on the Loddon, 1851–1912', *Journal of Australian Colonial History*, Vol. 6, 2004, p. 22; Colleen Morris, 'Chinese market gardens in Sydney', *Australian Garden History*, Vol. 12, No. 5, March/April 2001, pp. 5, 7; Oline

Richards, 'Chinese market gardening: a Western Australian perspective', *Australian Garden History*, Vol. 13, No. 1, July/August 2001, p. 19; Sandra Pullman, 'Along Melbourne's rivers and creeks', *Australian Garden History*, Vol. 12, No. 5, March/April 2001, p. 9; and McGowan, 'Chinese market gardens', n.p.

23. *Melbourne Age,* 24 February 1865, reproduced in *Sydney Morning Herald,* 2 March 1865, p. 2.
24. McGowan, 'Chinese market gardens', n.p.
25. Barry McGowan, 'Adaptation and organization: the history and heritage of the Chinese in the Riverina and Western New South Wales, Australia', *Chinese America: History and Perspectives*, Annual, 2007, p. 238.
26. John West, *Irrigation and Intense Culture*, Melbourne, 1889, p. 86, quoted in Frost, 'Migrants and technological transfer: Chinese farming in Australia, 1850–1920', *Australian Economic History Review*, Vol. 42, No. 2, 2002, p. 127.
27. Edwin Brady, *Australia Unlimited, Melbourne, 1919,* pp. 263, 438, 445, 470, 498.
28. Taylor argued that Australia could support only a limited population because of its limited resources and that efforts should be directed towards developing the better-watered areas of the south-east of the continent instead of the tropics and arid areas (J. M. Powell, 'Thomas Griffith-Taylor', *Australian Dictionary of Biography*, 1990, http://adb.anu.edu. au/biography/taylor-thomas-griffith-8765, accessed 30 April 2012).
29. Smaller-scale irrigation schemes for agriculture were developed in Central Otago and Canterbury in the late nineteenth century with private and government involvement, for example those capturing the waters of the braided, snow-fed rivers crossing the Canterbury Plains to the sea (Benjamin Evans, *A History of Agriculture, Production and Marketing in New Zealand*, Palmerston North, 1969, p. 13; and Martin Ward and Shona Russell, *Water Sharing Schemes: Insights from Canterbury and Otago,* Report for Foundation for Research, Science and Technology, Wellington, 2010, p. 4.)
30. *Northern Advocate*, Whangarei, 4 May 1920, p. 3.
31. Chinese market gardeners were particularly affected by increases in land values and land rents, as the majority were lessees rather than land owners.
32. Alex McLellan, 'Market gardening in Otaki', *Otaki Historical Society Historical Journal*, Vol. 5, 1982, p. 61; Anne Thorpe, 'Otaki's market gardens', *Otaki Historical Society Journal*, 2008, p.13.
33. *Bay of Plenty Times,* 24 November 1909, p. 2.
34. Donald T. Hunt, 'Market gardening in metropolitan Auckland', MA Honours Thesis, University of Auckland, Auckland, 1956, pp. 45–46.

35. Pamela Wai Shing, 'Locational and structural changes of market gardening in Pukekohe–Bombay–Patumahoe', MA Thesis, University of Auckland, Auckland, 1977, p. 15; and Thong Ling Lee, 'Chinese market gardening in the Auckland region', MA Thesis, University of Auckland, Auckland, 1974, pp. 14–15.
36. New Zealand's temperate climate restricted where the tropical and subtropical crops introduced by Polynesian settlers could be grown. The archaeological evidence for Māori gardening is most extensive in the northern half of the North Island, especially in coastal areas where frosts are rare or infrequent (Louise Furey, *Maori Gardening: An Archaeological Perspective*, Wellington, 2006, p. 21).
37. The 100 hectare Otuataua Stonefields reserve is one of only two surviving remnants of these once extensive gardens. Department of Conservation, 'Historic Otuataua Stonefields', http://www.doc.govt.nz/conservation/historic/by-region/auckland/central-and-south-auckland/otuataua-stonefields/, accessed 11 February 2013.
38. Lily Lee and Ruth Lam, *Sons of the Soil: Chinese Market Gardeners in New Zealand*, Pukekohe, 2012, p. 65.
39. *Dunstan Times*, 23 September 1870.
40. Lee and Lam, *Sons of the Soil*, pp. 65–69, 74–75.
41. As occurred in Otaki in New Zealand in October 1908, destroying acres of potatoes (*Evening Post*, 28 October 1908, p. 6).
42. Lee and Lam, *Sons of the Soil*, p. 432.
43. Ibid., pp. 213, 283–284, 294, 297.
44. Māori gardeners made a number of adaptations to traditional Polynesian horticultural practices in order to grow kumara in New Zealand's temperate climate, including the introduction of a winter storage phase and other techniques such as soil lightening, mulching, and constructing raised planting beds or mounds and wind breaks (Helen Leach, *1,000 Years of Gardening in New Zealand*, Auckland, 1984, pp. 58, 61, 72).
45. Lee and Lam, *Sons of the Soil*, pp. 412–413.
46. The most popular Māori puha dish is called 'boil up'. It consists of puha boiled with meat (pork, beef, mutton bird or often pork bones), potatoes and kumara (*Mindfood*, 'Traditional Maori recipes', 27 April 2012, http://www.mindfood.com/at-traditional-maori-recipes-global-tastes.seo, accessed 22 February 2013).
47. Lee and Lam, *Sons of the Soil*, pp. 289–290.
48. Kaaren Hiyama, *High Hopes in Hard Times: A History of Grey Lynn and Westmere*, Grey Lynn, 1991, p. 28.
49. Ian Jack and Katie Holmes surveyed and excavated a small section of the site in 1984. The site was occupied by a series of Chinese leaseholders between 1883 and 1934. Ah Toy, after whom the site is named, was the last of the

three leaseholders; he held the lease from 1900 until 1934 (Ian Jack and Katie Holmes, 'Ah Toy's garden: a Chinese market garden on the Palmer River goldfield, North Queensland', *Australian Journal of Historical Archaeology*, Vol. 2, 1984, pp. 51–52).

50. Ibid., p. 51.
51. Ibid., pp. 51–52.
52. Ibid., p. 52.
53. Ibid., pp. 55–56.
54. *Townsville Daily Bulletin*, 31 May 1932, p. 9.
55. Grant Vinning, *Select Markets for Sweet Potato, Taro and Yam*, Report for Rural Industries Research and Development Corporation, Kingston, ACT, May 2003, pp. 7, 37.
56. Queensland Department of Agriculture, Fisheries and Forestry, 'About taro', http://www.daff.qld.gov.au, accessed 27 March 2013.
57. Jack Golson, 'Aboriginal food plants: some ecological and cultural considerations', in D. J. Mulvaney and J. Golson (eds.), *Aboriginal Man and Environment in Australia,* Canberra, Australian National University Press, 1971, pp. 217, 222; and Doug Yen, 'The sweet potato in historical perspective', in Ruben Villareal and T. E. Griggs, *International Symposium on Sweet Potato, Taiwan, 23–27 March 1981*, AVRDC Publication 82–172, 1982, pp. 17–30.
58. Tim Denham, Mark Donohue and Sara Booth, 'Horticultural experimentation in northern Australia reconsidered', *Antiquity*, Vol. 83, 2009, p. 645.
59. Indentured labourers 'blackbirded' from the islands of Melanesia (predominantly Vanuatu, the Solomons, New Caledonia and Papua New Guinea) formed the major workforce in the Queensland sugar industry from the 1860s (Henry Reynolds, *North of Capricorn: The Untold Story of Australia's North*, Crows Nest, 2003, pp. 38–41).
60. *Queenslander*, 7 February 1885, p. 231.
61. *Bacchus Marsh Express,* 30 November 1878, p. 3.
62. *Townsville Daily Bulletin,* 6 September 1932, p. 7. The taro variety favoured by Chinese, and most commonly cultivated in Australia today, is a soft cooking type known as *bun-long* (Queensland Department of Agriculture, Fisheries and Forestry, 'About taro', http://www.daff.qld.gov.au, accessed 27 March 2013).
63. *Argus,* 25 September 1888, p. 43.
64. Leonie Ryder, 'Incorrigible colonist: ginger in Australia 1788–1950', PhD Thesis, University of Adelaide, Adelaide, 2010, pp. 89–91.
65. R. Bartlett, 'The cultivation of ginger', *Agricultural Gazette of New South Wales*, November 1922, p. 818, cited in Ryder, 'Incorrigible colonist', p. 135.
66. Ryder, 'Incorrigible colonist', p. 89.

67. *Queenslander,* 19 January 1878, p. 27.
68. *Argus,* 8 December 1886, p. 15.
69. J. M. Powell, *Plains of Promise, Rivers of Destiny: Water Management and the Development of Queensland 1824–1990,* Bowen Hills, 1991, p. 54.
70. Ibid., pp. 54–55.
71. Reynolds, *North of Capricorn*, p. 78.
72. Prior to Federation in 1901 at least three censuses of the European population in the Northern Territory were held, in 1881, 1891 and 1901. Aboriginal and Chinese people were not included (National Archives of Australia, 'Commonwealth Government records about the Northern Territory', http://guides.naa.gov.au/records-about-northern-territory/part2/chapter10/10.9.aspx, accessed 29 November 2016). See also Northern Territory Archives Service, 'Chinese People in the Northern Territory', https://dtc.nt.gov.au/arts-and-museums/northern-territory-archives-service/archives-subject-guides/chinese-people-in-nt, accessed 29 November 2016.
73. Ibid., p. 111; C. Y. Choi, *Chinese Migration and Settlement in Australia,* Sydney, 1975, p. 35.
74. Shi Kwong Ho, *From My Darwin Notebook,* Darwin, 1989, p. 43, quoted in Reynolds, *North of Capricorn,* p. 114.
75. *South Australian Register,* 26 January 1878, p. 4S. Palmerston was the name first chosen for the capital of the Northern Territory in 1864. It was renamed Darwin in 1911 after administration of the Northern Territory passed from South Australia to the Commonwealth of Australia.
76. *Northern Territory Times and Gazette*, 14 August 1903, p. 2.
77. *Northern Territory Times and Gazette*, 3 August 1906, p. 2.
78. *Northern Territory Times and Gazette*, 17 September 1897, p. 2.
79. Reynolds, *North of Capricorn*, p. 110.
80. *Northern Territory Times and Gazette*, 22 February 1879, p. 2.
81. McGowan, 'Chinese market gardens', 2005, n.p.
82. Geoffrey Svenson, 'Marginal people: the archaeology and the history of the Chinese at Milparinka', MA Thesis, University of Sydney, Sydney, 1994, pp. 7, 12.
83. Ibid., pp. 135, 138.
84. Report to Department of Mines, 1882, p. 99, quoted in Svenson, 'Marginal people', p. 40.
85. Svenson, 'Marginal people', p. 72.
86. *Sturt Recorder*, 8 March 1896, p. 2.
87. Svenson, 'Marginal people', p. 131.
88. Ibid., p. 93.
89. Ibid., pp. 93, 99, 100, 106.

90. Neville Ritchie, 'Traces of the past: archaeological insights into the New Zealand Chinese experience in southern New Zealand', in Manying Ip (ed.), *Unfolding History, Evolving Identity: The Chinese in New Zealand*, Auckland, 2003, p. 38.
91. Franklin Hiram King, *Farmers of Forty Centuries*, London, 1949, p. 36.
92. Svenson, 'Marginal people', pp. 87, 131.
93. Ibid., p. 84. See also King, *Farmers of Forty Centuries*, pp. 70, 199, 257–258.
94. King, *Farmers of Forty Centuries*, p. 343. The former province of Chihli was dissolved in 1928 and the new province of Hebei created, carving out the municipalities of Beijing and Tianjin (Wikipedia, 'Hebei', http://en.wikipedia.org/wiki/Hebei, accessed 3 June 2011).
95. Svenson, 'Marginal people', p. 143.
96. McGowan, 'Adaptation and organization', p. 238.
97. King, *Farmers of Forty Centuries*, pp. 173–175.
98. Ibid., p. 78.
99. McGowan, 'Chinese market gardens', 2005, n.p.
100. Ibid.
101. *Register*, Adelaide, 18 November 1913, p. 9.
102. Jennifer Martiniello, '"As Strands of Plaited Music," my Chinese-Aboriginal-Anglo heritage', in Penny Edwards and Yuanfang Shen (eds.), *Lost in the Whitewash: Aboriginal-Asian Encounters in Australia, 1901–2001*, Canberra, 2003, pp. 28–29.
103. Alexander Don, *Annual Inland Tour 1896–1897*, Dunedin, 1897, p. 480; see also *Christian Outlook*, 3 July 1897, p. 268.
104. Neville Ritchie, 'Traces of the past: archaeological insights into the New Zealand Chinese experience in southern New Zealand', in Manying Ip (ed.), *Unfolding History, Evolving Identity: The Chinese in New Zealand*, Auckland, 2003, pp. 32, 45.
105. Lee and Lam, *Sons of the Soil*, p. 21.
106. Ng, *Windows on a Chinese Past*, Vol. 1, p. 321.
107. Neville Ritchie, pers. comm., February 2011.
108. King, *Farmers of Forty Centuries*, pp. 67–68.
109. Jeffrey Fee, 'Idaho's Chinese Mountain Gardens', in Priscilla Wegars (ed.), *Hidden Heritage: Historical Archaeology of the Overseas Chinese*, Amityville, 1993, pp. 65–66.
110. Ng, *Windows on a Chinese Past*, Vol. 1, p. 323; and Lee and Lam, *Sons of the Soil*, p. 21.
111. When Phil George ate with the Chinese at Kyeburn, the menu included Chinese cabbage, Chinese radish, Chinese parsley, garlic and celery (Ng, *Windows on a Chinese Past*, Vol. 1, fn. 151b, p. 341).
112. Ibid., Vol. 2, p. 36.

113. Alexander Don, 'Chinese Mission Work in Otago: Nineteenth Annual Inland Tour, 1905–1906', Dunedin, 1906, p. 26.
114. *Outlook,* 14 April 1906, p. 13.
115. *Outlook,* 11 May 1901, p. 26.
116. Ng, *Windows on a Chinese Past,* Vol. 1, fn. 151a, p. 341.
117. Fruit growing was a long-established tradition in China and fruit trees had rich aesthetic and symbolic associations as well as economic value (Craig Clunas, *Fruitful Sites: Garden Culture in Ming Dynasty China,* London, 1996).
118. *Otago Witness,* 1 March 1894, p. 4; 4 March 1897, p. 26; and 24 February 1898, p. 26.
119. *Otago Witness,* 28 May 1896, p. 25.
120. *Otago Witness,* 11 November 1903, p. 27.
121. Lee and Lam, *Sons of the Soil,* p. 31.
122. *Otago Witness,* 11 November 1903, p. 27.
123. Alexander Don, NZ Presbyterian Chinese Mission Inland Tours XXIII and XXIV 1909–1911 and Westland Tour, 1911, reprinted from the *Outlook, Otago Daily Times,* Dunedin, 1911, p. 11.
124. Beattie, 'Hungry dragons: expanding the horizons of Chinese environmental history–Cantonese gold-miners in colonial New Zealand 1860s-1920s', *International Review of Environmental History,* Vol. 1, 2015, n.p.; William Heinz, *Bright Fine Gold: Stories of the New Zealand Goldfields* Wellington, 1974, p. 17.
125. Ng, *Windows on a Chinese Past,* Vol. 1, fn. 77, p. 167.
126. Neville Ritchie, 'Archaeology and history of the Chinese', p. 60.
127. Beattie, 'Hungry dragons', n.p.
128. John McCraw, *Mountain Water and River Gold: Stories of Gold Mining in the Alexandra District,* Dunedin, 2000.
129. Ibid., pp. 227–229.
130. Ibid., p. 229. The physical remains of Lye Bow's garden and orchard are listed on the register of the New Zealand Historic Places Trust as an Historic Area (NZHPT, Register No. 7547, New Zealand Historic Places Trust, 'Lye Bow's Historic Area' http://www.historic.org.nz/theregister/RegisterSearch/ RegisterResults.aspx?RID=7547, accessed 14 June 2011).
131. *Dunstan Times,* 14 October 1898, quoted in Ng, *Windows on a Chinese Past,* Vol. 1, p. 321. The Matakanui garden was still in existence in 1911, when Don reported that there were five men working there (*Outlook,* 1 August 1911, p. 14).
132. *Tuapeka Times,* 26 September 1874.
133. Lee and Lam, *Sons of the Soil,* p. 45.
134. Helen Leach, pers. comm., 4 February 2011.

135. Donald T. Hunt, 'Market Gardening in Metropolitan Auckland', MA Thesis, University of Auckland, Auckland, 1956, p. 54.
136. Lee and Lam, *Sons of the Soil*, pp. 89, 91.
137. *Taranaki Herald*, 9 June 1892, p. 2.
138. *Evening Post*, 20 November 1905, p. 8.
139. Ted Chung Gon, 'James Chung Gon', Chinese Heritage of Australian Federation, La Trobe University, www.chaf.lib.latrobe.edu.au, accessed 20 May 2012.
140. *Mercury*, Hobart, 21 September 1942, p. 3.
141. The Chung Gon family greengrocer business was still in operation in Launceston in 2014 (Andrew Piper, pers. comm., January 2014).
142. Jill Cassidy, 'Chung Gon, James (1854–1952)', Australian Dictionary of Biography, National Centre of Biography, Australian National University, http://adb.anu.edu.au/biography/chung-gon-james-9745/ text17213, accessed 3 November 2013.
143. *Mercury*, Hobart, 28 March 1949, p. 7.
144. *Mercury*, Hobart, 28 July 1953, p. 12.
145. *Mercury*, Hobart, 28 August 1942, p. 3 and 28 July 1953, p. 12.
146. *Mercury*, Hobart, 28 July 1953, p. 12.
147. Lesley Head, Pat Muir and Eva Hampel, 'Australian backyard gardens and the journey of migration', *Geographical Review*, Vol. 94, No. 3, July 2004, p. 326.

CHAPTER 4

Making a Living in New Gold Mountain

The prime motivation of Chinese immigrants to Australia and New Zealand was economic – to earn money to support their families in China. Possessing a strong work ethic and sense of responsibility for their families, their dream was to save enough to return to their family and friends in China as wealthy men. In common with other migrant communities they faced significant social and economic barriers in earning a living and relied heavily on social and economic support networks within their own community rather than the financial and social institutions of their adopted countries. Thus complex, interwoven social networks developed within and between Chinese communities across Australia and New Zealand, based on common ties of kin, village, county and language.

Historians of technology emphasise the importance of economic factors as crucial components of the successful adaptation of a given technology to a new environment.[1] As Edward Beatty points out, economic factors are part of an interrelated network of technical, economic, social, environmental and political influences that mediate the relationship between imported knowledge and the recipient society.[2] In a similar vein, Nathan Rosenberg dissects the complex processes by which technological change is generated. He points out that the diffusion of inventions and technological knowledge is not simply an economic phenomenon, determined by expected costs and benefits, but is rather a discontinuous, intermittent process of invention and reinvention, influenced by the receptivity and capacity of societies to assimilate new technologies.[3]

J. Boileau, *Chinese Market Gardening in Australia and New Zealand*,
Palgrave Studies in the History of Science and Technology,
DOI 10.1007/978-3-319-51871-8_4

Recent research into immigrant entrepreneurship and small business provides a useful framework for analysing the economic dimensions of Chinese market gardening. In his study of Chinese entrepreneurship in Australia Jock Collins draws on international studies of 'ethnic' entrepreneurs such as the study by Roger Waldinger and colleagues.[4] In Waldinger's model the business success of a particular migrant group is determined by how the specific characteristics of that group – their pre-migration experiences, history of migration and settlement, and post-migration experiences – interact with each other and with the socio-economic environment or opportunity structure of the 'host' country.[5] Collins' work has been criticised by Michael Williams, who points out that his nation-based perspective fails to take account of the importance of ongoing links with family and native place in China, focuses narrowly on the period before and after World War II, and concentrates on economic motivations and institutional racism in the host society as the main determinants of the lives of Chinese immigrants.[6] These criticisms are valid. When examining the course of Chinese market gardening in Australia and New Zealand it is necessary to use an extended timeframe and consider the diverse influences on the lives of Chinese market gardeners, including the vital importance of ongoing ties with China. For those who could afford it, return trips to China allowed them to take their earnings home, spend time with their extended family and form new bonds of kinship, marrying and fathering children. These social networks are explored in a later chapter.

However, Collins' characterisation of farmers as 'small business people on the land' is a useful concept.[7] Using it as a starting point, the following two chapters examine the economic factors which influenced the course of Chinese market gardening in Australia and New Zealand: the economic opportunities and resources which helped Chinese immigrants succeed in this line of small business and the economic barriers which they faced. The first chapter outlines the economic environments within which Chinese gardeners operated in Australia and New Zealand and the historical development of market gardening. The second chapter examines the economic resources and opportunities that market gardeners drew upon to meet the challenges posed by a complex and changing set of social and economic circumstances.

This chapter begins with a brief overview of the economic environments of Australia and New Zealand which Chinese immigrants encountered between the 1850s and the 1940s. It outlines the post-migration experiences of Chinese immigrants and the historical development of market gardening, beginning with the economic niche which Chinese market gardeners established on the goldfields then tracing the movement of Chinese market gardeners to towns and cities after the gold rushes and their place in the urban economy.

Settler Societies in the Southern Hemisphere

Australia and New Zealand are among the group of countries (including the United States and Canada) variously referred to by economic historians as the Anglo cluster, settler societies or regions of recent settlement.[8] These countries have a number of common characteristics: they share an Anglo-Saxon cultural heritage and the English language; they are products of British colonialism which spread across the world in the eighteenth and nineteenth centuries; and they are all developed 'Western' economies. Historically they were resource rich and labour poor, relying on large-scale immigration, capital inflow from Europe and the export of a narrow range of primary products (food and raw materials) for their economic development.[9] J. W. McCarty points out that the commercial cities of these countries became increasingly linked to the world market and the levelling effects of the expanding global economy in the nineteenth century.[10] As this chapter will show, the rapidly growing cities of Australia and New Zealand in the late nineteenth century provided an ideal economic niche for Chinese market gardeners supplying fresh food to urban populations.

Among settler societies Australia and New Zealand share particular commonalities. They were occupied by Europeans later than the other nations, during the early stages of the industrial revolution in Britain; pastoral industries emerged as the core of their economies; and their economic development was strongly influenced by their distance from European markets.[11] Their rapid economic growth during the second half of the nineteenth century was founded on the wool industry and fuelled by the discovery of gold and a favourable demographic structure created by large-scale immigration.[12] The development of refrigerated shipping in the 1890s transformed the agricultural industry in both countries and resulted in a boom in the export of meat and dairy produce.[13]

These common characteristics are reflected in similar patterns of economic growth. For example, David Greasley and Les Oxley find a close correspondence between real wages in Australia and New Zealand between 1873 and 1913 indicating that the trans-Tasman labour market was well integrated during this period and wage movements were shaped by common forces.[14] Many economists have noted the high levels of growth and high standards of living achieved in Australia and New Zealand between 1850 and 1913, apart from an economic downturn in the 1890s.[15] While estimates vary depending on the economic models and indicators employed (in particular how indices of prices, cost of living and expenditure relative to wages are arrived at and inter-country comparisons are taken into account) there is general agreement that Australia and New Zealand were more prosperous than Britain and the United States in the second half of the nineteenth century.[16] According to Schedvin, per capita incomes in Australia were higher than in the United States from as early as 1830, and in New Zealand from at least 1870, until the early twentieth century. Economic growth in Australia and New Zealand slowed in the early twentieth century, when both countries were in the early phases of industrialisation.[17] Incomes fell below the United States after 1902 in New Zealand and after 1912 in Australia.[18]

Given the dominance of pastoralism and broad-scale agriculture, market gardening was a minor part of the agricultural economies of Australia and New Zealand. This still holds true today.[19] Table 4.1 shows the acreages of land devoted to market gardens in New South Wales and Victoria from 1860 to 1930 and in Queensland from 1910

Table 4.1 Acreages of market gardens as a percentage of all land under crop: NSW, Victoria and Queensland, 1860–1930

Season	*NSW*	*VIC*	*QLD*
1860–1861	3.5% [8,707*]	1.7 % [7,298*]	–
1870–1871	4.5% [17,168*]	1.6% [14,856*]	–
1880–1881	2.9 % [18,130*]	1.1% [22,288*]	–
1890–1891	2.8% [24,276*]	1.2% [33,654*]	–
1900–1901	0.32% [8,000]	1.5% [57,496*]	–
1910–1911	0.28% [9,813]	0.27% [10,778]	0.34% [2,317]
1920–1921	0.22% [9,888]	0.27% [12,201]	0.25% [2,018]
1930–1931	0.1% [7,448]	0.3% [20,197]	0.07% [903]

*Figure includes orchards and gardens

Sources: Office of the Government Statist, Victoria, *Statistical Register of the Colony of Victoria*, Melbourne, *1861–1901;* New South Wales Government Statistician, Sydney, *New South Wales Statistical Register*, 1861–1901; and Australian Bureau of Statistics, Canberra, New South Wales, Victoria and Queensland *Year Books, 1910–1930.*

to 1930. The proportion of market gardens as a percentage of the total acreage under cultivation was never above 4.5 percent for the years recorded, even when orchards were included (in the years before 1900). However, despite its small size, the horticultural industry, and the market gardening industry within it, has historically played a significant role in providing fresh vegetables and fruit to local markets. It should be noted that the small acreages devoted to market gardening do not reflect the high yields that could be produced from them.[20] In the 1840s, before Chinese immigrants arrived in Australia in large numbers, European gardeners could make handsome profits from vegetables. In 1841 it was reported that one man near Melbourne made £500 in a year from just 3 acres.[21]

Perhaps a more valid indicator of the relative importance of market gardening is the numbers employed in the industry. In Australia, for example, market gardeners made up only 0.05 percent of the total agricultural workforce in 1911 and 0.03 percent in 1921.[22] Numbers were similarly small in New Zealand. Market gardeners comprised 0.05 percent of the total agricultural workforce in 1906 and 0.03 percent in 1926.[23]Another factor is the money economy versus the real economy; in other words the invisible production of market gardens not recorded in official statistics. It is probable that much of the production of market gardeners was not recorded in official statistics and also that the numbers working in the industry were under-enumerated due to the high mobility of the workforce. This would have been particularly true of Chinese market gardeners.[24]

Migrants from diverse non-English speaking backgrounds are highly represented in the horticultural industry today, particularly in Australia. This heritage extends back to the Chinese immigrants who played such a key role in the market gardening industry from the late nineteenth century, despite their declining numbers due to immigration restrictions.[25] Although New Zealand's immigration intake since the mid nineteenth century has been less diverse, the market gardening industry has also had a significant representation of non-English speakers, notably from China and India.[26] In Australia in 1901, Chinese market gardeners comprised 67.5 percent of all market gardeners in New South Wales and 43.5 percent of all market gardeners in Victoria (see Table 4.2). In New Zealand, Chinese market gardeners comprised between 38.3 percent and 52.9 percent of all market gardeners between 1896 and 1926, reaching a peak of 52.9 percent in 1926 (see Table 4.3).

Table 4.2 Market Gardeners in NSW and Victoria, 1891 and 1901

Year	*1891*		*1901*	
State	*NSW*	*VIC*	*NSW*	*VIC*
Total Chinese population	14,156	9,377	11,263	7,749
Number of Chinese market gardeners	3,841	2,095	3,563	2,051
Chinese market gardeners as a percentage of the Chinese workforce	27.0	22.3	31.6	27.0
Chinese market gardeners as a percentage of all market gardeners	70.3	*	67.5	43.5
European market gardeners as a percentage of the European workforce	0.14	*	0.14	0.22

* *Note*: No figure is available for European market gardeners in Victoria in 1891.
Sources: Based on data compiled by Yong from Censuses of NSW and Victoria, 1891 and 1901. C.F. Yong, *The New Gold Mountain*, Richmond S.A., 1977, p. 262, Table 2.

Table 4.3 Market gardeners in New Zealand, 1871–1936

Year	*Total Chinese population*	*Number of Chinese market gardeners*	*Chinese market gardeners as a percentage of the Chinese workforce*	*Chinese market gardeners as a percentage of all market gardeners*	*European market gardeners as a percentage of the European workforce*
1871	2,641	49			
1874	4,816	–	–	–	
1878	4,433	–	–	–	
1881	5,004	400			
1886	4,542	647			
1891	4,444	609	13.7		
1896	3,711	657	17.7	46.9	
1901	2,857	685	24.0	49.8	0.41
1906	2,570	791	30.8	38.3	0.50
1921	3,266	1,252	38.3	46.1	0.49
1926	3,374	1,144	33.9	52.9	
1936	2,943	914	31.0		

Source: Based on data in Sedgwick, 'Politics of survival', Christchurch, 1982, Appendices III, VIII, XVIII.

The transition of Chinese immigrants from gold mining to market gardening had a very similar history in Australia and New Zealand, influenced by the close economic, cultural and social ties between the two countries.[27]

Gardening on the Goldfields

Chinese market gardening had its origins in the gold rushes. Although they initially came in search of gold many Chinese men saw there was a more reliable income to be earned supplying vegetables to the thousands of gold seekers who flocked to the diggings and created temporary camps and fledgling townships. Some took up market gardening as a means of supplementing the income they earned from mining while others went on to become full-time market gardeners.

The discovery of gold in Victoria and New South Wales in 1851, and later in Queensland and Western Australia, transformed the economic fortunes of the Australian colonies and contributed to their political, social and cultural development. In Victoria, for example, the gold rushes sparked the spectacular growth of Melbourne and regional cities such as Ballarat and Bendigo, and the State's population grew from 77,000 to 541,000 in a single decade, from 1851 to 1861.[28] However, as Charles Fahey points out, the heady days of alluvial prospecting when men could make their fortunes from the readily accessible river gravels were short lived and after only a few years large amounts of capital and machinery were required to recover gold from leads deep underground. By the 1860s many miners were wage earners employed by mining companies.[29] Similarly in New Zealand the gold rushes which began in 1861 were already experiencing a slump by 1865. The estimated 18,800 European miners in Otago in early 1864 numbered only around 6,000 by the end of 1865, as they returned to Australia or followed the rush to newly discovered fields on the West Coast.[30] It was not until the 1890s that the dredge developed by Chinese entrepreneur and merchant Choie Sew Hoy (also known as Charles Sew Hoy) with his son Choie Kum Poy sparked a boom in dredging and revived the gold mining industry.[31]

It is difficult to estimate the precise earnings of individual miners, as they were understandably reluctant to advertise the extent of their finds, but it is clear from contemporary accounts that mining was a risky enterprise. In 1851 the *Sydney Morning Herald* reported on conditions at the Ophir diggings in New South Wales: 'Some are doing very well . . . and great numbers scarcely earning their rations. The gold digger's fortune cannot be more suitably expressed than by terming it a lottery.'[32]

The gold rushes created a high demand for fresh food and often a more reliable income could be made from supplying other miners with vegetables than from mining.[33] A recent study of census and economic data from the Californian goldfields has shown that, consistent with the historical literature, miners generally made negative or small positive gains while non-miners were much better off economically than if they had pursued similar occupations in their place of origin. This was true of both American-born and foreign-born miners. Thus merchants and other service providers reaped substantial profits from mining.[34]

Reduced earnings from mining after the heady early days of the gold rushes would have encouraged Chinese miners to diversify into other activities that generated a more reliable income. The high prices of goods and services on the goldfields and the high demand for fresh food, combined with the skills in horticulture which many brought from their homeland, made market gardening an attractive option. In March 1852 the Lieutenant Governor of Victoria, Charles Latrobe, reported: 'Meat (mutton) has always been procurable at a reasonable price, but every other article is purchased at an advance of from 50 to 100 percent upon ordinary prices'.[35] An intrepid young British woman, Ellen Clacy, who visited the Victorian goldfields in 1852–1853, observed:

> It is not only the diggers however, who make money at the Gold Fields. Carters, carpenters, storemen, wheelwrights, butchers, shoemakers, &c. usually in the long run make a fortune quicker than the diggers themselves, and certainly with less hard work or risk of life. They can always get £1 to £2 a day without rations whereas they may dig for weeks and get nothing.[36]

Chinese immigrants were quick to identify this niche market on the goldfields. Recalling the early days of the gold rush at Gabriel's Gully in New Zealand, a correspondent to the *Evening Post* in 1899 wrote that they had Chinese people to thank for supplying vegetables.[37] From 1864 there was an influx of hungry miners to the West Coast, a region with no established farms or market gardens. Chinese gardeners there quickly recognised the opportunity to fill the need for fresh vegetables.[38] In 1877 the *West Coast Times* reported:

> In the market garden business the Chinese as usual take up the field . . . were it not for their efforts the place would at present be supplied with garden stuff only in an intermittent and dear fashion.[39]

Given the uncertainty of returns from mining, the scarcity of fresh vegetables and the high cost of food on the diggings it is not surprising that miners in Australia and New Zealand grew some vegetables to feed themselves.[40] In September 1854 the Melbourne *Argus* reported: 'Gold fields diggers are abandoning the nomadic life, erecting more substantial dwellings and paying more attention to horticulture, which will lead to a more abundant supply of vegetables'.[41] Changes to mining regulations encouraged miners to establish a more settled existence. After 1857 the issue of a miner's right allowed the holder to establish a small residence and garden near their claim and by the early 1860s changes in mining technology were freeing up river flats for gardening.[42] These developments gave miners greater security of tenure and more land on which to grow vegetables for their own consumption and also for sale to others. In 1863 the journalist A. J. Patterson wrote of the changes that had occurred on the Victorian goldfields:

> On almost every field quartz raising and crushing on a large scale have superseded the primitive systems of six or eight years ago . . . licences to occupy land for residence and cultivation have been freely issued by the Government . . . water supply has been provided to some extent, railways and roads have been opened up; the remotest districts are easily accessible . . . to the purveyor of stores and the market gardener.[43]

It is probable that Chinese miners initially grew vegetables to supplement their income from mining and were involved in small-scale commercial activities, mainly supplying their own community. Some moved into full-time market gardening once it became clear there was a sufficient market in the wider community. As Waldinger et al. point out, the initial market for immigrant businesses often arises within the migrant community itself, as their needs and cultural preferences are best known and best served by members of their own group. Food is one of the major products involved. This commercial activity may then serve as a platform for expansion of the business into the open market.[44] Thus ready access to customers beyond the Chinese community was an important factor in the growth of Chinese market gardening on the goldfields.

Writers such as Warwick Frost and Anne Curthoys have suggested that there is little evidence of Chinese market gardening on the Victorian and New South Wales goldfields in the 1850s and that market gardening only became important after the end of the gold rushes in the mid to late 1860s.[45] Certainly in the mid 1850s there is little evidence of Chinese market gardening in Victoria. In 1857 a Select Committee on Chinese Immigration reported:

> The Chinese hitherto have in no one instance applied themselves to the cultivation of the land, nor indeed to any of the industrial pursuits of the colony, save that of digging for gold; their object being to acquire a sufficiency of means whereby to return to their own country.[46]

However, there is evidence for Chinese market gardening on the goldfields by the late 1850s, somewhat earlier than Frost and Curthoys contend.[47] In his report on the Victorian goldfields in 1862, A. J. Patterson notes the variety of occupations the Chinese had taken up in addition to mining. He suggests that for some time Chinese people had been growing both Chinese vegetables for their own community and European vegetables for the wider market:

> ...it is in market-gardening, apart from digging, that they have chiefly made themselves useful in the colony. Near Castlemaine, Sandhurst and Ballarat, they follow this art with great perseverance, and for a length of time they had almost the supply of these markets in their own hands. They cultivate Chinese cabbage, and the usual vegetables familiar to Europeans, and they dig and water, and manure with indomitable perseverance.[48]

So by at least the early 1860s it appears that Chinese market gardeners were well established in business on the Victorian goldfields and had few European competitors. Three years later, in 1865, Patterson reported:

> The class of Chinamen most useful to their European neighbours, next to miners, who are adding to the wealth of the country, are the gardeners. There are few townships in the colony the inhabitants of which are not mainly dependent on the Chinese for the constant and regular supply of fresh vegetables of the very finest quality. Such, at least, may be said to be the case on the gold-fields. Any idea of competing with the Chinese among our gold-fields townships, in the production and sale of vegetables, appears to be pretty generally abandoned.[49]

In 1866 the *Argus* reported on the skill of Chinese gardeners on the goldfields and suggested that their earnings exceeded those of their compatriots who were miners.

> There is a garden kept by five Chinamen near Mr. Patterson's slaughter yard... The ground is about three acres in extent, and the five Chinamen earn each man about £1 per week. This, to Europeans, may appear small, but it is an amount in excess of the average earned by their digging brethren, and more by about 200 percent, than these men could earn in their own country. If they choose they can save money.[50]

The Reverend W. Young ministered to the Chinese community in Victoria for many years. In a report in 1869, he records 813 market gardeners on the goldfields; the largest numbers were in Beechworth and Bendigo.[51] One Chinese market gardener in Victoria who amassed enough savings from growing vegetables to enable him to return to China was Gin Long, who was well known in Beechworth and the surrounding district. In 1867 the *Ovens and Murray Advertiser* reported that he was returning to China for about two years, and commented:

> During his residence here Gin Long amassed no inconsiderable amount of the precious metal, which had attracted him to the colony, not by gold mining but by digging the ground for the growth of vegetables. At several of the Agricultural and Horticultural Societies Exhibitions he was awarded prizes for his skill in vegetable raising and the superiority of the productions he vended, combined with his civility and probity, made him a general favourite with a very numerous class of customers.[52]

In New Zealand Chinese immigrants took up gardening from an early stage. Not all those who arrived from the Victorian diggings travelled directly to the Otago goldfields; some remained in Dunedin and leased land for market gardens. In August 1867, the *Otago Witness*, reporting the arrival of 236 Chinese from Melbourne on the steamers *Rangitoto* and *Otago*, stated:

> ...gradually they are being drafted off to different parts of the gold fields where those who preceded them have settled, and one or two drays daily are being dispatched with their goods and chattels. Not a few of their number seemed disposed to settle about the city, and in different places they have leased portions of ground and are already at work,

trenching, delving and otherwise preparing their sections for the cultivation of garden produce.[53]

The archaeological evidence from the New Zealand goldfields supports the model of immigrant business development proposed by Waldinger et al. This indicates that, like their countrymen on the Australian goldfields, Chinese miners in Central Otago initially grew vegetables for their own consumption in small plots adjacent to their huts. Neville Ritchie terms these 'house gardens'.[54] The camp Ritchie recorded at Cromwell consisted of around 30 huts surrounded by small garden plots, with pig pens, hen coops and fruit trees. A small stone lined spring provided water year round.[55] Thus the miners recreated in the far south of New Zealand a form of subsistence similar in many ways to that which they had left behind in southern China. They established small self-sufficient farms, where the raising of pigs and chickens provided both food and valuable manure for the vegetable plots.

Ritchie identifies two further categories of Chinese gardens in southern New Zealand which are indicative of the development of commercial market gardening. First, large garden areas (settlement gardens) were established adjacent to Chinese settlements, either owned communally or by one relatively affluent person such as a storekeeper, for example in Arrowtown, Cromwell and Lawrence. These settlements served as centres for the small groups of miners scattered through the gullies in the surrounding districts, with gardens, stores, grog shops, gaming rooms and brothels. Second, large market gardens operated on a full-time commercial basis were established in many towns in Otago by the 1880s.[56]

The documentary evidence supports the archaeological evidence that Chinese miners in Central Otago first established small vegetable gardens to feed themselves and their countrymen, and then branched out to supply the wider community. Drawing chiefly on the diaries of Alexander Don, James Ng and Ritchie suggest that this may have occurred at Wetherstones, Macetown, Conroy's Gully and Bald Hill Flat in Central Otago and Round Hill in Southland. They document large market gardens established by the 1870s in Beaumont, Butchers Gully, Clyde, Ophir, Matakanui, Arrowtown and Palmerston. Chinese gardeners and vegetable hawkers were also located in larger towns away from the goldfields including Milton, Gore, Oamaru, Invercargill and Dunedin.[57]

Chinese market gardeners selected crops which found a ready market among Europeans and which were suited to the New Zealand climate. European observers noted with interest that Chinese immigrants had introduced new vegetables from their native land, including the Chinese cabbage, Chinese turnip, bean sprouts, Chinese celery, garlic, parsley and a variety of radish, but as there was little demand for them from European consumers, they were generally grown only for Chinese consumption.[58] In 1865 the *Otago Witness* reported:

> The Chinese are the first market gardeners on the goldfields and they have already introduced several new vegetables previously not seen amongst us – one of them a species of beet known as 'the Chinese cabbage'.[59]

James Ng observes that the two most common Cantonese vegetables grown in New Zealand were Chinese cabbage or *bak toi* [bok choy] and bamboo mustard cabbage or *guy toi*. Other hardy Chinese vegetables cultivated by market gardeners were Chinese sugar peas, spring onions and kohlrabi. Less common vegetables grown in southern New Zealand included white radish, Chinese chives, coriander, box thorn and chrysanthemum greens. Chinese melons and gourds were grown in the North Island.[60]

It appears that the initial favourable reception of Chinese people on the New Zealand goldfields was due in part to their enterprise in supplying much needed fresh produce to the mining community at reasonable prices, and the fact that they were able to occupy an economic niche where they did not compete directly with Europeans. In May 1869 the *Dunstan Times* wrote of the 'immense value' of Chinese gold seekers. They had nowhere unduly competed with Europeans on the labour market, they 'filled in gaps' by taking over abandoned ground and kept up the value of mining property, they reduced the cost of vegetables with their gardens and 'generally acquitted themselves as respected members of society'.[61] The work of Chinese market gardeners in manuring and tilling also improved the quality of the soil for later European land users, a theme which recurs in later Chinese settlement in the North Island. In 1877 the *Tuapeka Times* commented on the survey of quarter acre sections in the township of Roxburgh:

> The soil is of extra good quality, having been occupied by a Chinese gardener for the last three or four years and the finest vegetables of all descriptions have been produced.[62]

After the gold rushes Chinese market gardeners followed the movement of the general population to towns and cities. The following section traces this movement, and the shift of Chinese immigrants into a range of occupations other than mining from the 1870s. Market gardening was one of the major occupations they took up. Thus Chinese immigrants were not only geographically mobile, they were also economically mobile, building on their existing businesses or changing occupations as they sought to increase their wealth.

After the Gold Was Gone

As Keir Reeves points out, the concept of Chinese as sojourners is problematic as it is based on a simplistic model of a single journey to the goldfields and return to China after the gold rushes. Many Chinese remained in their adopted countries, moving into other occupations and forming enduring communities.[63] Reeves' study of McLaren's Flat on the Mount Alexander diggings in Victoria traces the transition of Chinese activity from mining to other occupations, predominantly market gardening. He shows that many Chinese miners stayed on, either by choice or circumstance, developing strong networks with the Chinese community in Melbourne and other regional centres and becoming integrated into the European community.[64] Thus market gardening provided an economic foundation for the transition of Chinese immigrants from sojourners to settlers. C. Y. Choi makes a similar argument in his study of Chinese settlement in Australia, together with more recent researchers in immigrant entrepreneurship in Australia such as Jock Collins and Bruce Missingham.[65]

Historians of the Chinese in Australia have noted the significance of the post-gold rush shift into other occupations.[66] The general Australian population became increasingly urbanised in the late nineteenth century and this process was almost as rapid among the Chinese community, as

Table 4.4 Rural and urban Chinese populations, NSW, Victoria and Queensland, 1861–1901, compared to general population

	1861	*1871*	*1881*	*1891*	*1901*
Sydney	189	336	1,321	3,465	3,474
% of NSW	1.46 [27]	4.65 [27]	12.94 [30]	26.34 [35]	33.98 [37]
Melbourne	581	614	1,094	2,585	2,431
% of VIC	2.34 [23]	3.44 [26]	9.15 [31]	30.45 [41]	38.30 [40]
Brisbane	14	47	115	258	408
% of QLD	2.60 [20]	1.42 [13]	1.02 [14]	3.03 [24]	5.32 [24]

Note: The percentage figures refer to the numbers of Chinese in each capital city as a percentage of the Chinese population in the colony. Figures in square brackets refer to the percentage of the general population residing in the capital city.
Sources: Commonwealth Censuses and Yearbooks. Adapted from Choi, *Chinese Migration and Settlement*, Table 2.3, p. 28; and J.W. McCarty and C.B. Schedvin (eds), *Australian Capital Cities*, Sydney, 1978, Table 2.3, p. 22.

they sought employment opportunities in the cities. Table 4.4 shows the percentages of the Chinese populations of New South Wales, Victoria and Queensland living in capital cities between 1861 and 1901. The increase in the urban Chinese population is particularly marked between 1881 and 1901. By 1901, when the colonies federated to form the Commonwealth of Australia, the percentages of the Chinese populations of Victoria and New South Wales living in the capital cities of each state were approaching the figures for the general population. The urbanisation of the Chinese community continued after Federation. While the overall Chinese population decreased due to immigration restrictions, the proportion of Chinese residing in metropolitan areas increased from 32 percent in 1911 to 41.4 percent in 1933. By 1947 58.9 percent of Chinese lived in cities, a larger proportion than the general population (50.72 percent).[67]

The cities provided employment opportunities for Chinese people and also the security and community support provided by benevolent associations and clubs.[68] Market gardening was an economically and socially viable occupation for urban Chinese, as the demand for fresh food increased dramatically with the growth of urban populations. There remained sizeable Chinese populations dispersed across rural areas, in small groups on the fringes of country towns, or in larger groups of 100 or more in some regional centres.[69] Market gardening was one of the major occupations of Chinese people in rural areas, in addition to store keeping and work on pastoral stations.

Table 4.5 New Zealand: relative populations North and South Island, 1864–1916

Year	*% Population North Island*	*% Population South Island*
1864	37.91	62.09
1867	36.55	63.45
1871	37.78	62.22
1891	44.92	55.08
1906	53.65	46.35
1916	59.22	40.78

Source: New Zealand Government Statistician, *Results of a Census of New Zealand*, Wellington, 1916, p. 45.

Similarly in New Zealand the pattern of Chinese settlement mirrored that of the European population. The gold rushes stimulated rapid population growth in the South Island. Dunedin was the country's leading commercial centre during the 1860s and a major centre for the Chinese community.[70] But from the late 1870s the mining industry was in decline and there was a marked shift of the population from the South Island to the North Island. By the late-nineteenth century Dunedin was eclipsed by the growing cities in the North Island.[71] As Table 4.5 shows, by 1916 almost 60 percent of the total population was in the North Island.[72] Former Chinese miners followed these population movements, seeking business opportunities and markets in the burgeoning towns of the North Island.[73] Many service towns grew up to cater for the booming dairy and sheep rearing industry following the advent of refrigerated shipping in the 1880s, and this rural expansion continued into the first decades of the twentieth century.[74]

As occurred in Australia the Chinese community in New Zealand became increasingly urbanised in the late nineteenth and early twentieth centuries, to a greater extent than the general population. In 1921 69 percent of the Chinese population lived in urban centres, compared to 56 percent of the general population.[75] By 1926 almost 71 percent of the Chinese population lived in cities and they were concentrated in the two major centres, Wellington (34 percent) and Auckland (32 percent).[76] Alexander Don's *Roll of Chinese* charts the movement of Chinese from the South to the North Island and the occupational shift from mining to market gardening.[77] His data clearly shows the emergence of centres of Chinese market gardening in the North Island, notably in Palmerston

North, Wanganui, Otaki, Gisborne and Auckland. In the South Island, Dunedin remained a major centre of Chinese market gardening with smaller clusters in Christchurch, Greymouth, Invercargill, Oamaru and Nelson.

The occupational shift that accompanied the rural-urban shift of Chinese communities in Australia and New Zealand involved a variety of occupations, predominantly store keeping, laundries and market gardening. In Australia, cabinet making was another popular occupation. However, market gardening remained a major occupation in both rural and urban areas. As Table 4.6 shows, market gardening was the major livelihood of Chinese people in Victoria, New South Wales and Queensland in 1891 and 1901, representing about one-third of the total Chinese workforce. The percentage of the Chinese workforce who were market gardeners increased between 1891 and 1901 in both New South Wales and Victoria. Chinese dominated the occupation of market gardening during this period, particularly in New South Wales. As Table 4.2 shows, they represented over 70 percent of all market gardeners in the

Table 4.6 Major occupations of Chinese males in NSW, Victoria and Queensland, 1891 and 1901 (percentage of all Chinese breadwinners)

	1891			*1901*		
OCCUPATIONS	*NSW*	*VIC*	*QLD*	*NSW*	*VIC*	*QLD*
Miners	14.83	27.48	10.45	10.22	21.17	7.76
Market gardeners	29.26	26.51	30.53	35.76	33.02	28.29
Pastoral workers	4.74	1.21	1.76	4.71	.44	.62
Other agriculture	13.84	6.68	17.98	3.54	8.41	21.95
General labourers	5.48	7.35	6.93	5.88	1.45	2.95
Domestic servants	6.03	4.85	8.25	5.95	1.63	8.75
Hotel/boarding house workers	2.13	.85	3.06	2.94	1.97	3.65
Merchants	.43	.38	.28	.72	.90	.34
Greengrocers	2.42	1.17	2.39	6.52	4.12	3.46
Storekeepers	2.77	6.20	6.04	2.91	4.74	7.03
Hawkers	3.07	3.64	1.49	3.65	2.55	3.04
Clerks	1.64	.92	.35	2.27	.82	.45
Cabinet makers	2.64	3.10	.36	6.64	10.12	.68
Laundry	.02	.93	.19	.68	4.41	-
Total no of breadwinners	13,127	7,937	8,399	9,968	6,123	8,468

Source: Based on data compiled by Choi from 1981 and 1901 Censuses of NSW, Victoria and Queensland (Choi, *Chinese Migration and Settlement in Australia*, Sydney, 1975, Table 2.4, pp. 30–31).

State in 1881 and over 67 percent in 1891. In Melbourne in 1905, 44 percent of the 4,600 market gardeners in the city were Chinese.[78] The economic importance of market gardening for the Chinese community was maintained in 1911 when, according to Commonwealth census figures, market gardeners represented just over 29 percent of the total Chinese population in Australia.[79] The 1921 census provides no information on numbers of Chinese market gardeners but Yong suggests that they would have represented at least 20 percent of the total Chinese population in Australia.[80]

A similar pattern can be seen in the occupational structure of the much smaller Chinese community in New Zealand. Table 4.7 shows that the rise of market gardening as the major Chinese occupation in New Zealand occurred later than in Australia, during the first two decades of the twentieth century, reaching almost 44 percent of the Chinese workforce by 1921. The statistics show a marked decrease in mining as an occupation between 1901 and 1921, and a corresponding shift into other occupations, predominantly market gardening. The 1936 census recorded 914 Chinese market gardeners, one-third of the total Chinese workforce.[81]

It is significant that the dominance of Chinese in market gardening was maintained despite the decline of the Chinese populations in Australia and

Table 4.7 Major Chinese occupations as a percentage of the Chinese workforce in New Zealand, 1871–1936

Year	*Mining*	*Market gardening*	*Store keeping*	*Laundry*
1871	84.7	1.2	2.4	-
1874	84.0	2.1	1.8	-
1878	76.7	6.3	2.4	-
1881	77.1	8.0	2.3	-
1886	69.0	14.2	3.1	-
1891	68.0	13.7	3.1	.09
1896	58.5	17.7	5.8	.83
1901	46.0	24.0	7.3	3.0
1921	1.8	43.9	22.9	11.4
1926	0.8	33.9	22.4	10.6
1936	0.8	42.5	25.6	7.3

Source: Adapted from Sedgwick, 'Politics of Survival', Christchurch, 1982 Appendix VIII.

New Zealand in the late nineteenth and early twentieth centuries, as a result of restrictive immigration policies in both countries. There are a number of explanations for this phenomenon. Sociological studies of migrant groups in Australia highlight the tendency of immigrants to congregate in 'niche' occupations where they complement rather than compete with members of the dominant society. In his study of Chinese entrepreneurs in Australia Collins argues that the institutional racism directed towards Chinese immigrants in the form of restrictive immigration policies and discriminatory employment practices, supported by the trade union movement, denied them access to many avenues of work. While some former miners found employment as contract labourers on pastoral properties, many turned to self-employment as market gardeners or store keepers.[82]

In this analysis the attraction of self-employment in small business is as much to provide shelter from racist hostility in the workplace as to provide a greater income. Indeed, Collins argues that the move into small business, particularly market gardening, store keeping and furniture making, was critical to the economic survival of individual families and the Chinese community as a whole.[83] Yong similarly emphasises the barriers which prevented Chinese immigrants from establishing large businesses such as banks or department stores in Australia: restrictive immigration laws, racist attitudes and an unpredictable business environment. He argues that although Chinese merchants had the capital and experience to start new businesses in Australia they preferred to invest in the familiar and less hostile environment of their homeland.[84]

In common with writers on Chinese entrepreneurship in Australia, Charles Sedgwick argues that the stable small businesses which Chinese immigrants in New Zealand established after the gold rushes formed the foundation for the emergence and survival of the community. Their market gardens, fruit and vegetable stores and laundries provided them with a relatively steady income from a predictable clientele, and the possibility of a more settled lifestyle with more regular savings and return visits to China.[85] Certainly market gardening involved establishing a longer-term relationship with the land compared to mining as crops had to be tended daily, all year round, in order to provide a constant return.

Like other migrant groups around the world Chinese immigrants in Australia and New Zealand tend to cluster in the small business sector which has low barriers to entry and a high labour to capital ratio. As

Waldinger et al. argue, barriers to upward mobility are a powerful incentive to enter small business.[86] In her study of the Chinese community in Australia Christine Inglis highlights some of these barriers, in particular the difficulties new arrivals have in finding employment if they do not speak fluent English or their skills are either not recognised or not usable in their new environment.[87]

There were further reasons why market gardening became the major small business enterprise Chinese immigrants entered after the gold rushes. Firstly, Europeans recognised and valued their horticultural skills, although their successes were more often attributed to their frugality and 'plodding industry' than their technological expertise. For example in 1901 the *Brisbane Courier* made this comment about Chinese market gardeners in Charters Towers: 'White gardeners did not have same patient application as Chinese in cultivating the soil in dry weather and as vegetables were a necessity, the Chinese had to return'.[88] Secondly, market gardening represented a very small part of the agricultural sector and was regarded as a labour-intensive, low-status occupation which most Europeans were not keen to enter. When asked why Europeans did not take up vegetable growing even though the soil was fertile, a farmer in Fielding, New Zealand, replied: 'Farmers are too lazy, also the times are too good for them to devote their energies to what they consider misplaced energy'.[89]

The literature on immigrant entrepreneurship emphasises the importance of niche markets within immigrant communities as the seedbed for business enterprises in the wider economy. Chinese hawkers and small fruit and vegetable stalls became an integral part of fruit and vegetable distribution, acting as the link between producers and the public.[90] Large Chinese wholesale fruit and vegetable distribution firms were established in Sydney and Melbourne between 1910 and 1920, and attained a virtual monopoly of the business at that time.[91] Yong contrasts the situation in Australia with that in America, arguing that Chinese market gardening never became a significant occupation in America because labourers had a wider range of other employment opportunities including road and railway building, forest clearing, fruit growing and picking, fishing, cooking and laundry work.[92] However, Chinese people also entered a relatively wide range of occupations in Australia, as Table 4.6 shows. There are more complex economic and social reasons for the dominance of Chinese in market gardening in Australia and New Zealand.

Although market gardening remained the major occupation in the Australian-Chinese community, Choi documents the growing importance of cabinet making, laundry and wholesale fruit and vegetable marketing businesses in Sydney and Melbourne by the early twentieth century.[93] However, in contrast to market gardening where they faced relatively little competition from Europeans, Chinese in the furniture-making industry faced strong competition and there was much agitation against them in New South Wales and Victoria.[94] During the debate in the Victorian Parliament in 1885 on the Factories, Workroom and Shops Bill, which contained measures to counter the dominance of Chinese people in the furniture trade, Mr Winter was one of the few members who spoke in support of the Chinese, highlighting the contribution made by market gardeners. He said: 'there would have often been a disastrous vegetable famine here and in other parts of Australia if it had not been for the Chinese. They should not be handicapped in any way.'[95] Discriminatory legislation was introduced in Victoria, New South Wales and later Western Australia to curtail Chinese business operations other than market gardening.[96] As a result Chinese furniture-making enterprises declined rapidly, while market gardens declined much more slowly.[97]

There is evidence of concerns about Chinese market gardeners competing with Europeans, particularly during periods of economic downturn and high unemployment. However, these anti-Chinese sentiments were often tempered by the belief that any measures taken against them would impact on the supply of fresh produce and that most consumers would continue to buy Chinese-grown vegetables. Measures taken specifically against Chinese market gardeners were not widely taken up and did not extend to restrictive legislation. In New Zealand in 1890, for example, when a meeting of the Trades Council resolved to ask unions to refrain from dealing with Chinese and encourage small farmers to form an association for the supply of vegetables, these initiatives were not widely supported.[98]

In 1895–1896 the Anti-Chinese League in Wellington proposed to establish a European market garden in competition with Chinese which would provide work for the unemployed. A delegation met with the Minister for Lands to request a grant of land and a European farmer, Frank Moore, offered land rent-free. However, it appears that this venture did not proceed.[99] In Christchurch there were movements to boycott Chinese-grown vegetables spearheaded by the Market Gardeners and Fruiterers Association in Christchurch in 1895 and anti-Chinese demonstrations in 1912, but these were not widespread.[100] There were also localised and short-lived anti-Chinese movements in other centres of Chinese market gardening, for example in Ohakune in 1907 and Otaki in 1909.[101] In Auckland in 1926, a meeting of the New Zealand Natives Association was reconvened, largely in response to concerns about the dominance of Chinese and Indian market gardeners.[102]

In Australia in August 1887, a meeting of European market gardeners in the Cheltenham district of Victoria resolved not to supply Chinese hawkers with vegetables but it was reported that this initiative had been only partially carried out.[103] In the same month an anti-Chinese meeting held in Brisbane resolved to boycott Chinese-grown vegetables, but a journalist reporting on the meeting commented that if they were put into effect the town of some 30,000 would overnight be deprived of vegetables.[104]

The decline of Chinese market gardening in New South Wales and Victoria appears to have begun after 1910 and accelerated in the 1930s. Yong attributes this to three factors. First, the rapid decrease in the Chinese population following the Immigration Restriction Act 1901 necessarily contributed to the decline in numbers of market gardeners. Second, southern European migrants from Italy, Greece, Malta and Yugoslavia began to enter market gardening in the 1920s and 1930s, in areas such as Werribee near Melbourne and Blacktown, Fairfield, Cabramatta, Ryde and Nepean near Sydney. They also settled in rural areas during this period and took up market gardening.[105] Finally, as cities expanded, European land owners resumed the land they leased to Chinese market gardeners in order to subdivide for residential and industrial uses.[106] This occurred in Brighton in Melbourne and in Rose Bay, Kingsford and Rockdale in Sydney.[107] In the early twentieth century local government authorities also resumed market gardening land for public recreation reserves and sporting facilities such as golf courses, for example in the Sydney suburbs of Willoughby and Manly, and in South Perth.[108]

In New Zealand Chinese dominated the market gardening industry for longer than in Australia, reaching a peak in the 1960s.[109] In 1957 it was estimated that Chinese market gardeners supplied over 75 percent of the domestic production of green leafy vegetables.[110] This was probably due to the regeneration of the Chinese community that occurred with the arrival of wives and children as war refugees between 1939 and 1947, and the fact that New Zealand did not experience such a large influx of post-war migrants from Europe as Australia. Lily Lee and Ruth Lam's study indicates peak periods for

Chinese market gardening in Otaki, Hawke's Bay and Mangere in the 1960s and 1970s, Gisborne in the 1950s and 1960s and Ohakune between the 1930s and 1970s.[111] In the South Island the major peaks were in Dunedin between the 1950s and 1970s and in Oamaru in the 1960s.[112]

This chapter has traced the journey of Chinese immigrants from the goldfields where they first saw the opportunity to grow vegetables to feed hungry miners, to cities and towns across Australia and New Zealand where they became major suppliers of fresh food to urban populations. They occupied this important economic niche in Australia well into the twentieth century, and for even longer in New Zealand. Geographically mobile and also economically mobile, they followed general population movements and markets, and were adept at identifying niche markets, particularly in the growing cities. The following chapter explores how Chinese market gardeners met the challenges posed by a complex and changing set of social and economic circumstances, including restrictive immigration policies, population growth, urbanisation, changes in land use and changes in retailing.

Notes

1. For example, Darwin Stapleton, *The Transfer of Early Industrial Technologies to America*, Kansas, 1987, p. 16; Edward Beatty, 'Approaches to technology transfer in history and the case of nineteenth century Mexico', *Comparative Technology Transfer and Society*, Vol. 1, No. 2, August 2003, p. 183; and V.W. Ruttan and Yujiro Hiyami, 'Technology transfer and agricultural development', *Technology and Culture*, Vol. 14, No. 2, Part 1, April 1973, p. 121.
2. Beatty, 'Approaches to technology transfer', p. 183.
3. Nathan Rosenberg, *Inside the Black Box: Technology and Economics*, Cambridge, 1982, pp. 191, 193, 245.
4. Roger Waldinger, H. Aldrich, R. Ward and Associates, *Ethnic Entrepreneurs: Immigrant Business in Industrial Societies*, Newbury Park, CA, 1990. The use of the term 'ethnic' in the American literature sets up a dichotomy between Anglo-Americans as the majority culture and people of other ethnicities. The term 'immigrant entrepreneurs' is preferred in this study.
5. Waldinger et al., *Ethnic Entrepreneurs*, p. 48.
6. Michael Williams, 'Destination qiaoxiang', PhD Thesis, University of Hong Kong, Hong Kong, 2002, pp. 16–18.

7. Jock Collins, Katherine Gibson, Caroline Alcorso, Stephen Castles and David Tait, *A Shop Full of Dreams: Ethnic Small Business in Australia*, Leichardt, 1995, p. 58. See also Bruce Missingham, Jacqui Dibden and Chris Cocklin, 'A multicultural countryside? Ethnic communities in rural Australia', *Rural Society*, Vol. 16, No. 2, 2006, n.p.
8. One of the first historians to take this comparative approach was Ragnar Nurske (*Equilibrium and Growth in the World Economy*, Harvard, 1961), quoted in J. W. McCarty, 'Australian capital cities in the nineteenth century', *Australian Economic History Review*, Vol. X, No. 1, March 1970, p. 109. See also: J. W. McCarty, 'Australia as a region of settlement in the nineteenth century', *Australian Economic History Review*, Vol. XIII, 1973, pp. 148–167; Neal Ashkanasy et al., 'The Anglo cluster: legacy of the British empire', *Journal of World Business*, Vol. 37, Issue 1, Spring 2002, n.p.; and C. B. Schedvin, 'Staples and regions of Pax Britannica', *Economic History Review*, Vol. 43, Issue 4, November 1990, pp. 533–539.
9. Ashkanasy et al., 'The Anglo cluster', n.p.; Schedvin, 'Staples and regions', pp. 533–539; and McCarty, 'Australia as a region of settlement', pp. 148–167.
10. McCarty, 'Australian capital cities', p. 109.
11. Geoffrey Blainey's classic study *The Tyranny of Distance: How Distance Shaped Australia's History* (Sydney, 2001), highlights the crucial influence that Australia's isolation from Europe had in shaping the nation's history.
12. Schedvin, 'Staples and regions', p. 536. Economic historians agree that the globalising forces of trade and migration, together with local productivity performance, played key roles in wages movements but they are divided on their assessment of magnitude and the relationships between migration and real wages (David Greasley et al., 'Real wages in Australia and Canada, 1870–1913: globalisation and productivity', *Australian Economic History Review*, Vol. 40, Issue 2, July 2000, pp. 185–186).
13. David Greasley and Les Oxley, 'Refrigeration and distribution: New Zealand prices and real wages 1873–1939', *Australian Economic History Review*, Vol. 45, Issue 1, March 2005, pp. 23–44; and Greasley and Oxley, 'Globalization and real wages in New Zealand 1873–1913', *Explorations in Economic History*, Vol. 41, Issue 1, January 2004, pp. 26–47.
14. Greasley and Oxley, 'Globalization and real wages', p. 35.
15. Schedvin, 'Staples and regions', p. 536.
16. Ibid., p. 534; Mark Thomas, 'A substantial Australian superiority? Anglo-Australian comparisons of consumption and income in the late nineteenth century', *Australian Economic History Review*, Vol. 35, No. 2, 1995, pp. 10–38; and Greasley and Oxley, 'Refrigeration and distribution', pp. 23–44.
17. Discussing the possible explanations for these growth patterns, Schedvin argues that Australia and New Zealand were caught in a 'staple trap', which

hampered attempts to diversify the agricultural economy in the early twentieth century (Schedvin, 'Staples and regions', pp. 553, 557).

18. Schedvin, 'Staples and regions', p. 541.
19. Francis Parker, 'Making periurban farmers on the fringe matter', Proceedings of the State of Australian Cities Conference, Sydney, 2007, p. 260.
20. For example the average value of production per acre for market gardens in New South Wales increased steadily from £26.3s in 1902–1904 to £56 in 1920–1921 (New South Wales Legislative Council, Report on the Agricultural Industry in New South Wales, Sydney, 1923, p. 17). This can be compared to the average value of the total Australian wheat crop in 1906–1907 of £1.12s.9d per acre (George Handley Knibbs, *Official Yearbook of the Commonwealth of Australia,* No. 1, 1908, Melbourne, 1908, p. 306). In 1913–1914 this figure was £2.0s.5d per acre (Knibbs, *Official Year Book of the Commonwealth of Australia*, No. 8, 1915, Melbourne, 1915, p. 317).
21. Richard Broome, *Coburg: Between Two Creeks*, Coburg, 2001, p. 49.
22. Australian Bureau of Statistics, Commonwealth Censuses, 1911 and 1921, 2112.0, http://www.abs.gov.au/websitedbs/censushome.nsf/home/historicaldata?, accessed 3 November 2013.
23. Office of Registrar General, *New Zealand Statistics: Population Census taken for the night of 29th April, 1906*, Wellington, 1907, pp. 362, 365, 383; Census and Statistics Office, *Dominion of New Zealand Population Census 1926*, Vol. IX, Industrial and Occupational Distribution, Wellington, 1930.
24. Economists are working on the problem of modelling the hidden or shadow economy. Aggregated shadow economy data is available for Australia and New Zealand from 1990 to 2012. However, the issue of the unrecorded contribution of migrant groups such as the Chinese to the economies of Australia and New Zealand in the late nineteenth and early twentieth centuries remains an area for further research (Professor Friedrich Schneider, pers. comm.).
25. Ibid.
26. Pamela Wai Shing, 'Locational and structural changes of market gardening in Pukekohe–Bombay–Patumahoe', MA Thesis, University of Auckland, 1977, pp. 22–23; and Sally Miller, 'European immigrant communities and New Zealand: comparative reflections on a little known history', in Olavi Koivukang and Charles Westin (eds.), *Scandinavian and European Migration to Australia and New Zealand*, Proceedings of the conference held in Turku Finland, June 1998, Turku, 1998, pp. 234–235.
27. Keir Reeves, for example, explores the cultural continuities between the experiences of the Chinese on the goldfields of Australia and New Zealand (Keir Reeves, 'Tracking the dragon down under', Chinese cultural

connections in gold rush Australia and Aotearoa New Zealand', *Graduate Journal of Asia-Pacific Studies*, Vol. 3, No. 1, 2005, pp. 49–66).
28. Joanne Monk, 'The diggers in the trenches: a history of market gardens in Victoria 1835–1939', *Australian Garden History*, Vol. 4, No. 1, July/August 1992, p. 3.
29. Charles Fahey, 'Gold and land', in Deborah Gare and David Ritter (eds.), *Making Australian History: Perspectives on the Past*, South Melbourne, 2008, pp. 206–208.
30. James Ng, *Windows on a Chinese Past,* Vol. 1, Dunedin North, 1993, p. 123.
31. This was the first dredge in New Zealand. With its central ladder of buckets and shallow draught it could be used to work river beds, beaches and flats. With modifications it was a prototype for gold and tin dredges around the world (ibid, p. 318).
32. *Sydney Morning Herald*, 6 June 1851, p. 2.
33. Food prices in non-mining areas also rose sharply, as farmers and rural workers deserted their farms for the diggings (Michael Symons, *One Continuous Picnic: A Gastronomic History of Australia*, Melbourne, p. 67).
34. Karen Clay and Randall Jones, 'Migrating to riches? The evidence of the California gold rush', *Journal of Economic History*, Vol. 68, No. 4, 2008, pp. 997–998.
35. Charles Latrobe, Latrobe to Grey, 2 March 1852, 'Further papers relative to the recent discovery of gold in Australia', pp. 169–172, Victorian Parliamentary Papers 1852–1853, LXIV, p. 1607.
36. E. Clacy, *A Lady's Visit to the Gold Diggings of Australia in 1852–1853*, London, 1853, pp. 88–89.
37. *Evening Post*, Wellington, 23 December 1899, p. 3.
38. Julia Bradshaw, *Golden Prospects: Chinese on the West Coast of New Zealand*, Greymouth, 2009, p. 70.
39. *West Coast Times*, 9 February 1877.
40. There is considerable evidence for this on the Central Otago goldfields, documented in Alexander Don's reports of his annual inland tours. See, for example, *The New Zealand Presbyterian*, 1 May 1889, pp. 204–205, and *Chinese Mission Work in New Zealand: Annual Inland Tour 1900–1901*, Dunedin, 1901, pp. 21, 29.
41. *Argus,* Melbourne, 8 September 1854, p. 5.
42. Zvonkica Stanin, 'From Li Chun to Yong Kit: A Market Garden on the Loddon, 1851–1912', *Journal of Australian Colonial History,* Vol. 6, 2004, p. 23.
43. J. A. Patterson, *The Gold Fields of Victoria in 1862*, Melbourne, 1863, pp. 2–3.
44. Waldinger et al., *Ethnic Entrepreneurs*, pp. 21–22.

45. Warwick Frost, 'Migrants and technological transfer: Chinese farming in Australia, 1850–1920', *Australian Economic History Review,* Vol. 42, No. 2, 2002, p. 117; and Ann Curthoys, 'Men of all nations except Chinamen: Europeans and Chinese on the gold-fields of New South Wales', in I. Coalman, A. Cook, A. Reeves (eds.), *Gold: Forgotten Histories and Objects in Australia*, Cambridge, 2001, pp. 115–116.
46. Legislative Council of Victoria, *Report of Select Committee on the Subject of Chinese Immigration*, Votes and Proceedings of Legislative Council of Victoria, 1856–1857, Vol. 2, pp. iii–iv.
47. In 1859 the *Bendigo Advertiser* reported that Chinese cabbage was being grown for sale in the Chinese camp at Spring Gully, near Bendigo (*Bendigo Advertiser*, 26 March 1859, p. 2).
48. Patterson, *Gold Fields of Victoria*, p. 138.
49. *Sydney Morning Herald,* 2 March 1865, p. 2, from the *Melbourne Age*, 24 February, 1865.
50. *Argus,* 14 July 1866, p. 640.
51. W. Young, Report on the Condition of the Chinese Population of Victoria, Victorian Parliamentary Papers, Melbourne, 1869, pp. 5–14. There were also numbers of Chinese engaged in pastoral and agricultural pursuits such as shearing and harvesting, as shown in Table 4.6.
52. *Bruce Herald,* 11 September 1867, p. 3, Intercolonial News Report from *Ovens and Murray Advertiser.*
53. *Otago Witness,* 24 August 1867, p. 13.
54. Neville Ritchie, 'Archaeology and history of the Chinese in southern New Zealand during the nineteenth century: a study of acculturation, adaptation, and change', PhD Thesis, University of Otago, 1986, p. 640.
55. Neville Ritchie, 'Traces of the past: archaeological insights into the New Zealand Chinese experience in southern New Zealand', in Manying Ip (ed.), *Unfolding History, Evolving Identity: The Chinese in New Zealand*, Auckland, 2003, p. 35.
56. Ibid.; Ritchie, 'Archaeology and history of the Chinese', pp. 640–641.
57. Ng, *Windows on a Chinese Past*, Vol. 1, p. 321; and Ritchie, 'Archaeology and history of the Chinese', p. 641. In 1874 for example, the *Tuapeka Times* reported that six or eight Chinese had established a market garden in Lawrence, reportedly working the ground very finely and lavishly working in horse manure (*Tuapeka Times*, 21 October 1874, p. 2). In the same year the newspaper reported that a party of Chinese had started a garden by the town boundary in Wetherstones (*Tuapeka Times,* 26 September 1874, p. 2).
58. Ng, *Windows on a Chinese Past*, Vol. 1, p. 323.
59. *Otago Witness,* 30 September 1865, pp. 13–14.
60. Ng, *Windows on a Chinese Past*, Vol. 2, p. 55, fn. 35b.
61. *Dunstan Times*, 14 May 1869.

62. *Tuapeka Times,* 26 May 1877, p. 3.
63. Reeves, 'Tracking the dragon down under', p. 53; and Keir Reeves and Benjamin Mountford, 'Sojourning and settling: locating Australian history', *Australian Historical Studies*, Vol. 42, No. 1, 2011, p. 118.
64. Keir Reeves, 'Historical neglect of an enduring Chinese community', *Traffic*, Issue 3, July 2003, pp. 53–77.
65. C.Y. Choi, *Chinese Migration and Settlement in Australia,* Sydney, 1975, p. 33; Jock Collins, 'Chinese entrepreneurs: the Chinese diaspora in Australia', *International Journal of Entrepreneurial Behaviour and Research*, Vol. 8, No. 1/2, 2002, p. 115; and Bruce Missingham, Jacqui Dibden and Chris Cocklin, 'A multicultural countryside? Ethnic communities in rural Australia', *Rural Society*, Vol. 16, No. 2, 2006, n.p..
66. Choi, *Chinese Migration and Settlement,* pp. 28, 52; Michael Williams, *Chinese Settlement in New South Wales*, Sydney, 1999, p. 6; and Barry McGowan, 'Chinese market gardens in Southern and Western NSW', *Australian Humanities Review*, Issue 36, 2005, n.p..
67. Choi, *Chinese Migration and Settlement,* p. 52; and Roland Wilson, *1947 Census Bulletin No 1: Population of States and Territories, Urban and Rural Population*, Canberra, 1947, Table 6, p. 7.
68. Choi, *Chinese Migration and Settlement*, p. 33.
69. Paul Jones, *Chinese-Australian Journeys: Records on Travel, Migration and Settlement, 1860–1975*, Canberra, 2005, p. 16.
70. By 1889 there were an estimated 290 Chinese living in Dunedin and suburbs, of whom 110 were gardeners (Niti Pawakapan, 'The Chinese in Dunedin between the 1920s and the 1930s', MA Thesis, University of Otago, Dunedin, 1987, p. 37).
71. The population movement to the North Island was promoted by the move of the capital from Auckland to Wellington in 1876 and the establishment of a central government, replacing the former provincial system (Charles Sedgwick, 'The politics of survival: a social history of the Chinese in New Zealand', PhD Thesis, University of Canterbury, Christchurch, 1982, pp. 6, 73).
72. New Zealand Government Statistician's Office, *Results of a Census of New Zealand*, Wellington, 1916, p. 45.
73. Sedgwick, 'Politics of survival', pp. 73–74, 154.
74. Ibid., p. 74. In 1901 there were only 752 Chinese in the North Island out of a total Chinese population of 2,857. By 1936 the total Chinese population in New Zealand was 2,580, and 2,058 were resident in the North Island (Ng, *Windows on a Chinese Past*, Vol. 3, p. 160).
75. New Zealand Government Statistician's Office, Dominion of New Zealand Population Census, Wellington, 1921, p. 103.
76. New Zealand Government Statistician's Office, Dominion of New Zealand Population Census, 1926, Wellington, 1926, Vol. 4, p. 4 and Vol. 5, p. 3.

77. Alexander Don, *Roll of Chinese, 1883–1929,* reproduced as Vol. 4 of James Ng, *Windows on a Chinese Past,* Dunedin, 1999.
78. Joanne Monk, 'The diggers in the trenches: a history of market gardens in Victoria 1835–1939', *Australian Garden History*, Vol. 4, No. 1, July/ August 1992, p. 4.
79. C.F. Yong, *The New Gold Mountain*, Richmond SA, 1977, Table 3, p. 262.
80. Ibid., p. 36.
81. Ng Bickleen Fong, *The Chinese in New Zealand*, Hong Kong, 1959, p. 84.
82. Collins, 'Chinese entrepreneurs', pp. 114–115.
83. Ibid., p. 117.
84. Yong, *New Gold Mountain*, pp. 53–54.
85. Sedgwick, 'Politics of survival', p. 316.
86. Waldinger et al., *Ethnic Entrepreneurs*, p. 32.
87. Christine Inglis, 'Chinese in Australia', *International Migration Review*, Vol. 6, No. 3, Autumn, 1972, pp. 269–270.
88. *Brisbane Courier,* 25 October 1901, p. 2.
89. *Fielding Star*, 15 January 1908, p. 2.
90. Collins, 'Chinese Entrepreneurs', p. 117.
91. By 1910 four large Chinese firms dominated the banana trade in Sydney and Melbourne: Wing Sang, Wing On, Tiy Sang and Tim Young (Choi, *Chinese Migration and Settlement*, pp. 53, 89).
92. Yong, *New Gold Mountain*, p. 36.
93. Choi, *Chinese Migration and Settlement*, p. 32.
94. Ibid., p. 53.
95. *Argus*, Melbourne, 10 December 1885, p. 4.
96. The Factories and Shops Act, NSW 1896, Victoria 1896; see Choi, *Chinese Migration and Settlement,* p. 53; Williams, *Chinese Settlement,* p. 44. Factory Act, Western Australia, 1904; see Anne Atkinson, 'The socio-economic aspects of the Chinese community in Perth 1900–1920', *Journal of the Royal Western Australian Historical Society*, Vol. 9, No. 2, 1984, p. 97.
97. John Connell and Angela Ip, 'The Chinese in Sydney: from Chinatown to suburbia', *Asian Profile*, Vol. 9, No. 4, August 1981, p. 293.
98. *Wanganui Herald*, 28 July 1890, p. 2.
99. *Evening Post*, 8 August 1895, p. 4; 22 August 1895, p. 4; 13 May 1896, p. 2; 14 May 1896, p. 4; and 11 July 1896, p. 5.
100. *Star,* Christchurch, 16 May 1895, p. 2; and *Ashburton Guardian,* 26 February 1912, p. 2.
101. *Observer,* Auckland, 28 December 1907, p. 15; and *Otago Witness*, 13 January 1909, p. 53.
102. *Evening Post,* 25 January 1926, p. 4.
103. *Argus*, Melbourne, 25 August 1887, p. 5.
104. *Brisbane Courier*, 19 August 1887, p. 10.

105. Missingham et. al., 'A multicultural countryside?', n.p.; Frances Parker, 'The safe use of farm chemicals by market gardeners of non-English speaking background: developing an effective extension strategy for the Sydney basin', Sydney, 2000, pp. 15–16.
106. Yong, *New Gold Mountain*, p. 38.
107. Weston Bate, *A History of Brighton*, Melbourne, 1983, p. 390; and James Jervis, *The History of Woollahra, a Record of Events from 1788 to 1960 and a Centenary of Local Government*, Sydney 1965, pp. 68–69.
108. Ian Rannard, *The Forgotten Gardens: The Story of the Last Market Gardens in Willoughby and Northbridge, NSW*, Douglas Park, 2005, pp. 7, 46; Terry Metherill, 'Faster: Manly in the 1920s', unpublished ms., City of Ryde Library Service, p. 7; and Cecil Florey, *Peninsula City: A Social History of the City of South Perth*, Perth, 1995, p. 246.
109. Lily Lee and Ruth Lam, *Sons of the Soil: Chinese Market Gardeners in New Zealand*, Pukekohe, 2012, p. 508.
110. NZPD Vol. 314, 23 October 1957, p. 2371, quoted in Nigel Murphy, 'A history of the horticultural industry', unpublished ms., 2010, p. 65.
111. Lee and Lam, *Sons of the Soil*, pp. 149, 233, 268, 284, 383.
112. Ibid., pp. 45, 83.

CHAPTER 5

Small Businessmen on the Land

This chapter examines the economic resources and opportunities that market gardeners drew upon to meet the challenges they faced in their adopted countries. As noted in the previous chapter migrants tend to cluster in small businesses with low capital but high labour input. Market gardening required relatively little capital to establish, but the traditional intensive agricultural techniques used by Chinese gardeners meant that the year-round work of preparing the soil, planting, watering, hoeing, manuring, harvesting and preparing produce for sale was labour intensive. Anti-Chinese attitudes, legal restrictions on Chinese immigrants owning land, their limited capital and high mobility meant that most market gardens were leased.

Through their social and business networks which were based on common ties of kinship and dialect and allegiance to village and district in China, Chinese market gardeners developed distinctive forms of labour organisation and sales and marketing strategies which enabled them to earn a livelihood and adapt to complex and changing economic and social conditions. These work practices and ways of doing business had their basis in traditional Chinese culture, for example sharing capital and creating small family enterprises that were shareheld (*gongsi*), and the central concept of *guanxi*, cultivating personalised networks of influence based on mutual obligations and reciprocity. This chapter examines the key economic factors influencing the course of Chinese market gardening in Australia and New Zealand: access to capital, labour, land and markets.

J. Boileau, *Chinese Market Gardening in Australia and New Zealand*,
Palgrave Studies in the History of Science and Technology,
DOI 10.1007/978-3-319-51871-8_5

Capital

Most Chinese immigrants to Australia and New Zealand arrived indebted to relatives, labour agents or local merchants for the cost of their passage and payment of the poll tax. Compensating for their limited personal capital, Chinese market gardeners generally worked in partnerships, several men sharing the lease of a market garden and sharing operating costs and profits. For example in 1913 Ben Hing employed eight men, all as partners, on his 6-acre market garden in Kogarah, Sydney. The profits were divided equally, any partner not working on the garden contributing the equivalent of a weekly wage (around £1.10s) each week, which was added to the total revenue before it was divided.[1]

As most Chinese market gardens were leased, the major capital inputs were rent, seeds, gardening implements, fertiliser and the cost of transporting produce to market. To make most efficient use of their carts many market gardeners in cities such as Sydney and Melbourne carried back loads of manure and straw to their gardens after their early morning trip to deliver their produce to the central markets.[2] Giving evidence to the Royal Commission into food supplies and prices in 1913, European gardener Henry Tasker estimated the annual costs per acre of seed, manure, bone dust and cartage to be approximately £13.5s per acre per year.[3]

The size of partnerships varied, depending on the size of the holding. The men working together generally came from the same clan, village or district in China. In New Zealand the records of Alexander Don in the early twentieth century show clusters of gardens operated by groups of men from the same district. They were often related by kinship or marriage. Similarly, in Australia, there were strong clan associations on many market gardens. For example, families living in villages in the Go Yui (Gao Yau) district in Guangdong regularly sent their sons to work on market gardens in Alexandria in Sydney. A clan association, the Yiu Ming Hung Fook Tong, was established there to protect their interests; the organisation still exists today.[4]

These flexible business arrangements enabled individual gardeners to draw on the economic and social resources provided by networks of kinship and friendship, and share risks. They also accommodated the high mobility of the overseas Chinese community, as when one partner moved elsewhere or returned to China (often for several years) his place could easily be taken by another.[5] In their studies of the Chinese community in New Zealand, Sedgwick and Pawakapan describe how employees on market gardens had to wait their turn for a partnership, until they had accumulated enough capital and found someone retiring or returning to China. Alternatively they could strike out on their own. Self-employment was the goal of most Chinese immigrants, conferring stability and status. In his study of Chinese market gardeners in the Sydney suburb of Willoughby, Ian Rannard highlights their self-reliance and preference for running their own businesses.[6] This is still true of Chinese market gardeners working in Sydney today.[7]

Waldinger et al. note the reliance of migrant entrepreneurs on community networks for raising capital, recruiting labour, and information on business opportunities, local regulations, and suppliers.[8] Studies of migrant small farmers in Australia similarly highlight the importance of chain migration and social and kinship networks in their economic success.[9] Chinese market gardeners typically invested most of their capital into their income producing garden. A list of the debts and assets of a Chinese market gardener in Wellington, New Zealand in the early 1900s, who rented 19 acres of land in the suburb of Miramar at £4 per acre a year, gives a snapshot of his financial situation. Ming Hong's assets lay chiefly in the crops he had in the ground. These were valued around £353. Against this he had debts of £107, including £80 for two horses and two carts and the balance for horse feed.[10]

Market gardeners invested relatively little capital in their own accommodation. Writing about the Chinese market gardens in Dunedin in the early twentieth century, Stedman contrasts the neatly laid out and carefully tended vegetable plots with the ramshackle huts occupied by the gardeners.[11] While many European observers commented on the poor living conditions of Chinese market gardeners, they were making a rational economic decision as to where they should allocate their limited finances. The facts that the majority of Chinese market gardeners leased their land and were single men were additional disincentives to improving their accommodation. Their priority was to earn money to support wives and families in China.

It is difficult to assess the returns individual Chinese made from market gardening, as this data was either not collected in official statistics or aggregated. For example, in Australia before 1901 the statistics on acreages and value of production of market gardens in New South Wales and Victoria were aggregated with orchards and no data is available from Queensland. The income of market gardeners was also highly variable depending on the season and market prices, which depended on supply and demand. However, some data is available from government inquiries. In his evidence to the Royal Commission of Inquiry into food supplies and prices in New South Wales in 1913 the market gardener Ben Hing stated:

> Income from gardens varies. When there is a good crop you make more money. I could not tell you how much per week per man. At times you do not make over 30s a week. With a good season you might make £6 a week.[12]

It is possible to make some broad comparisons between earnings from market gardening and earnings in other occupations. According to some of Pawakapan's informants in New Zealand, in the 1920s market garden workers in Dunedin earned higher incomes from market gardening compared to other occupations, although gardening involved longer hours and hard physical labour. Market gardening in Dunedin was expanding and workers could earn £4 a week plus keep, compared to the average wage of an assistant in a fruit and vegetable shop of £2.10s to £3 a week. It was said that a frugal labourer could save enough after three or four years to lease his own small garden.[13] Data provided by Sedgwick, from a small sample of 84 men who were applying to the government to bring wives from China in 1954, suggest that in New Zealand in the 1950s the incomes of Chinese market gardeners were towards the higher end of the income range of Chinese in a variety of occupations. Their incomes ranged from £516–524 per annum compared to £364–324 for fruit shop owners, £208–520 for shop assistants and £364 for laundrymen.[14]

The limited economic data indicates that not all Chinese market gardeners were poor, barely eking out a living from their small plots of land. While prices, and therefore returns, fluctuated, many were able to make a reasonable living with incomes equivalent to the earnings of European general labourers. This is supported by the evidence given by witnesses to the 1913 Inquiry into food supplies and prices in New South Wales. Henry Tasker commented: 'They [Chinese market gardeners] live sparely but they have their luxuries when they are doing well'.[15]

Chinese immigrants were upwardly mobile, always seeking to better their financial position. Market gardening acted as a springboard for entry into other small businesses such as retail fruit and vegetable stores. It provided gardeners with sufficient capital to expand into other businesses, and some went on to become successful entrepreneurs. One was the Auckland market gardener Ah Chee (Chan Dah Chee). When he first arrived in New Zealand in the 1870s he worked as an itinerant hawker. In 1882 he leased a 7.25 acre market garden in the inner suburb of Parnell, which he operated until 1920. Ah Chee named his garden *Kong Foong Yuen,* 'the garden of prosperity', and it lived up to its name. In the 1890s he established a second market garden in Epsom with three partners, Ah Chong, Ming Ling and Ah Hing. He purchased another 35 acres in Avondale in 1905, and 13 acres in Patumahoe in 1925. His success in market gardening enabled Ah Chee to move into a range of business enterprises including fruit and vegetable retailing, banana and ginger plantations in Fiji, importing Chinese groceries and merchandise as well as exporting dried wood ear fungus and cured rabbit skins to China.[16]

In 2008 an excavation of the site of Ah Chee's market garden in Parnell, where he lived with his family, was carried out. This was the first archaeological investigation of a Chinese market garden site in New Zealand. The artefacts recovered support the recollections of his descendants that Ah Chee and his family had a comfortable, if not lavish, lifestyle. In addition to the Chinese ceramics commonly found in overseas Chinese archaeological sites (celadon rice bowls, porcelain serving bowls and spoons and rice wine jars), the assemblage included fine European teaware – two matching bone china cups and saucers in a Doulton willow pattern dated between 1891 and 1902, a side plate with a gilt banded design, and a Rockingham style teapot. This interpretation is further supported by evidence of a gas supply and a water closet in Ah Chee's house, both luxury conveniences at the time.[17] However, there is not necessarily a direct relationship between high cost artefacts and the social status of the occupants of a site, as the work of Grace Karskens and Jane Lydon in the Rocks in Sydney demonstrates.[18]

The European ceramics embody multiple layers of meaning. When contrasted with the other Chinese domestic ware found on the site, they reflect class divisions within the Chinese community and the characteristic work organisation of larger market gardens, with divisions between partners and managers and ordinary workers. Ah Chee housed many members of his extended family and assisted other Chinese immigrants by providing them with work and lodging on the gardens.[19] The European ceramics found on the site of the Chee family's house counter the general impression of frugality presented by the assemblage. They suggest a middle-class existence and aspirations to upward social mobility.[20] They also illustrate the status and acceptance Ah Chee and his family enjoyed in European social circles and his ability to operate in two worlds. They conjure up notions of gentility, respectability, middle-class manners and social expectations when entertaining Europeans. In March 1894 the *Observer* reported that Lady Glasgow (wife of the Governor, David Boyle, 7th Earl of Glasgow) and her daughters paid a visit to Ah Chee at his home in the gardens:

> On a recent Monday afternoon Lady Glasgow sent a note to her greengrocer (Ah Chee) that she and her daughters would pay him a visit at his home in Mechanics Bay. At the time appointed the ladies duly arrived, and were entertained by Mrs Ah Chee. The ladies Boyle played and sang, partook of afternoon tea, fruit etc, and the whole party (yellow and white) had a good time.[21]

There were other successful market gardeners in New Zealand regional centres such as Wanganui, Hamilton, Gisborne and Oamaru. One was Ngan Fore (Ngan Seung Fore) who operated several market gardens around Wanganui by 1920 and employed up to 60 men. His main garden at Aramoho covered over 40 acres. He could always be counted on to provide lodging and a job to his newly arrived countrymen. In 1917 he built a fine home for his family at Aromoho, complete with a flagpole flying the Chinese flag.[22]

Chinese market gardeners also prospered in regional centres in Australia, for example in Bendigo and Goulburn. Bendigo, which owed its prosperity to the gold rushes, supported several large market gardens of up to 30 acres, employing many people. They endured up to World War II and supplied a large surrounding area with fresh produce.[23] In Goulburn in around 1900, William James Nomchong, who had been a partner in a bank in Braidwood with the well-known Sydney businessman Quong Tart, established a successful market garden on 14 acres bordering the Mulwaree River and recruited numerous workers from China. His son Lionel recalls:

> My father... he made a lot of money from the market garden side of it, he had other people working for him, and he had managers in his market garden, he was doing things in Sydney to get him money and so on, buying property and selling it, and he did the same in Goulburn on his visits to and fro... But the start of him making big money was from market gardening.[24]

LABOUR

Chinese market gardeners brought with them from China distinctive patterns of business management and labour relations which set them apart from Europeans.[25] As traditional Chinese gardening techniques were so labour intensive, market gardens in Australia and New Zealand depended on a reliable supply of labour from China through kin and clan networks. Many writers on the overseas Chinese in Australia and New Zealand note the process of chain migration as sons, nephews, brothers or cousins followed other family members and the fundamental importance of kinship, clan and district ties to their survival in a new land.[26] Williams, for example, emphasises the strong identification of people from the Pearl River delta region by district, village and dialect; it was on this basis that they organised their migration to Australia and the businesses and societies they established in their adopted country.[27] There were also close social, economic and political connections between the Chinese communities in Australia and New Zealand.[28] Sedgwick notes that many Chinese travelling to New Zealand or returning to China stopped in Sydney *en route* and stayed with contacts there from the same village or district.[29]

The pattern of flexible business partnerships characteristic of Chinese market gardens was established when Chinese first migrated to the Australian goldfields. In 1865 the *Melbourne Age* reported on Chinese market gardens in the Castlemaine district of Victoria, commenting on the division of labour within gardens and the apparent cooperation between market gardens to avoid competition:

> There are a great many of those vegetable gardens in various parts of the Castlemaine district, being so distributed that it would seem as if it had been arranged that enough of vegetables should be reared by Chinese in each locality, to supply the whole of each neighbourhood. The number of Chinamen employed in each is generally about from four to six, and all appear to have an equal interest in the business. Each has allotted to him the particular duties which he is to perform. One hawks the vegetables and collects manure, another is employed chiefly in weeding, a third in watering, and a fourth in digging, preparing the plots, and sowing the seeds. These arrangements once made are rigidly adhered to, and are universally carried out in the most amicable spirit.[30]

The working environments of Europeans and Chinese were very different. From the late nineteenth century, European work culture was based on the principle of a fair hourly wage and an eight-hour day while Chinese work culture was based on profit sharing which rewarded long hours of labour.[31] For example, in his evidence to the 1913 inquiry into food supplies and prices, Henry Tasker commented that Chinese market gardeners on the share system worked harder and for much longer hours than Europeans employed on wages.[32]

The artefacts recovered in the few archaeological investigations of Chinese market garden sites in Australia and New Zealand are testimony to the hard manual labour involved in creating and maintaining market gardens. During the excavation of Ah Chee's market garden in Auckland, a variety of agricultural implements were recovered including spades, shovels, hoes, rakes, pitchforks, a slasher, files for sharpening blades and the remains of the metal rim and handle of a wooden bucket. A large number of brown bulbous shaped rice spirit bottles, *ng ka py*, were also excavated. According to the wife of one of Ah Chee's descendants, market gardeners who suffered from rheumatism, the result of long hours of physical labour in all weathers, drank rice spirits to ease their aches and pains.[33] However, there is ample archaeological evidence that the Chinese miners who came to New Zealand during the gold rushes drank copious quantities of rice spirits and other liquor as a social activity.[34]

Workers on market gardens had a frugal lifestyle and generally lived communally. In her archaeological study of the Loddon River in Victoria, Stanin contrasts the simple one-room dwellings of Chinese gardeners, used for communal sleeping and cooking, with those of European settlers where the presence of families and the accumulation of possessions led to expansion and division of spaces.[35] In Sydney in the late nineteenth century, Chinese gardeners in Randwick lived in simple corrugated iron huts divided into two rooms, one for communal sleeping and one for storing carts, tools and liquid manure.[36] Lionel Nomchong recalls the living conditions of the workers on his grandfather's market garden in Goulburn, established around 1910:

> They had accommodation, dormitory rooms in the big sheds . . . you would go down and there would be a hallway with rooms going off each side. There was a big dining room with chairs all around it and at the end of the dining room, further down, was a big kitchen and there was a chopping block there made from the trunk of a tree and then near that was a great big open fire with big bars across it and chains hanging down from inside and there was always a great big kettle of water boiling all the time.[37]

In a community made up predominantly of single men, communal living and working arrangements created a sense of companionship, reinforced by communal food preparation, eating and drinking, and fostered the preservation of Chinese customs and beliefs.[38] Sharing food and accommodation costs also made economic sense.

There was clear segmentation within the occupation of Chinese market gardening and upward occupational mobility resulting in some successful market gardeners going on to become merchants and entrepreneurs. As Jane Lydon observes in her study of the Chinese community in the Rocks in Sydney, the Chinese community was not as cohesive as the word suggests: ‘it embodied class divisions related to wealth and status within the Confucian social order’.[39] Men with sufficient capital would take on a lease and employ their countrymen, creating a distinction between owners or partners and general labourers. Workers on market gardens who accumulated sufficient savings could go on to become partners. The evidence to the 1892 Royal Commission into Alleged Chinese Gambling and Immorality indicates that some Chinese merchants in Sydney took over market gardens and brought out Chinese workers on contract. Pow Chee, for example, an interpreter and school teacher, described how the doctor On Lee engaged 30 Chinese workers to work on his gardens for five years, paying them £1 a week plus food and lodging. He deducted the amount of the poll tax from their wages over the course of their contract.[40]

Labour relations on Chinese market gardens were managed according to traditional kin and clan loyalties and were patriarchal. The data in Alexander Don’s *Roll of Chinese* shows that partners and workers on market gardens in New Zealand tended to come from the same district in China and were clustered in particular cities. For example, market gardens in Palmerston North and Wanganui were operated by men from Poon Yu county and market gardens in Fielding by men from Jung Seng (Zengcheng). Immigration records show that numerous Chinese workers at Ah Chee’s gardens in Parnell, many of whom were his kinsmen, relied on him for housing and employment.[41] Similar conditions existed on other large market gardens in New Zealand and Australia, such as those operated by Ngan Fore in Wanganui and William Nomchong in Goulburn.[42]

Despite their paternalistic living and working environments, workers on market gardens sometimes banded together to assert their rights. In a Wellington newspaper in 1913, William Ah Chee, son of Ah Chee, pointed out that Chinese employers also had their labour troubles. Workers petitioned their employers for higher wages and if their demands were not met they went out on strike; they had done so two or three times in the last year. However, Chee stated they usually settled their troubles before they went very far, 'as the result of conferences between the parties concerned'.[43] He claimed he paid some workers in his gardens as much as £3 and £4 per week, including keep. These higher paid workers would probably have been managers; it would have been in Chee's interests to point out the generous wages he paid his workers. The majority were probably paid considerably less and were also in a position of indebtedness to him which would have limited their bargaining power. As they worked in partnerships, Chinese market gardeners operated largely outside the systems of state arbitration governing wages and working conditions which were introduced in Australia and New Zealand from the late nineteenth century.[44]

The flexible organisation of Chinese market gardening enterprises in Australia and New Zealand helped them to adapt to changing social and economic climates and labour market conditions. The major factor influencing the labour supply on Chinese market gardens, and therefore labour costs, was the operation of restrictive immigration policies in Australia and New Zealand from the late nineteenth century, resulting in sharply declining Chinese populations. As Williams points out, the immigration restrictions introduced by New South Wales in 1888 and the Commonwealth in 1901 increased the wages of Chinese gardeners and cabinet makers due to their improved bargaining position.[45] In his evidence to the 1892 Royal Commission into Alleged Chinese Gambling and Immorality, the market gardener Quin Young commented: 'When I first started business I paid wages of 14s. week. I have to pay more now, because we cannot get men to work, there are not so many Chinese now.'[46] Giving evidence to the 1913 Inquiry into food supplies and prices, the market gardener Ben Hing alluded to the ongoing problem of labour shortages and the resultant pressure on wages:

> I am making less out of the garden than ten years ago, everything is dear, very few Chinese are coming in now, Chinese workers are scarce...You have to keep him and find him in everything.[47]

In Australia concessions to immigration restrictions eased the labour supply problem to some extent.[48] Exemptions were granted for assistants to, or substitutes for, market gardeners to enter Australia under the *Immigration Restriction Act,* enabling market gardens to continue into the 1940s.

Chinese market gardeners faced less direct competition from Europeans than their compatriots in the furniture making industry, as Europeans were less inclined to take on the heavy manual labour and long hours involved in market gardening and governments recognised that they were providing a valuable service to the community.[49] Immigration restrictions were relaxed to admit men engaged in less competitive occupations such as merchants, market gardeners and later, cooks. However, these arrivals did not keep pace with the retirement of ageing Chinese market gardeners and market gardens in urban areas were increasingly overtaken by industrial and residential development. After World War II few Chinese market gardens survived in Sydney and Melbourne. Younger market gardeners and new arrivals from China moved into the expanding restaurant industry, while other newly arrived migrants from Europe and later South-East Asia took up market gardening on the urban fringes.[50]

To meet labour shortages caused by immigration restrictions, market gardeners in Australia and New Zealand turned to local labour and employed Europeans, for example in Mangere, Auckland in the 1930s and in Narrabri, New South Wales in the 1950s.[51] They also employed Aboriginal people in rural New South Wales and Queensland, and Māori in New Zealand.[52] These working relationships are explored further in a later chapter.

In contrast to European market gardeners, Chinese gardeners in New Zealand continued to rely on family labour into the 1970s. One of the reasons they gave was that growing vegetables required care and skill; they felt that hired labourers did not generally take great care in their work and would need constant supervision.[53] This shows an interesting continuity in the care and individual attention which Chinese market gardeners have traditionally given to their crops. The reliance on family labour is still evident on some of the few remaining Chinese market gardens operating in Sydney, such as the 5.5 acre garden in Botany farmed by Gordon Ha. Having taken over from his father Wally Ha who came to Australia in 1952 to work on his uncle's garden in La Perouse, Gordon Ha operates the farm with his brother and cousin and runs a small business specialising in Asian vegetables and herbs.[54]

Land

In common with other settler nations with large agricultural sectors, land was the major source of wealth in Australia and New Zealand.[55] As Powell observes, it was also a prime source of political influence. Processes of acquiring title to land and the direction and pace of the spread of settlement were tightly controlled by governments and politicians through legislation, with varying degrees of success.[56] Geoffrey Bolton emphasises the economic motivation for Britain's initial settlement of New South Wales and the important role which private land companies played in the economic development of both Australia and New Zealand in the early nineteenth century.[57]

However, Chinese immigrants were not part of the European agrarian ideal. Until well into the twentieth century they were restricted from owning land by a complex web of institutional racism and were relegated to working as labourers on European pastoral properties or as leaseholders cultivating small plots of land owned by Europeans. In New Zealand the only period when Chinese were not legally able to buy land was during World War II, between 1942 and 1945.[58] However, until this time the majority of Chinese market gardeners leased their plots. Most were single men, mobile and lacked the capital to invest in land. In addition, many European land owners may have been reluctant to sell their land to Chinese, although they would consider leasing it for short periods. More Chinese market gardeners began to purchase land after World War II. By this time many had accumulated enough capital for a deposit, having supplied vegetables for the armed forces during the war years, and more financial assistance was available from produce marketing firms and the State Advances Corporation (the forerunner of Housing New Zealand). Many growers had also been able to bring their wives and children from China, encouraging them to settle down on their own land.[59] Surveying Chinese market gardeners in Auckland in the 1970s, Lee found that growers aspired to owning their own land. They saw farming leasehold land as a stepping stone to buying land once they had enough capital.[60]

Similar factors in Australia restricted the ability of Chinese market gardeners to purchase land. Chinese people were unable to own land unless they were naturalised Australian citizens and only some 5,000 were naturalised under the laws of the Australian colonies prior to 1901.[61] After Federation legislative restrictions varied from state to state. In 1912, Queensland enacted legislation allowing 'aliens' to acquire leases of less than 5 acres for up to 21 years; those leasing over 5 acres were required to pass a dictation test in a prescribed language. In New South Wales legislation enacted in 1913 and 1916 prevented aliens from acquiring certain land holdings direct from the Crown. In South Australia after 1914, Asians were not permitted to lease Crown land in irrigation areas.[62]

Sizes of market gardens were highly variable according to their location and the number of partners involved. In New Zealand few figures are available for the sizes of gardens on the goldfields, but Ng suggests they were initially only small, becoming larger as the populations of the gold mining settlements grew.[63] By the early 1900s, when the numbers of Chinese involved in market gardening exceeded those in mining, extensive gardens employing large groups of men were established in North Island population centres such as Otaki and Wanganui.[64] Chinese market gardens in Otaki ranged from 10 to 30 acres, and there was a 41-acre garden outside Wanganui operated by Ngan Fore.[65] By 1930 there was a 50-acre garden outside Fielding, trading under the name Yee Lee and Co.[66]

Market gardens in inner city areas tended to be smaller than those on the outskirts. In Dunedin in the early twentieth century, most Chinese market gardens were between 3 and 8 acres.[67] In Auckland, early Chinese market gardens between the 1870s and 1890s ranged from 3 to 7 acres in inner city areas such as Ponsonby, Parnell and Arch Hill to between 26 and 41 acres in semi-rural areas on the outskirts such as Mangere, Avondale and Epsom.[68] By the 1930s there were large gardens of 127, 95, 70 and 50 acres in Panmure (the average was around 30 acres) and 70 and 64 acres in Mangere.[69]

The acreages of market gardens in Australia were also variable. Market gardens on the Victorian goldfields in the 1860s ranged from 1 to 6 acres, with larger gardens on the outskirts of larger towns such as Bendigo. The limited quantitative data available suggest a similar pattern to New Zealand – Chinese market gardens established in inner city areas by the late nineteenth century tended to be smaller than those on the outskirts of cities or country towns. In Sydney in 1900, there were 85 Chinese market gardens, averaging 3.9 acres.[70] In the same year, gardens in the Melbourne suburb of Brighton averaged 9.7 acres. In Brisbane in 1889, market gardens ranged from 1 to 10 acres.[71] In rural areas between 1902 and 1928, there were large gardens of 30 acres in Warwick and 40 acres in Samford (both in Queensland), and gardens of 20 to 30 acres outside Bendigo in Victoria.[72]

As most Chinese market gardeners leased land, their wealth lay in the improvements they made to it and the crops they produced. But it was often European land owners who benefited from the labour of Chinese market gardeners, who transformed uncleared, 'unproductive' land into fertile acres suitable for agriculture or pasture.[73] The environmental impact which market gardening had on the land has been noted in earlier chapters. In New Zealand, the Chinese market gardeners who arrived in the Ohakune district in the central North Island from the early 1900s took up leases on land that had been logged for the booming timber industry and grew vegetables in the fertile volcanic soil among the stumps. With limited tools and the aid of a horse, if they could afford one, they cleared the bush, burnt it off and removed the stumps. As they cleared they gradually increased the area under crops.[74]

European landowners usually leased their land rent-free to Chinese gardeners for two to five years, on the condition that they returned the land in pasture, which could then be stocked with sheep and cattle. The gardeners moved on to another block of uncleared land and repeated the process. Chinese gardeners played a major role in developing the market garden industry in Ohakune, which remains an important market gardening centre, particularly for root crops.[75] Unlike most European farmers, who regarded market gardening as a supplement to mixed farming and used haphazard methods, Chinese farmers derived their sole income from what they produced and applied the traditional intensive cultivation techniques of their homeland.[76] Similarly, the Chinese gardeners who leased land from Europeans in Dunedin in the early twentieth century improved the fertility and productivity of the land and increased its value. Their leases were generally only word of mouth agreements and could be terminated at any time.[77]

An item published in the Sydney Chinese language newspaper the *Tung Wah News* in July 1899 providing advice to market gardeners gives an insight into concerns about the terms of their leases and security of tenure. Gardeners were advised that any lease they signed should include the size of the land, who paid the rates, the amount of rent and should not make the tenant responsible for repairs. It also suggested that the added value Chinese gardeners gave to uncultivated land could be used as a bargaining tool to negotiate a period rent-free.[78]

Productive land for vegetable growing could be worth ten times as much as unimproved land. Giving evidence to the New South Wales inquiry into food supplies and prices in 1913, Thomas Elliot, a market gardener in Bowral, observed that improved and drained land ideal for growing cabbages was worth £50 an acre compared to unimproved land which was only worth £4 or £5 an acre.[79] In New Zealand in the early twentieth century, uncleared land on the outskirts of the growing towns of the North Island could be leased for 7 shillings an acre in the first year, increasing to £1 to £2 in the second year and perhaps £4 or £5 in the third year as it was progressively improved.[80] Chinese market gardeners themselves could also benefit from the increased capital value of market garden land which they had developed through their own labour. In the 1920s a Dunedin market gardener returning to China for an extended visit could sell his share in the lease of an established 5-acre garden that had been worked for a number of years to a countryman for £200 to £300, or £40 to £60 an acre.[81]

Land values of market gardens were also determined by their location in relation to markets and population centres and their relative value for competing land uses. In 1875 Johann Heinrich Von Thunen developed a general theory of urban growth and a rationale for the location of market gardens on the fringes of urban centres, arguing that the arrangement of land uses around a city is based on the notion of economic rent and competition for land. As urban land uses realise higher rents, rural land uses are progressively displaced by residential, commercial and industrial uses.[82] Improved communications and refrigerated storage and transport have extended the area from which cities can draw supplies of fruit and vegetables, leading some geographers to question the relevance of von Thunen's model.[83] However, as Derek Smith argues, it is still relevant in Australia where centres of consumption are widely dispersed.[84]

Market gardeners had to adapt to changes in land use over time, particularly in cities. Faced with rising land values and corresponding increases in rents and rates as urban populations grew and suburbs expanded into former rural areas, many had to move their operations to the urban fringes where land values were lower or take up another occupation altogether. By the early twentieth century market gardens which had been established in the inner suburbs of cities in Australia and New Zealand from the 1870s and 1880s were being encroached upon by residential and industrial development. In the era before urban planning there were tensions between the competing demands of industrial and residential uses, and a social geography of class and ethnicity emerged as working-class and 'ethnic' enclaves developed.[85] Growing cities also needed space for waste disposal, recreation and transport networks. In Melbourne market gardens began to disappear from inner suburbs after World War I. In 1919 the last remaining bastion of Chinese market gardening in Hawthorn, the 40-acre Urquart Estate, which had been cultivated since the 1880s, was sold and subdivided for housing.[86] In the Sydney suburbs of Botany, Alexandria and Willoughby Chinese market gardening was declining by the 1920s as former agricultural land was subdivided for housing estates.[87] The decrease in the number of Chinese gardens in Sydney's inner suburbs due to rising land rents was noted in the *Tung Wah Times* in 1928.[88]

In New Zealand, as Dunedin expanded Chinese market gardeners moved from suburbs close to the city centre to outer areas such as the Kaikorai valley and the Taieri plain where land was cheaper and they could afford larger plots. As land became more valuable for housing and industry than food production, European landlords first increased rents, then as suburban sprawl gathered pace, terminated leases and subdivided their land for housing.[89] A retired market gardener interviewed by Pawakapan described how his father began with a small garden in Forbury in 1920, then moved to the Kaikorai valley. After he took over from his father he moved to North Taieri, Outram, then finally to Momona. Each time he expanded the size of his holding as he saved enough money to lease more land.[90] Similarly in Auckland, market gardeners in Mangere, Avondale and Mount Wellington moved to areas on the outskirts of the city such as Pukekohe, Bombay and Kumeu as inner city land values rose (refer to Figure 5).[91]

As an alternative to moving their operations Chinese market gardeners could intensify them, perhaps on a smaller holding, by increasing inputs of capital to compete with alternative urban land uses. In the Auckland suburb of Mangere from the 1960s, some Chinese market gardeners (including the Luen brothers Percy, Harry and George; Jack Chong, Sam Lee, David Yee, Poy Lee and Sherman Lowe) established glasshouses on small holdings of 1 to 2 acres. Chinese growers still produce tomatoes under glass in the area today, with cucumbers as a rotation crop. Glasshouse growing also gained popularity among gardeners in Avondale, Otahuhu and Onehunga in the post-war period. By using automated irrigation and heating equipment they achieved higher yields per acre and lower labour costs.[92]

Markets and Prices

In contrast to the high population densities of southern China in the nineteenth century the populations of Australia and New Zealand were small and scattered over large distances, particularly in Australia. As vegetables were perishable, market gardens needed to be located close to markets so that produce could be harvested and sold as quickly as possible, ideally the same day. Locating close to markets also saved freight costs and time lost tilling the soil.[93] Thus from the late nineteenth century Chinese market gardeners established market gardens in cities to serve the rapidly growing urban populations.

Apart from a period during World War II when the Australian and New Zealand governments introduced controls over the prices and supply of vegetables, as part of a program of centralising control of their economies, vegetables were sold under a free market.[94] Prices fluctuated widely. In 1904, for example, a report in the *Agricultural Gazette of New South Wales* noted that prices for vegetables could rise or fall as much as 300 percent due to sudden gluts and shortages.[95] Thus the major challenges for European and Chinese market gardeners alike were to manage gluts and shortages in specific crops and fluctuating prices, and try to ensure an even supply of produce year round.

Chinese market gardeners applied their traditional horticultural skills, such as the use of hot beds and transplanting, to hasten the growth of young seedlings and extend their growing season. Their aim was to produce a crop, and thus a monetary return, as quickly as possible.[96] They typically cultivated a wide variety of vegetable crops with different growing seasons, whereas European market gardeners tended to concentrate on larger scale cultivation of one or two crops, for example potatoes and onions.[97] This helped ensure an even supply of produce year round. Although Chinese market gardeners focused almost exclusively on edible crops, in some cases they grew flowers as a profitable sideline, for example in the Sydney suburb of Willoughby from the 1920s and 1930s. Ian Rannard recalls the gardens along Sailors Bay Creek up until the 1950s, with their orderly terraces of vegetables in shades of green, offset by the warm pinks, reds, yellows and oranges of poppies, dahlias and carnations.[98]

In Australia, keen-eyed Europeans often commented on the orderliness of Chinese market gardens and the variety of crops at different stages of growth. As early as 1865 the *Melbourne Age* commented:

> In no respect are the wisdom, foresight, and accurate calculation of the Chinese gardener more wonderfully displayed than in his arrangement of his crops, so that he has always got some description ready for sale, and yet never has a glut, or is compelled to sell immature vegetables, or allow them to become overgrown. He sows and reaps something every day all the year round, with him it is always seed time and harvest. In the garden are to be

> seen vegetables in all stages of development, and as soon as one plant is taken from the ground new seed is put in its place. The soil is never allowed to remain a day idle, but it is too liberally treated ever to get exhausted.[99]

However, Chinese market gardeners were not always successful in managing gluts and shortages, particularly in regional centres with smaller markets. Ernest Sue Fong recalls the Chinese market gardens in Inverell, New South Wales in the 1930s, where there were two large gardens, one employing 28 men, the other 20 men.

> There was Sam Sing and Hop Sing... I used to go down and take a lot of fertiliser down to them... The trouble was when they got a good crop, everybody else got a good crop and they'd get a glut and they'd get no money out of it... It just got tougher and tougher and they dwindled away.[100]

Even in Otaki, close to the large Wellington market, Chinese market gardeners were sometimes hit by a glut of cabbages, beans or other seasonal vegetables, flooding the market. In 1909 a Wellington newspaper reported:

> The Chinese have found that with present prices for their produce it does not pay to send them from Otaki to Wellington. Numbers of them are ploughing the rotting vegetables into the ground.[101]

Through intensive cultivation, frequent watering and manuring Chinese market gardeners were able to produce more to the acre. This helped them to keep their prices relatively low and still make a profit. Europeans often commented on the astonishing amounts Chinese gardeners could produce from a small area of land. Writing in the 1930s Depression, Alex Bennett advised readers on how to make a living from an acre and suggested they visit a Chinese garden as an example of how to get the utmost out of a small plot of land. He advised careful cultivation, saying that the aim was to get the highest possible return from a given area in the shortest possible time. Drawing on European kitchen gardening traditions and also referring to traditional Chinese agricultural methods, he promoted intensive hand cultivation, interplanting, thorough manuring, weeding and watering and giving the growing plants individual attention. His motto was 'an acre means no idle hours'.[102]

The economic returns from market gardening varied greatly according to market conditions, the vagaries of the weather, the location of the garden and the number of men working it. In his history of Brighton, Bate observes that market prices vitally affected gardeners. He notes that the shift to vegetable production in the area after 1854 coincided with the gold rushes and can be explained by rising demand and rising prices. He argues that even though vegetables were an unstable commodity they were no worse a business proposition than other commodities, citing evidence for similar or greater fluctuations in the prices of wheat, bread and dairy produce between 1849 and 1854.[103] The Chinese market gardeners who settled in Brighton in the 1870s and 1880s faced relatively little opposition from European market gardeners until the Depression when they showed how well they were insulated against low prices. Bate comments that they could even make a profit out of incidental crops like parsley, garlic and salad vegetables which European gardeners found uneconomic.[104] C. M. Yuan points out that Chinese could not access social services during the 1930s Depression and many lived in poverty, but within their community, market gardeners remained relatively prosperous, providing consumers with fresh produce.[105] Some gardeners even gave work to unemployed Europeans, for example in Mangere in Auckland where they employed them to clear rocks and create stone walls and gave them bread and a bowl of soup for lunch.[106]

In an address to the Birkenhead and Northcote Fruitgrowers Association in Auckland in 1899, Allan Wilson noted the surprisingly large returns Chinese market gardeners gained from some lines of vegetables, for example a good crop of lettuce could yield £300 per acre and cabbages £50 to £100 an acre. However, he warned that the highest prices were rarely obtained, and much depended on catching the market at the best time. Often prices fell so low it did not pay to market the produce, so it was well to watch the market closely and try to always have a variety of crops ready for sale at a reasonable price.[107] An indication of the keen interest which Chinese market gardeners in Sydney had in the market prices for their produce is the regular publication of vegetable prices in the Chinese language *Tung Wah News*.[108]

During droughts the irrigation skills of Chinese market gardeners often enabled them to produce vegetables where Europeans' crops either failed or they did not attempt to grow them. For example, in Botany in Sydney, an area of sand dunes and swamps with plentiful water, Chinese gardeners could reap the benefits of high prices at times when other areas were stricken by drought and vegetables were scarce. One local historian commented 'when others pray for rain, these market gardeners pray no less fervently for a continuation of dry weather'.[109]

In May 1902, during a drought, an officer of the Department of Agriculture visited the Chinese market gardens around Brisbane, citing them as examples of the efficacy of irrigation. In his report he commented 'in fact, the longer the dry weather continues the better for the Chinese gardeners, as their products bring a higher price'. He estimated the annual value of vegetables produced by the gardeners to be between £150 and £200 per acre.[110] Over 20 years later the *Brisbane Courier* reported in a similar vein on the successes of Chinese gardeners in Samford, west of Brisbane, who obtained good prices for their produce during the drought of 1927, thanks to their extensive irrigation system.[111]

The evidence for the cost and availability of vegetables and fruit in the late nineteenth and early twentieth centuries in Australia is contradictory. Some contemporary sources point to their cheapness and ready availability, while others paint a picture of scarcity and high cost. This probably indicates unreliable supply with alternating gluts and shortages, and uneven distribution. As Andrea Gaynor points out in her study of gardening in Australian cities the availability of cheap vegetables from Chinese market gardeners made urban residents less inclined to grow their own in their backyards. Some suburbs were well served by Chinese market gardeners, while others were not.[112] Enid Ross, who grew up in Newcastle in the early 1900s, recalled:

> We didn't make gardens in those days because the fruit and vegetables were so cheap it wasn't worthwhile. We did have fruit trees in the yard . . . and that was all we had growing. Parsley perhaps, and mint is something that's needed for the addition to food, but tomatoes were only four pence a dozen and the Chinaman with his baskets would come . . . and he would have quite a variety of vegetables, cabbage – eight pence, perhaps cheaper sometimes, cauliflower and the like, and celery, all freshly picked out of his garden and a soup bunch made up . . . Well, it wouldn't be worth making a garden for that.[113]

The returns of Chinese market gardeners were also affected by the economic downturns experienced in Australia and New Zealand in the early 1890s and again in the Depression of the 1930s. During times of economic hardship anti-Chinese feeling and fear of competition increased and from the late 1880s there were renewed calls from trade unions for an end to Chinese immigration.[114] More people turned to growing their own vegetables and were less willing to pay the prices the gardeners charged. For example, in 1894 in Milparinka local residents complained about the high prices charged by Chinese market gardeners when wages had been reduced all round.[115] During the 1930s Depression, like the rest of the community, Chinese market gardeners struggled to make a living and the hard-won savings they had accumulated to send to their families in China or make a return trip to see them dwindled. However, as discussed in a later chapter, they were insulated to some extent by the fact that they could grow their own food, and for those in urban areas, by their support networks in the Chinese community. Not eligible for government relief work or food handouts, they were forced to depend on their kin and countrymen and create their own form of social security. In some cases they provided employment to others, including Aboriginal and Māori people, and they extended credit to their regular customers. George Lee Kim, whose father had a market garden in Bendigo, recalls:

> When the Depression hit it hit everyone hard…He [my father] still delivered vegetables to the households. He didn't worry so much if he got paid, he just went on the same old way, if they owed they owed, if they paid that was good. Life went on; he never actually refused credit to anyone.[116]

In the Northern Territory, Eddie Quong recalls that during the Depression his parents held a soup kitchen at their market garden outside Pine Creek every Sunday.

> We had to grow our own food during the Depression. We used to have people coming to our garden two and a half miles out of town, wealthy people in town who became friends of ours and they came to our soup kitchen every Sunday. My mother would have a soup kitchen with all the vegetables and big stews, and the baker used to bring some bread and everybody would have one big hot meal a day.[117]

In his study of the Chinese community in Dunedin, Pawakapan argues that although Chinese gardeners were self-employed, they were exploited by the demands of consumers for cheap vegetables.[118] But there is evidence of Chinese market gardeners and vegetable hawkers exercising their economic power, even though they were hedged in by the prejudices of the wider community and discriminatory legislation. The fears and prejudices of consumers influenced their purchasing choices. In 1881 there was a smallpox scare in New South Wales, and Chinese market gardeners and vegetable hawkers were the scapegoats – householders refused to buy their produce for fear of contamination and they had to bear the taunts and the sticks and stones of larrikins. In Newcastle the Chinese community responded by banding together and refusing to sell vegetables direct to householders for one month. They sold their produce to European greengrocers instead, forcing householders to pay double the price for their vegetables. The *Sydney Morning Herald* noted that Newcastle householders were keen to welcome back the Chinese hawkers, commenting 'the pocket is a very sensitive object when touched'.[119]

There are other examples of Chinese market gardeners taking collective action. In 1890 Chinese gardeners in Mackay, Queensland, reacted promptly to the introduction of a local bylaw imposing a special tax on vegetable hawkers. They refused to do business, cutting off supplies of fresh vegetables and returning the townspeople to a monotonous diet of tea, damper and mutton or beef. The bylaw was soon repealed.[120] In New Zealand in September 1907, Chinese market gardeners in Wanganui met and decided to increase the prices of their vegetables by 50 percent. It was later reported that Chinese market gardeners in Dunedin wanted to follow their example but 'could not trust each other to hold to the terms of the proposed bond'.[121] Yet in 1908 a Wellington reporter commented on the relatively uniform prices charged by Chinese fruit and vegetable retailers as evidence of 'a subtle understanding among them'.[122] These comments reflect the prevailing prejudices against Chinese people.

Marketing Strategies

Through their social and business networks within the Chinese community, market gardeners had access to a variety of avenues for distributing and marketing their produce. These flexible marketing strategies enabled them to adapt to changes in the retail sector. A characteristic feature of the Chinese immigrant economy was the integration of the various occupations involved in food supply: market gardening; hawking produce door-to-door; fruit and vegetable stores; and wholesale fruit and vegetable marketing.[123] As Sedgwick notes, in New Zealand in the late nineteenth and early twentieth centuries the Chinese community was linked occupationally, with growers, hawkers and retailers all handling the same commodities and controlling distribution.[124] Similarly, in Australia, many merchants acquired their capital by working in market gardens or as vegetable dealers at the markets before becoming partners in a store.[125] This integration minimised the length of the supply chain and meant that Chinese market gardeners were able to provide superior quality and freshness as well as reliable supplies. After World War II when the tastes of Australians and New Zealanders became more cosmopolitan, the increasing popularity of Chinese restaurants created a further outlet for Chinese market gardens and slowed their decline.[126]

It is difficult to establish what proportion of Chinese market gardens had direct connections to Chinese retail outlets. In her study of Chinese market gardening in Perth between 1900 and 1920, Anne Atkinson finds only limited evidence for direct connections between Chinese greengrocers and market gardens; most sales were through hawking or wholesale produce markets.[127] However, there are examples of combined market garden and retail enterprises in Queensland and New South Wales, such as the garden and store operated by Willie Mar and his son (also named Willie Mar) in Winton from 1927 to 2000. The Sydney based firm Tong Sing and Co., operated two market gardens on the Gwydir River outside Moree from the 1890s to the 1950s.[128] Opening a greengrocery store was a natural avenue of business expansion for successful market gardening enterprises, particularly where market gardeners were able to bring wives and children from China to help in the business. The experience of Chung Kwong is typical of many Chinese immigrants to New Zealand. In the 1920s he ran a garden and fruit shop in the North Island town of Otaki with the assistance of his wife. After returning to China for three years so the children could receive a Chinese education, the family returned to New Zealand in 1932 and moved to Wairoa where they again ran a fruit shop and market garden.[129] Lee and Lam record many other successful businesses combining market gardens and fruit and vegetable stores, particularly in smaller provincial towns.[130]

Chinese hawkers working on their own account sold the produce of European and Chinese market gardeners door-to-door. In 1887 a regional Victorian newspaper reported that European market gardeners in Melbourne were jealous of the Chinese hawkers who supplied suburban householders with such regularity, and also the Chinese market gardeners who sold directly to consumers.[131] Gardeners who sold their produce directly to householders rather than through city markets obtained better returns. Where there were sufficient numbers in a market gardening partnership work roles were divided between the partners. Generally it was men with English language skills who dealt with customers.[132]

In later years, Chinese market gardeners in cities and country towns sold their wares door-to-door to householders, for example in Dunedin in the early 1900s.[133] The Golden Threads Project in New South Wales records local residents' memories of Chinese market gardeners selling their produce from horses and carts, hand carts or baskets carried on a shoulder pole in many country towns, including Eugowra, Deniliquin, Dubbo, Gilgandra, Inverell, Parkes and Wagga.[134] Direct marketing enabled market gardeners to keep their prices low and still realise a profit, while consumers benefited by having a regular supply of fresh vegetables brought to their door. As a rural Victorian newspaper commented in 1887:

> For years residents in the suburbs of Melbourne have depended for their supply of vegetables on the Chinese hawkers, who wait on their customers with unerring regularity. The suburban resident has no need to seek for the article. There are regular days of the week on which every variety of the choicest vegetable is brought to his door by the industrious Chinaman.[135]

With the establishment of central markets in cities such as Sydney, Melbourne, Perth, Wellington and Auckland by the early twentieth century, the auction system of selling produce gained popularity.[136] It is difficult to determine how many Chinese market gardeners sold their produce directly to consumers and how many sold via the auction system, due to the lack of quantitative data. Probably many continued to do both. With the auction system came middlemen, both Chinese and European. In Sydney and Melbourne, for example, Chinese men who had English language skills worked as vegetable dealers at the city markets and gardeners with limited English would sell their entire loads to them.[137] Thus by the late nineteenth century market gardeners had a number of avenues for selling their produce. Smaller enterprises sold direct to customers through hawking and door-to-door sales, at city produce markets or to Chinese or European greengrocers. In Dunedin in the 1920s and 1930s, for example, gardeners took their crops to the produce markets on Tuesdays and Thursdays and delivered the remainder directly to retailers.[138]

Larger enterprises operated by Chinese merchants produced sufficient volume to enable them to supply organisations such as clubs, hotels, shipping companies and hospitals under contract. For example, in

Sydney in the 1890s Quin Young had contracts to supply the Grosvenor, Australia, Royal and Metropolitan hotels and clubs such as the Reform Club, the Union Club and the Australian Club. Young employed hawkers to sell direct to customers, owned two horses and carts and also sold produce though his shop in Castlereagh Street.[139] In 1913 the entrepreneur Ben Hing was a major partner in a large garden in Kogarah worked by eight men. He paid his partners wholesale market prices then sold the produce, mainly on order, to hotels, shipping companies and Sydney Hospital. He also sold produce though his shop in central Sydney.[140] In Dunedin, historian James Ng suggests that the volume of vegetables produced by the large number of Chinese market gardeners in the 1880s and 1890s would have been well in excess of the needs of the Dunedin population, indicating that they also supplied the ships that visited the port for repairs and reprovisioning.[141]

Market gardening enterprises were flexible enough to adapt to changes in retailing, particularly in cities. Increased regulation by local governments of matters such as public health and traffic flow in the early twentieth century brought about a shift away from hawking and door-to-door sales towards retail stores and the auction system. In Sydney, for example, the number of suburban fruit shops increased significantly between 1900 and 1930 as a result of regulations limiting the operation of street hawkers and barrows.[142] However, there is evidence that Chinese market gardeners, their carts laden with produce, still plied their trade in the suburbs of Melbourne as late as the 1940s. In 1946 the *Argus* reported that there was a good chance of home deliveries resuming despite problems of fuel shortages after the war.[143]

Another marketing option for Chinese growers was selling directly to the public from roadside stalls adjacent to their gardens. This became economically viable from the 1950s and 1960s, with increases in private car ownership and improvements in the road system. In New Zealand, for example, growers such as George Yee in Taradale, Hawke's Bay, Ho Sue Shee in Mangere, Auckland, and Jim Shack Tai in Bombay, south of Auckland, all ran roadside vegetable shops in conjunction with their gardens from the late 1950s and early 1960s.[144]

The auction system was the major method of produce marketing in New Zealand from the 1950s to the late 1980s, when large supermarket chains began to dominate, buying produce in bulk on contract. The auction system was generally supported by Chinese growers as they believed it was a fair way of selling their produce, placed growers on a more equal footing and rewarded their efforts.[145] Most felt that prices were more realistic through the auction system, despite fluctuations according to supply and demand. Looking back over his years of experience in the industry, Jeffery Turner of Turners and Growers, one of the oldest established wholesale produce firms in New Zealand, puts it succinctly:

> The auction system has many advantages – it allows the law of supply and demand to have full play and it rewards the goods of the best quality, condition and presentation with the highest prices. It encourages the growers to aim for the best.[146]

Turner also comments on the responsibility of the auctioneer, balancing the needs of growers and retail buyers:

> Auctioning was a lot of fun but it was also tricky because you knew you had a very responsible position, you were selling the livelihood of those growers. Hard months and hard days of battling the weather to get their crops in and it was all over in a few seconds...[147]

Conclusion

Taking as their starting point the concept of Chinese market gardeners as small business people on the land, the last two chapters have explored the economic resources and opportunities which Chinese immigrants to Australia and New Zealand were able to draw upon to successfully apply traditional Chinese horticultural techniques and business practices in new lands. They faced many challenges in Australia and New Zealand: barriers of language and culture, racist attitudes, restrictive immigration policies, discriminatory employment practices and an unpredictable business environment. These barriers limited their occupational and social mobility and encouraged them to enter niche occupations such as market gardening, with low capital but high labour inputs. Chinese market gardeners also faced economic constraints that were characteristic of the market gardening industry generally: the small size of the industry compared to other agricultural pursuits; the need to be located close to markets; rising land prices and changing land uses in cities; the vagaries of weather which determined crop yields; seasonal fluctuations in supply and demand and therefore in prices; and changes in retailing.

Chinese market gardeners developed distinctive models of business organisation and labour relations as well as sales and marketing strategies which were based on common ties of kinship and allegiance to village and district in China and traditional work practices and ways of doing business. These included flexible business partnerships, capital sharing, often paternalistic labour relations, a work culture that rewarded long hours of labour and flexible marketing strategies integrating the various occupations involved in food supply. The social and business networks within the Chinese community played a central role, providing sources of capital and labour, information on local business opportunities, laws and regulations, and avenues for distribution and marketing.

The flexible business partnerships adopted by Chinese market gardening enterprises enabled them to adapt to a complex and changing set of social and economic circumstances, including labour shortages due to the significant decline in Chinese immigration as a result of restrictive immigration policies, population growth, urbanisation and changes in retailing.[148] Despite these constraints Chinese market gardeners were able to apply their traditional skills in intensive horticulture and irrigation to maximise their returns from small acreages and manage gluts and shortages. The limited economic data available in the sources surveyed indicate that many market gardeners made an adequate living and some went on to become successful merchants and entrepreneurs. Thus market gardening was an important springboard for upward economic mobility and expansion into other businesses, particularly retail fruit and vegetable shops.

Chinese market gardeners remained important suppliers of fresh food in towns and cities across Australia well into the twentieth century, and for even longer in New Zealand. There were a number of reasons for their predominance in market gardening. Chinese gardeners followed general population movements and markets, and were adept at identifying niche markets, particularly in the growing cities. Europeans recognised the horticultural skills of Chinese immigrants, which were valued in the Australian and New Zealand economies. In addition, market gardening was regarded as a labour-intensive, low-status occupation which most Europeans were not keen to enter, reducing direct competition between Europeans and Chinese in the industry.

The evidence presented in these chapters also counters the stereotype of Chinese immigrants as sojourners. If market gardeners were to conduct a successful commercial enterprise they needed to make a longer-term commitment to living in their adopted country and establish links to the land. Market gardening was therefore a crucial economic foundation for the transition of Chinese immigrants from sojourners to settlers. Australia and New Zealand were leading the world in per capita incomes for most of the period from 1860 to 1914. The rapid urbanisation of the populations of Australia and New Zealand and the high standards of living they enjoyed underpinned the growth of market gardening.

Developments in technology, transport, communications and industry, which flowed from population growth and urban expansion, also supported the growth of market gardening. The complementary industries associated with early market gardening included tanneries, boiling down works, piggeries and dairies. These factors are explored in the following chapter.

Notes

1. New South Wales Legislative Assembly, *Royal Commission of Inquiry as to Food Supplies and Prices 1913*, Sectional Report on the Supply of Fruit and Vegetables, Sydney, 1914, Sydney, 1914, p. 1.
2. Weston Bate, *A History of Brighton*, Melbourne, 1983, p. 166; and Randwick Municipal Council, *Randwick: A Social History*, Sydney, 1985, p. 159.
3. New South Wales Legislative Assembly, *Royal Commission of Inquiry as to Food Supplies*, pp. 69–70.
4. Shirley Fitzgerald, *Red Tape Gold Scissors: The Story of Sydney's Chinese*, Sydney, 1996, pp. 97, 162.
5. Williams records that Billy Gay stated that he knew many gardeners who would 'go back for 12 months whenever they had saved £100' (interview with Billy Gay, 19 March 1998, Tape 2, A 115, quoted in Michael Williams, *Chinese Settlement in New South Wales*, Sydney, 1999, p. 44).
6. Ian Rannard, *The Forgotten Gardens: The Story of the Last Market Gardens in Willoughby and Northbridge, NSW*, Douglas Park, 2005, p. 78.
7. Author interviews with Gordon Ha, Sydney, 27 July 2011 and Wayn Chew, Sydney, 3 August 2011, recordings held by author.
8. Roger Waldinger, H. Aldrich, R. Ward and Associates, *Ethnic Entrepreneurs: Immigrant Business in Industrial Societies,* Newbury Park, CA, 1990, p. 36.
9. See Bruce Missingham, Jacqui Dibden and Chris Cocklin, 'A multicultural countryside? Ethnic communities in rural Australia', *Rural Society*, Vol. 16, No. 2, 2006, pp. 131–150; and Frances Parker, 'The safe use of farm chemicals by market gardeners of non-English speaking background: developing an effective extension strategy for the Sydney basin', Sydney, 2000.

10. His only assets were three pigs valued at £10, garden implements worth £15, and the vegetables in the ground, mostly cabbages, which were valued at £245 (*Evening Post*, 26 June 1906, p. 6).
11. Geoffrey Stedman, 'The South Dunedin Flat: a study in urbanisation 1849–1965', MA Thesis, University of Otago, Dunedin, 1966, p. 132.
12. New South Wales Legislative Assembly, *Royal Commission of Inquiry as to Food Supplies*, p. 2.
13. Niti Pawakapan, 'The Chinese in Dunedin between the 1920s and the 1930s', MA Thesis, University of Otago, Dunedin, 1987, p. 41.
14. Charles Sedgwick, 'The politics of survival: a social history of the Chinese in New Zealand', PhD Thesis, University of Canterbury, Christchurch, 1982, p. 473.
15. New South Wales Legislative Assembly, *Royal Commission of Inquiry as to Food Supplies*, pp. 69, 96.
16. Hans Dieter Bader and Janice Adamson, 'Kong Foong Yuen, The Garden of Prosperity: final report on the archaeological excavations at Carlaw Park, Auckland', Auckland, March 2010, pp. 31–44.
17. Ibid., pp. 163–164, 221–223.
18. From her analysis of the artefacts recovered in this predominantly working-class district, Karskens identifies a culture of comfort, domesticity and respectability which cuts across boundaries of class and social status (Grace Karskens, 'Divergent Visions: The Archaeology of Working People's Culture', in *The Cumberland/Gloucester Streets site, The Rocks: Archaeological Investigation Report,* Redfern, 1999, pp. 159–184). See also Jane Lydon, *Many Inventions: The Chinese in the Rocks 1890–1930*, Clayton, Victoria, 1999, pp. 14, 180–191.
19. Bader and Adamson, 'Kong Foong Yuen', p. 35.
20. Ibid, p. 221.
21. *Observer*, Auckland, 31 March 1894, p. 6.
22. James Ng, *Windows on a Chinese Past*, Vol. 3, Dunedin, 1999, pp. 76, 303–304; and Lily Lee and Ruth Lam, *Sons of the Soil: Chinese Market Gardeners in New Zealand,* Pukekohe, 2012, pp. 207–208.
23. George Lee Kim's grandfather Billy Lee King came to Australia in the 1870s as a gold miner and established a large market garden in Bendigo (George Lee Kim interviewed by Paul McGregor, 17 May 1994, Tape 1, National Library of Australia Post-War Chinese Australians Oral History Project, TRC 3522/4).
24. Lionel Nomchong interviewed by Matthew Higgins, 3 March 1995, National Library of Australia Post-War Chinese Australians Oral History Project, TRC 3196.
25. Similar patterns can be seen in the operation of Chinese stores. See Janis Wilton, *Hong Yuen: A Country Store and its People,* Armidale, 1989.

26. Ng, *Windows on a Chinese Past*, Vol. 1, p. 89; Sedgwick, 'Politics of survival', pp. 64, 445; C. F. Yong, *The New Gold Mountain*, Richmond SA, 1977, p. 54; C. Y. Choi, *Chinese Migration and Settlement in Australia*, Sydney, 1975, pp. 11–13; Lydon, *Many Inventions*, p. 66; and Williams, *Chinese Settlement*, p. 11.
27. Williams, *Chinese Settlement*, p. 11.
28. Keir Reeves, 'Tracking the dragon down under: Chinese cultural connections in gold rush Australia and Aotearoa New Zealand', *Graduate Journal of Asia-Pacific Studies*, Vol. 3, No. 1, 2005, pp. 49–66.
29. Sedgwick, 'Politics of survival', p. 319. Of the 6,580 Chinese who entered New Zealand between 1874 and 1896, 64.38 percent came from Australia (Mohommad Taher, 'Asians in New Zealand: a geographical review and interpretation', PhD Thesis, University of Auckland, 1965, pp. 41–42).
30. *Sydney Morning Herald*, 2 March 1865, p. 2, from the *Melbourne Age*, 24 February 1865.
31. From the 1880s the Australian and New Zealand governments introduced labour legislation aimed at encouraging unionism and the settlement of industrial disputes by conciliation and arbitration (G. T. Bloomfield, *New Zealand: A Handbook of Historical Statistics*, Boston, 1984, p. 127; and Ken Buckley and Ted Wheelwright, *No Paradise for Workers: Capitalism and the Common People in Australia 1788–1914*, Melbourne, 1988, pp. 231–232).
32. New South Wales Legislative Assembly, *Royal Commission of Inquiry as to Food Supplies*, p. 69.
33. Bader and Adamson, 'Kong Foong Yuen', pp. 166, 221.
34. Andrew Piper, pers. comm., December 2011.
35. Zvonkica Stanin, 'From Li Chun to Yong Kit: A Market Garden on the Loddon, 1851–1912', *Journal of Australian Colonial History*, Vol. 6, 2004, pp. 27–28.
36. Colleen Morris, 'Chinese market gardens in Sydney', *Australian Garden History*, Vol. 12, No. 5, March/April 2001, p. 7.
37. Lionel Nomchong interviewed by Matthew Higgins, 3 March 1995, National Library of Australia Post-War Chinese Australians Oral History Project, TRC 3196, transcript p. 16.
38. Lydon, *Many Inventions*, p. 173.
39. Ibid., p. 73.
40. By comparison the average wage in New South Wales in 1892 was £1.12s a week, without food and lodging. New South Wales Legislative Assembly, *Report of the Royal Commission into Alleged Chinese Gambling and Immorality*, Votes and Proceedings of the Legislative Assembly of New South Wales, *1891–1892*, Vol. 8, Sydney, 1892, p. 160. See also Yong, *New Gold Mountain*, p. 36.

41. Bader and Adamson, 'Kong Foong Yuen', p. 35; and Ng, *Windows on a Chinese Past*, Vol. 3, p. 235.
42. Lee and Lam, *Sons of the Soil*, pp. 207–209; and Lionel Nomchong interviewed by Matthew Higgins, 3 March 1995, National Library of Australia Post-War Chinese Australians Oral History Project, TRC 3196.
43. *Evening Post*, 21 May 1913, p. 3.
44. These systems were indicative of the rise of the welfare state and increasing government intervention in the labour market. Colin Forster,'Unemployment and minimum wages in Australia, 1900–1930', *Journal of Economic History*, Vol. 45, Issue 2, 1985, pp. 383–388; Neal Ashkanasy et al., 'The Anglo cluster: legacy of the British empire', *Journal of World Business*, Vol. 37, Issue 1, Spring 2002, n.p.; and Robin Walker and Dave Roberts, *Scarcity to Surfeit: A History of Food and Nutrition in New South Wales*, Kensington, 1988, pp. 62–63.
45. Williams, *Chinese Settlement*, p. 38.
46. New South Wales Legislative Assembly, *Report of Royal Commission into Alleged Chinese Gambling*, p. 418.
47. Ibid., p. 1.
48. Choi, *Chinese Migration and Settlement*, p. 41; and John Connell and Angela Ip, 'The Chinese in Sydney: from Chinatown to suburbia', *Asian Profile*, Vol. 9, No. 4, August 1981, p. 293.
49. In New South Wales in 1921, for example, the Minister for Labour refused the request of European market gardeners to apply the eight-hour day system to Chinese market gardeners (Yong, *New Gold Mountain*, p. 37).
50. Choi, *Chinese Migration and Settlement*, p. 53, and Williams, *Chinese Settlement*, p. 44.
51. Mangere Historical Society, *Mangere Chronicles*, Mangere, 1990, p. 21; and John Brooks, 'The last of the Chinese gardeners at Narrabri', unpublished ms., 1998, Golden Threads Papers, Armidale.
52. Ah Yook and Carole Gass interviewed by Joe Eisenberg, August 1999, Golden Threads Papers, Armidale; Henry Reynolds, *North of Capricorn, The Untold Story of Australia's North*, Crows Nest, NSW 2003, p. 76; and R. Wornall-Smith, 'The employment of Maori women by Asiatic market gardeners: an aspect of NZ race relations', *University of Auckland Historical Society Annual*, 1970, pp. 1–11.
53. Thong Ling Lee, 'Chinese market gardening in the Auckland Region', MA Thesis, University of Auckland, 1974, pp. 37, 38.
54. Gordon Ha interviewed by Joanna Boileau, Botany, 27 July 2011.
55. Martin Shanahan and John Wilson, 'Measuring inequality trends in colonial Australia using factor-price ratios: the importance of boundaries', *Australian Economic History Review*, Vol. 47, No. 1, 2007, p. 7. See also

J. M. Powell, *Mirrors of the New World*, Folkestone/Hamden Connecticut, 1977, pp. 35–36, 71.

56. Powell, *Mirrors of the New World*, pp. 71–75.
57. Geoffrey Bolton, 'Money: trade, investment and economic nationalism', in Deryk Schreuder and Stuart Ward (eds.), *Australia's Empire*, Oxford, 2008, pp. 212–213.
58. Lee and Lam, *Sons of the Soil*, pp. 17, 529.
59. Ibid., p. 383.
60. Ibid., p. 16.
61. The pre-Federation Chinese population in Australia fluctuated between 38,258 in 1861 and 29,627 in 1901 (Choi, *Chinese Migration and Settlement*, p. 22). Under the Naturalisation Act 1903 persons deemed naturalised in the colonies prior to the passing of the Act were deemed naturalised under Commonwealth law. However, few people were granted naturalisation between 1904 and 1956 (Paul Jones, *Chinese-Australian Journeys: Records on Travel, Migration and Settlement, 1860–1975*, Canberra, 2005, p. 119).
62. Banana Industry Preservation Act 1921, Queensland; NSW Crown Lands Consolidation Act 1913 and Closer Settlement (Amendment) Act 1916, New South Wales; and South Australia Irrigation and Reclaimed Lands Act 1914, South Australia.
63. Sedgwick, 'Politics of survival', p. 318; Ng, *Windows on a Chinese Past*, Vol. 1, p. 321; and Neville Ritchie, 'Traces of the past: archaeological insights into the New Zealand Chinese experience in southern New Zealand', in Manying Ip (ed.), *Unfolding History, Evolving Identity: The Chinese in New Zealand*, Auckland, 2003, p. 35.
64. Sedgwick, 'Politics of survival', p. 318.
65. Ng, *Windows on a Chinese Past*, Vol. 1, p. 321.
66. This was run by a partnership of six men from Zengcheng (Jung Seng) and had six employees (Lee and Lam, *Sons of the Soil*, p. 195).
67. Pawakapan, 'Chinese in Dunedin', pp. 37–38; and Stedman, 'The South Dunedin Flat', p. 133.
68. Lee and Lam, *Sons of the Soil*, pp. 319–324.
69. Ibid., pp. 342, 348, 371.
70. New South Wales Legislative Assembly, *Royal Commission of Inquiry as to Food Supplies*, p. xxxv.
71. *Brisbane Courier*, 30 September 1889, p. 7.
72. *Brisbane Courier*, 6 August 1902, p. 2 and 4 February 1928, p. 13; and George Lee Kim interviewed by Paul McGregor, 17 May 1994, Australia-China Oral History Project, National Library of Australia, Canberra, ORAL TRC 3522/4/4.

73. Cathie May has documented the important role Chinese played in land clearing in North Queensland (Cathie May, *Topsawyers: The Chinese in Cairns 1870 to 1920*, Townsville, 1984). See also Warwick Frost, 'Migrants and technological transfer: Chinese farming in Australia, 1850–1920', *Australian Economic History Review,* Vol. 42, No. 2, 2002, pp. 113–131.
74. Merrilyn George, *Ohakune: Opening to a New World*, Wanganui, 1990, p. 218; and Lee and Lam, *Sons of the Soil*, pp. 219–221.
75. Lee and Lam, *Sons of the Soil*, pp. 219–221; and Nigel Murphy, 'A history of the horticultural industry', unpublished ms., 2010, pp. 3, 11.
76. George, *Ohakune*, pp. 218, 222.
77. Pawakapan, 'Chinese in Dunedin', p. 39.
78. *Tung Wah News*, 19 July 1899, p. 2.
79. New South Wales Legislative Assembly, *Royal Commission of Inquiry as to Food Supplies*, p. 72.
80. Sedgwick, 'Politics of survival', p. 322.
81. Pawakapan, 'Chinese in Dunedin', p. 40.
82. Derek Smith, 'Market gardening at Adelaide's urban fringe', *Economic Geography*, Vol. 42, No. 1, January 1966, p. 19. See also J. W. McCarty, 'Australian capital cities in the nineteenth century', *Australian Economic History Review*, Vol. X, No. 1, March 1970, p. 109; and Lee, 'Chinese market gardening', p. 13.
83. For example, Andreas Grotewold, 'Von Thunen in retrospect', *Economic Geographer*, Vol. 35, 1959, pp. 346–355.
84. Smith, 'Market gardening at Adelaide', p. 19; and R. G. Golledge, 'Sydney's metropolitan fringe: a study in urban-rural relations', *The Australian Geographer*, Vol. 7, 1960, pp. 243–255.
85. For example, in the Cooks River area of Sydney. See Ian Tyrrell, 'Cooks River and environmental history' adapted from 2004 Botany Bay Forum University of New South Wales, http://iantyrrell.wordpress.com/cooks-river/, accessed 30 October 2012.
86. Sandra Pullman, 'Along Melbourne's rivers and creeks', *Australian Garden History*, Vol. 12, No. 5, March/April 2001, p. 10.
87. New South Wales Legislative Assembly, *Royal Commission as to Food Supplies*, p. 73; and Karla Whitmore, 'Willoughby's Chinese market gardens', unpublished ms., courtesy of author, 2004, p. 4.
88. *Tung Wah Times*, 21 April 1928, p. 7-b.
89. Stedman, 'The South Dunedin Flat', p. 137. Despite the forces of suburbanisation and industrialisation, an enclave of Chinese market gardeners remained in the Macandrew Road area of south Dunedin until the late 1930s, when the land was resumed for the building of a government school (ibid., p. 206).
90. Pawakapan, 'Chinese in Dunedin', p. 39.

91. Donald Hunt, 'Market gardening in metropolitan Auckland', MA Thesis, University of New Zealand, Auckland, 1956, p. 61; and Val Payne, *Celebrating Mangere Bridge*, Mangere, 2005, p. 27.
92. Lee, 'Chinese market gardening', p. 30. See also Lee and Lam, *Sons of the Soil*, pp. 341, 402–403.
93. This was still the case in the 1970s. Surveying Chinese market gardeners in Auckland, Lee found that a major factor in the location of their gardens was proximity to the expanding market of Auckland and lower transport costs (Lee, 'Chinese market gardening', pp. 9, 12).
94. Murphy, 'History of the horticultural industry', pp. 16, 37; and Marnie Haig-Muir and Roy Hay, 'The economy at war', in Joan Beaumont (ed.), *Australia's War 1939–1945*, St Leonards, 1996, pp. 111–114.
95. W. H. Clarke, 'The fruit and vegetable industries', *Agricultural Gazette of NSW*, December 2, 1904, p. 1204.
96. This dissuaded many early Chinese market gardeners from planting orchards; the trees took too long to grow to maturity and produce a return. See comment in *Sydney Morning Herald*, 2 March 1865, p. 2.
97. Lee, 'Chinese market gardening', pp. 1, 67; Lee and Lam, *Sons of the Soil*, p. 14; and Taher, 'Asians in New Zealand', p. 186.
98. Rannard, *The Forgotten Gardens*, pp. 7, 22.
99. *Sydney Morning Herald*, 2 March 1865, p. 2, from the *Melbourne Age*, 24 February 1865.
100. Janis Wilton, *Golden Threads: The Chinese in Regional New South Wales 1850–1950*, Sydney, 2004, p. 27.
101. *Evening Post*, 25 March 1909, p. 8.
102. Alex Bennett, *A Living from an Acre: How to Defeat the Depression*, Sydney, 1932, pp. 17–19, 74–75.
103. The prices of potatoes rose from £7 per ton in 1849 to £23 per ton in 1854 and carrots from 1.5 pence to 1 shilling in the same period. Bate, *History of Brighton*, p. 168.
104. *Southern Cross*, 22 September 1900, p. 2, quoted in Bate, *History of Brighton*, p. 357.
105. C. M. Yuan, 'Chinese in White Australia 1901–1950', in James Jupp (ed.), *The Australian People: An Encyclopedia of the Nation, its People and their Origins*, Cambridge, England, 2001, p. 306.
106. Mangere Historical Society, *Mangere Chronicles*, Mangere, 1990, p. 21.
107. *Auckland Star*, 28 June 1899, p. 6.
108. For example, see *Tung Wah News*, 13 July 1898, p. 4; 31 August 1898, p. 4; and 17 September 1898, p. 4.
109. Randwick Municipal Council, *Randwick: A Social History*, Kensington, 1985, p. 159.
110. *Brisbane Courier*, 3 May 1902, p. 15.

111. *Brisbane Courier*, 4 February 1928, p. 13.
112. Andrea Gaynor, *Harvest of the Suburbs: An Environmental History of Growing Food in Australian Cities*, Perth, 2006, pp. 40–41.
113. Enid Ross interviewed by Warwick Eather, 18 and 22 February 1987, New South Wales Bicentennial Oral History Collection, Mitchell Library, MLMSS 5163/Box 02, quoted in Gaynor, *Harvest of the Suburbs*, p. 41.
114. Sedgwick, 'Politics of survival', p. 155.
115. *Sturt Recorder*, 19 January 1894, p. 2.
116. George Lee Kim interviewed by Paul McGregor, 17 May 1994, National Library of Australia, Oral TRC 3522/4/4, Tape 2.
117. Eddie Quong interviewed by Diana Geise, 1993, National Library of Australia, TRC 3005.
118. Pawakapan, 'Chinese in Dunedin' p. 98.
119. *Sydney Morning Herald*, 30 June 1881, p. 6.
120. *Brisbane Courier*, 15 July 1890, p. 7.
121. *Hawera Normanby Star*, 30 September 1907, p. 7; and *Poverty Bay Herald*, 11 October 1907, p. 6.
122. *Fielding Star*, 8 May 1908, p. 4.
123. Between 1910 and 1920 Chinese wholesale fruit and vegetable firms virtually monopolised the business in New South Wales and Victoria. Choi, *Chinese Migration and Settlement*, p. 53.
124. Sedgwick, 'Politics of survival', p. 320.
125. Williams, *Chinese Settlement*, p. 26.
126. Yuan, 'Chinese in White Australia', p. 306.
127. Anne Atkinson, 'Chinese market gardening in the Perth metropolitan region, 1900–1920', *Western Geographer*, Vol. 8, 1984, p. 47. See also Yong, *New Gold Mountain*, p. 36.
128. Sandi Robb, 'Mar and Mar: a tale of two Chinese gardeners at Winton', in Kevin Wong Hoy and Kevin Rains (eds.), *Rediscovered Past: Valuing Chinese Across the North*, North Melbourne, 2012, pp. 36–44; and Kath Mahaffey, *Pioneers of the North West Plains*, Moree, 1982, p. 34.
129. Ann Thorpe et al., 'Otaki's market gardens', *Otaki Historical Society Historical Journal*, Vol. 30, 2008, p. 19.
130. They include S.C. Lee and Bing (from 1902) and Wong Sik Hum (from 1913) in Wanganui; Wong Cho Nam in Masterton in the 1930s and 1940s; and King Brothers in Ashburton from 1947 (Lee and Lam, *Sons of the Soil*, pp. 100, 139, 205–206).
131. *Alexandra and Yea Standard, Gobur, Thornton and Acheron Express*, 16 September 1887, p. 50.
132. Williams, *Chinese Settlement*, p. 44.
133. Pawakapan, 'Chinese in Dunedin', p. 26.

134. Golden Threads: The Chinese in Regional New South Wales Database, 1999. Records: S13, Notes, p. 1; S17, Notes, p. 2; and S10, Notes, p. 1. See also Marion Dormer, *Dubbo, City on the Plains, 1901–1988*, Dubbo, 1988, p. 105; and Marion Dormer and Joan Starr, *Settlers on the Marthaguy*, Dubbo, 1979, p. 99.
135. *Alexandra and Yea Standard, Gobur, Thornton and Acheron Express*, 16 September 1887, p. 50.
136. Yong, *New Gold Mountain*, p. 37. Public markets were built in Auckland in 1906 and Wellington in 1910 (Murphy, 'History of the horticultural industry', p. 17). In Australia public markets were established in Sydney in Campbell Street (Belmore Markets, 1869), then the Haymarket (New Belmore Markets, 1890s, Municipal Markets 1909–1914) and in North Melbourne (Victoria Markets, 1878). (Melbourne Wholesale Fruit, Vegetable & Flower Market, 'History', http://www.melbournemarkets.com.au/about_history.asp, accessed 12 September 2011; and City of Sydney Council, 'History of Sydney City Council', http://www.cityofsydney.nsw.gov.au/__data/assets/pdffile/0003/65550/hs_chos_history_of_council_1001.pdf, accessed 12 September 2011). In Perth produce markets were established in 1899 and in 1906 between Roe and James streets (Atkinson, 'Chinese market gardening' p. 480).
137. Williams, *Chinese Settlement*, pp. 44, 17.
138. Lee and Lam, *Sons of the Soil*, p. 43.
139. NSW Legislative Assembly, *Report of the Royal Commission into Alleged Chinese Gambling*, p. 418.
140. NSW Legislative Assembly, *Royal Commission of Inquiry as to Food Supplies*, p. 1.
141. James Ng, pers. comm., November 2011. See also Lee and Lam, *Sons of the Soil*, p. 41.
142. Walker and Roberts, *Scarcity to Surfeit*, p. 137.
143. *Argus*, 21 February 1946, p. 10.
144. Lee and Lam, *Sons of the Soil*, pp. 263, 397, 451.
145. See, for example, William Young interviewed by Lily Lee, 16 November 2006, quoted in Lee and Lam, *Sons of the Soil*, p. 193; and Allan Fong interviewed by Ruth Lam, 24 September 2007, quoted in Lee and Lam, *Sons of the Soil*, p. 501.
146. Jeffery Turner, 'Fifty years of progress in the fruit and vegetable industry', Address to the Annual Conference of New Zealand Commercial Chinese Growers Federation, 17 July 1992, quoted in Lee and Lam, *Sons of the Soil*, p. 18.
147. Jeffery Turner interviewed by Ruth Lam, 15 May 2008, Lee and Lam, *Sons of the Soil*, p. 362.

148. For example, the *Royal Commission of Inquiry as to Food Supplies and Prices 1913* in New South Wales found that the supply of fresh vegetables was not keeping pace with demand due to the rapid increase in the urban population, the expansion of suburban building areas absorbing land formerly devoted to market gardens and labour shortages due to the almost complete cessation of Chinese migration. The report also highlighted poor road and rail transport facilities, with delays and rough handling causing significant losses to growers and reduction in supplies (NSW Legislative Assembly, *Royal Commission of Inquiry as to Food Supplies*, pp. xxx, xxxvii).

CHAPTER 6

New Ideas in New Lands

Europeans settled Australia and New Zealand in the late eighteenth and early nineteenth centuries, during the height of the industrial revolution in Britain. In these isolated colonies they modified imported European technologies or improvised local technologies with European origins in order to cope with local conditions. Historians of technology have emphasised the unique set of conditions which existed in Australia and New Zealand in the nineteenth century, in particular the small, scattered populations in both countries and their reliance on the export of primary produce to Britain, and the vast size and aridity of the Australian continent.[1] Chinese immigrants who settled in Australia and New Zealand brought with them their own skills and technologies, particularly in horticulture and water management, and they too adapted and modified these technologies to suit local conditions.

Processes of technological change and exchange are complex and multifaceted, the shape and direction they take determined by a multitude of factors including historical movements of people and ideas; economic variables, particularly profitability; and developments in complementary industries or as some writers term them, supportive industries.[2] Most importantly, it is the mobility of people, technologies, skills and ideas accompanying networks of migration that is central to processes of technological change. Some of the industries that were crucial in supplying Chinese market gardeners with the production inputs they required were producers of fertilisers, pesticides, seeds and equipment such as pumps,

J. Boileau, *Chinese Market Gardening in Australia and New Zealand*,
Palgrave Studies in the History of Science and Technology,
DOI 10.1007/978-3-319-51871-8_6

irrigation plant and cultivating tools. Technological developments in transport facilities, in particular the expansion of railways and tramways, were also crucial in expanding the market reach of market gardeners.

This chapter explores the interplay between the continuity of the traditional horticultural practices and technologies which Chinese immigrants brought to Australia and New Zealand, and their adaptation of these technologies to new social and environmental conditions. The first section discusses the contribution that Chinese immigrants made to the horticultural industry and some of the economic and social factors involved in technological change. The second section discusses change and innovation in market gardening, illustrating the many ways in which Chinese market gardeners adapted their agricultural technology to new environments and adopted European technological developments which were appropriate to their intensive cultivation methods. It also discusses the complementary industries which supported technological change in Chinese market gardening, including transport, fertiliser, pesticide, motor vehicle and agricultural equipment industries and seed producers.

There were complex interactions between the social, economic and technological changes that occurred in agriculture and industry in Britain from the early-eighteenth century and laid the foundations for the industrial revolution.[3] There are ongoing debates among scholars on the nature of these relationships, such as the extent to which the size of the agricultural workforce declined, releasing labour to industry, and the precise chronology and speed of agricultural change, depending upon what economic indices are used.[4] Some historians such as Robert Allen and Norman McCarol emphasise the importance of structural changes in agriculture rather than changes in the overall size of the workforce.[5] However, there is general consensus that changes in agriculture laid the foundation of the industrial 'revolution' long before any significant increase in industrial output. The economic data, based on a variety of indices, shows that there was a steady increase in agricultural productivity between 1700 and 1850 which supported the growing population and facilitated the expansion of industry and commerce.[6]

This increase in productivity was the result of a number of interrelated factors, in particular: enclosures of common land and the rise of large estates which were more efficient to manage; greater security given to tenant farmers; advances in plant and animal breeding, resulting in improvements in the quality of crops and livestock; and the application of fertilisers and machinery.[7] In turn, industrial and technological advances promoted mechanisation in agriculture, saving both time and labour and increasing productivity.

In Von Tunzelmann's terms there were both process and product innovations in agriculture – new methods of doing things and new activities or outputs. New 'gadgets' to make the work of ploughing, planting, weeding, fertilising and harvesting easier were developed, along with new crop varieties and new commercial fertilisers.[8] All these developments had a direct impact on the progress of European settlement in Australia and New Zealand from the late eighteenth century, as they were transferred and adapted to suit local conditions.

The scenario most relevant to the situation of Chinese immigrants in Australia and New Zealand is one of creative settlers in a hostile environment, similar to that adopted by the economic historian Colin White. In White's scenario for Australia, new settlers pioneered a range of innovations in fields as diverse as food canning, refrigeration, wind power and irrigation. These were not so-called macro-inventions but a multitude of micro-inventions that transformed big ideas into practical innovations.[9] Scholars of the industrial revolution in Britain such as Von Tunzelmann also emphasise the importance of micro-inventions in the technological advances that occurred in agriculture and industry from the mid eighteenth century.[10]

White highlights the flow of people, ideas and skills throughout the British Empire in the nineteenth century and the diverse multinational input into technological developments in colonial Australia. He points out that the most dynamic industries, technologically and economically, were in the primary sector: agriculture, pastoralism and mining. Initially most innovations arose from the experimentation of individuals closest to the particular problem and from practical experience and technical ingenuity rather than scientific understanding and research.[11] As Joe Powell observes in Australia, and Stapleton in America, this was a process of trial and error as immigrants adapted their farming practices and crops to foreign environments.[12] Similarly, historians such as Jan Todd and Ann Moyal note the important role played by individual innovators and researchers in nineteenth-century Australia and the tardiness of governments in funding experimentation and applied research.[13]

Chinese immigrants were among these innovators. However, their contribution to agriculture in Australia and New Zealand has been largely overlooked, with the exception of the work of a few scholars such as Barry McGowan, Warwick Frost and Jock Collins in Australia and Lily Lee, Ruth Lam and Nigel Murphy in New Zealand.[14] As Frost observes, most specialised histories of Australian agriculture do not mention Chinese farmers and most historians of the Chinese in Australia focus on social and political aspects, making only brief reference to their skills in agriculture, their horticultural methods and work organisation.[15] There is a similar gap in the literature on the history of agriculture in New Zealand, due largely to an emphasis on broad-scale farming at the expense of small-scale horticulture.[16] Lily Lee and Ruth Lam's recent study of Chinese market gardening in New Zealand does much to fill this gap.[17]

As outlined in Chapter 2, the Chinese immigrants who made the long sea journey from Guangdong to Australia and New Zealand had behind them a long tradition of innovation and change in agriculture, combined with practical experience of farming. The literature on Chinese migration to Australia provides little detail on the characteristics of Chinese farming, tending to depict it as a set of traditional practices that was transferred to Australia fully formed during the gold rushes. However, Frost shows that Chinese agricultural practices evolved over time and that Chinese immigrants who took up farming and market gardening in Australia continued to adapt and innovate in their new environments.[18] He proposes a broad sequence for the development of Chinese farming in Australia, characterising the period 1850–1880 as one of adjustment, in which Chinese immigrants mainly worked as farm labourers under European supervision or in small-scale market gardening which wrought little technological change. In Frost's schema the period 1880 to 1900 was the golden period of Chinese agriculture, when they dominated intensive agriculture and became specialists in market gardening. The period 1900–1920 was a period of decline, when the scale and range of Chinese agriculture shrank significantly.[19] This is true in the case of more specialised Chinese horticulture, for example growing maize, bananas and rice in Queensland and tobacco in New South Wales and Victoria.

However, Frost's schema (which he acknowledges may be an oversimplification) is challenged by more detailed study of the development of Chinese market gardening. As the evidence presented in this chapter will show, Chinese market gardening underwent technological change and adaptation from the time the first Chinese immigrants began to grow crops on the goldfields. It continued to thrive in cities and towns across Australia until the late 1930s, and in New Zealand into the 1970s. Influenced by technological developments in the wider society, significant changes in Chinese market gardening occurred between the 1920s and 1950s.

Just as European immigrants transplanted established European agricultural practices to new and sometimes hostile environments, Chinese immigrants brought with them traditional Chinese agricultural methods that had been developed over centuries in their homeland. Both groups adapted their methods to the new environments they encountered and met with varying degrees of success. European farming methods, based primarily on growing cereal crops and raising sheep and cattle, met with economic success but at the cost of significant damage to the environment in both countries.[20] At the same time these industries triggered many significant innovations and adaptations of imported technology, for example in Australia the stripper harvester invented by John Ridley in 1843 and the stump jump plough in 1876.[21]

Although many features of traditional Chinese horticulture were remarkably enduring in Australia and New Zealand, Chinese market gardeners adapted their agricultural technology to new environments and adopted European technological advances which were appropriate to their intensive cultivation methods. The following section explores the complex mix of conservatism and innovation which characterised Chinese market gardening in Australia and New Zealand, and some of the mechanisms by which ideas and innovations diffused within the Chinese community and between the wider community and the Chinese community.

Conservatism, Innovation and the Diffusion of Ideas

As historians of technology observe, the diffusion of technology is never homogenous over time and space; it involves failures, delays and slowdowns, and constant adaptations and translations to meet the specific needs and circumstances of individuals.[22] The pace and direction of the diffusion of technology is also determined by individual cultural differences and historical experiences of migration. Thus the pace at which Chinese market gardeners innovated, adapted and adopted new technologies varied according to their individual circumstances.

Nineteenth-century attitudes towards Chinese immigrants were contradictory. Europeans characterised them as conservative, clinging to 'primitive' methods and ideally suited to manual labour, involving hours of patient toil. At the same time they acknowledged, to varying degrees, the ingenuity and entrepreneurial skills of Chinese people as well as their long tradition of technological advances. An article published in New Zealand in 1916 depicts Chinese gardeners in Otaki as conservative and resistant to change:

> But, with all their patience and all their hard work, the Chinese are primitive in their methods. If they were less primitive there would be less drudgery. However, there it is, and they . . . persistently and stubbornly refuse to adopt European methods . . . But, that same Celestial knows how to attend to his business. He always obtains results.[23]

Many later writers have also depicted Chinese market gardeners as conservative and resistant to change. For example, writing about Chinese market gardeners in South Dunedin in the early twentieth century, Geoffrey Stedman observes that they relied on traditional horticultural methods and were little influenced by their new environment.[24]

However, the evidence presented in this chapter provides many examples of Chinese market gardeners adopting European technological developments and adapting them to meet their own needs. These were both process and product innovations, ranging from irrigation pumps and agricultural implements to fertilisers and improved seed varieties. To explain the processes by which technological change was adopted, this chapter applies the analytical framework of the diffusion of innovation. Using this approach we can see Chinese market gardeners as active agents of technological change, interrogating new knowledge and technologies on the basis of their own experience and knowledge of local conditions and reinventing them to meet their specific needs.

There were individual differences within both the European and Chinese communities in the degree to which market gardeners were willing to adopt new ideas and experiment. Some market gardeners could be conservative and resistant to change.[25] However, Lee and Lam's study of Chinese market gardening in New Zealand reveals a number of Chinese market gardeners who were particularly innovative and creative, keeping up with the latest technological developments in New Zealand and overseas and leading by example.[26] Everett Rogers highlights the important role played by change agents in the diffusion of innovation, in particular opinion leaders in the community.[27] Similarly, scholars writing about the so-called 'Green Revolution' and technological change in agriculture emphasise the creativity of individual farmers and their ability to experiment, adapt and innovate as well as act as role models in the local community.[28]

There are complex social and economic factors influencing the take-up of technology beyond individual differences in willingness to take risks and experiment. As Gordon Clark argues, explanations of differences in the speed of adoption of innovation in terms of individual behavioural characteristics are overly simplistic. Those categorised as conservative or 'late adopters' may be unable to afford the capital investment involved, may have other priorities for their capital or the innovation itself may be marginally suited on technical grounds.[29] In his study of the Chinese community in Dunedin between 1910 and 1920, Niti Pawakapan observes that Chinese people were particularly conservative in financial matters. They worked for years to save enough to return to China to visit family or to bring relatives to New Zealand, and would defer the purchase of a truck or tractor.[30] Thus, characteristic of transnational communities, a priority of overseas Chinese was maintaining ties with their kin and home village in China and supporting them financially.[31]

The owners of larger, more successful market gardening operations were more likely to have the capital to invest in advances in technology, tractors and motor vehicles. The gardener Georgie Ah Ling ran a small one or two-man business in Donald, Victoria. He was conservative and made little investment in new technology. However, he did make some changes over the years. Living in a small town he was relatively isolated from other Chinese people after his partner died, but his social contacts with the local European community encouraged him to adopt some new technology. He replaced his covered wagon with a flat-bed lorry, which was still horse-drawn, and used a petrol engine to pump water from the reservoir. Although he dreamed of returning to China a prosperous man one day, Ah Ling remained in Australia, faithfully sending most of his modest earnings home to support his wife and children.[32]

There were also clear economic reasons for delaying investing in technology. Most market gardens were leased before World War II, discouraging gardeners from making large capital investments in their businesses and the small scale of their operations, with intensive cropping, was not amenable to mechanisation. In his study of market gardening in the Auckland region in the 1950s, Donald Hunt contrasts the farming methods of European and Chinese growers. European market gardeners, who had larger acreages than Chinese (up to 20 acres) and grew a smaller range of crops, generally used a half-horse power tractor, plough, disc scarifier, moulding machine, topdresser and farm trailer.[33] In contrast, Chinese gardeners, who had smaller acreages, particularly closer to the city centre, and grew a wide variety of crops with narrow spaces between rows, used only small hand-operated rotary hoes.[34] Thus, the intensive agricultural methods which had their origins in nineteenth-century China made it difficult for Chinese gardeners to use horse-drawn implements or tractors.

Another factor in technological change was labour supply. In the nineteenth century there was a generally reliable supply of labour on Chinese market gardens, which did not encourage mechanisation. As discussed in the previous chapter, there were increasing labour shortages in the late nineteenth century and early twentieth century after immigration restrictions were introduced in Australia and New Zealand. There were also generational changes. By the 1950s the Australian or New Zealand-born children of Chinese market gardeners were attending university and entering professional occupations. As the use of family labour became less important, mechanisation increased.[35]

Later generations of Chinese immigrants had different aspirations to their parents and grandparents. Harry Chong Choy came to Sydney in 1949, following in the footsteps of his father and grandfather. His father was a partner in a market garden in Rockdale with five other Chinese, and Choy joined his father there. But Harry Choy's ambition was to go into business, not till the land. He recalled:

> I was ambitious and worked hard. I came to see two things as essential to making a good life in Australia: speaking English and driving a car. Three years after I arrived, I bought a car. I paid 250 pounds for a Ford. Seeing the car, all my friends said: 'You are really over-extended!' I spent almost all my money saved from three years of hard work. It was the price of half a house at that time. There were only four other fellow countrymen of Gaoyao/ Gouyiu origin who owned a car. Even my father was against me buying the car, asking me: 'Why do you buy a car?' I told him: 'You would not be able to enjoy independence or own a business without owning a car.'[36]

The younger generation who did follow their parents or other relatives into market gardening learned the business from them. In New Zealand for example, the large Wong Sing & Co. partnership in Gisborne was an important training ground for younger kinsmen who arrived from China in the 1940s, and was recognised as the centre for Toi Shan-Sun Wui clan activities. The partnership later developed into several separate family-run gardens.[37] The younger generation played a key role in technological change, often encouraging their elders to adopt the latest developments. This was particularly the case in New Zealand, where significant Chinese involvement in market gardening continued longer than in Australia. Sue Chong, whose father Jack had a market garden in Taranaki with her grandfather Harry, remembers the arguments between them over mechanisation in the 1930s. Her father was all for buying the latest model tractor but her grandfather could not see the point of it.[38] Bill Wong learned the market gardening business from his father Bing Wong in Hastings after he left school in 1956. When he took over 10 years later he told his father he had new ideas and wanted to try new things. He recalls:

> Dad was a little bit old fashioned and didn't like to change much. I thought that if I take over, I'd like to look to the future. If you don't, you don't get ahead.[39]

Another avenue for the diffusion of innovation in agriculture within the Chinese community was via the ongoing links that Chinese settlers in Australia and New Zealand maintained with their homeland. Agriculture in China was also modernising and there was a fertile exchange of ideas between China and overseas Chinese in Australia and New Zealand from the late nineteenth century via the return visits of Chinese immigrants to China and visits of Chinese government officials, intellectuals and writers as well as Chinese language media.[40] In New Zealand, the Chinese consul Yung Liang Huang visited the west coast of the South Island in 1910. He updated the local Chinese community on the modernisation of their homeland and advised them not to be too conservative and to adopt English ways and methods of work when in an English country.[41]

European technological developments and ideas diffused from the European community through a variety of avenues. Exchanges between European and Chinese market gardeners and the establishment of cooperative organisations such as the Otaki District Commercial Gardeners Association in New Zealand, established in 1942, played an important role. They disseminated information about the latest seed varieties, fertilisers and pesticides and technology such as rotary hoes, tractors and irrigation systems.[42] It was also in the interests of vegetable processing firms, such as Watties in New Zealand, to encourage growers to produce crops suited to their production needs and develop new varieties to meet changing consumer preferences.

The auction system rewarded growers who experimented and produced improved crop varieties with higher prices. Produce marketing firms in New Zealand such as Produce Markets and Turners and Growers promoted technological change and innovation in market gardening, providing Chinese growers with capital to invest in additional land and purchase equipment such as tractors to make their operations more efficient and with advice on new crop varieties.[43] Repayments were deducted from the sales of the produce they supplied. In the 1950s there was strong competition between the auction companies and therefore a great desire to assist growers. Dick Young established a market garden in Waimauku, north of Auckland, in 1960 when he was only 18 years old. He recalls the assistance he received from Turners and Growers when he was starting out. Willing to take a business risk, they gave him a loan to purchase seed, sprays and fertiliser to grow lettuces. As his business grew Young obtained further loans to purchase machinery such as tractors.[44]

Media advertising also played a role in technological change, disseminating information about the latest developments in agricultural equipment. In Australia in the early 1900s, the engineering firm Tangyes Ltd advertised 'force pumps' and steam pumps in the Chinese language *Tung Wah Times*.[45] There are many factors involved in the decisions of advertisers to maintain a presence in a given publication, including the economic climate, the response from readers and the advertising opportunities offered by rival publications, but the firm's advertisements in a Chinese language newspaper are an indication that they saw a potential market for their products among Chinese market gardeners and that gardeners were open to adopting new technologies. In the early years of Chinese language newspapers in the 1890s, advertisers were almost exclusively Western businesses.[46] The *Tung Wah Times* also published articles about new types of irrigation pumps for gardens and new ways to use them, a further indication of readers' interest in technological developments.[47] Articles appeared on the importance of technology in agriculture in 1900, on an exhibition of manufacturing technology in Perth in 1910 and on agricultural research in New South Wales in 1926.[48]

The following section examines in more detail change and innovation in the development of Chinese market gardening in Australia and New Zealand. It demonstrates that while gardeners maintained many features of traditional Chinese intensive agricultural practices, they readily adopted

European technological advances which were appropriate to the scale and work organisation of their enterprises and where necessary, adapted them to meet their own needs.

Inventors and Innovators

The major contribution of Chinese immigrants to horticulture in Australia and New Zealand lay in the intensive cultivation and irrigation methods they used to produce vegetable crops. There was considerable continuity in the layout of gardens and the labour-intensive methods involved in preparing the soil, manuring, weeding and caring for individual plants. However, there was also ongoing adaptation of Chinese agricultural practices, for example in the cultivation of European crops to meet the demands of European consumers. Gardeners integrated advances in European technology into their farming systems where these were appropriate to the scale and work organisation of their enterprises. These innovations were a series of micro-inventions, both in processes and products. They included innovations in: gardening implements; motor vehicles and tractors; irrigation equipment; fertilisers; pest control methods; growing methods; and new seed and crop varieties. There were also changes in the built structures on gardens to accommodate these technological changes.

Gardening Implements

The basic implements of the nineteenth-century market gardener were hand tools: spade, hoe, fork, rake and watering can. These simple tools were the product of experience handed down over generations and were remarkably stable in design, enduring well into the twentieth century. They changed little over the decades. This lack of technological change was a product of the small-scale land holdings and labour-intensive cultivation methods which characterised market gardening. As Bing Wong, whose grandfather was a market gardener in Auckland, recalls: 'in New Zealand they usually just had simple tools, a shovel, a hoe and a basket, they use labour, they hoe and dig and they grow the crops'.[49]

In 1937 the Melbourne *Argus* newspaper published a story about two aged Chinese market gardeners returning to their homeland. Although the story presents a stereotyped image of Chinese, it illustrates the longevity of traditional hand implements and the attachment market gardeners had to their faithful tools, well worn and repaired over the years. Charlie Jung and Ah Chuey, who for decades had been market gardeners in Victoria, managed to save enough money to return to China. Boarding their ship, the two men were followed up the gangway by two red-capped porters carrying a strange assortment of luggage:

> Clothing and curios and presents were packed in huge wicker baskets slung on a wooden yoke. And bringing up the rear came the strangest articles of all – a number of ancient garden hoes, their original handles long since replaced by twisted saplings, and two huge and battered watering-cans of an antiquated pattern. Charlie Jung explained the reason for these strange pieces of luggage. They had used them in the gardens for more than 30 years, and had proved them trusty and reliable implements. When the two old men decided to return to Canton they offered the hoes and watering-cans for sale. There were no bidders, and rather than leave them in Australia they decided to carry them back to Canton with them.[50]

Another illustration of the longevity of the hoe comes from the description of the establishment of a 4.5-acre market garden on the banks of the Barwon River at Geelong in 1870, quoted in Chapter 2. The work of preparing the ground was entirely done 'with fork-shaped hoes, and heavy long-headed common hoes'.[51] Half a century later similar heavy three-pronged hoes were still in use (Fig. 6.1). George Lee Kim, whose grandfather had a market garden in Bendigo in the 1920s, describes how he used a digging fork like a hoe with three prongs, a long handle and a heavy metal head. It had to be heavy, to break up the ground and was used as a hammer to pound the lumps of dirt into small particles. Kim recalls:

> It was clayey soil . . . They used to have to dig the land with hoe and digging fork, they had to dig it out, break it up, to get down to the fine tilth that seedlings could survive in. That was very hard work, they were into middle age. I used to help them sometimes; I wasn't strong enough to lift this hoe, you had to be fairly strong to break this ground open and crush it up.[52]

As discussed in an earlier chapter, the careful preparation of soil before planting, breaking it up into a fine tilth, is a hallmark of traditional Chinese agricultural practices. The implements used to achieve this were adapted in new environments, saving time and labour and increasing efficiency. In the early twentieth century in New Zealand, for example, Chinese and European market gardeners in Otaki used a horse-drawn clod crusher, a

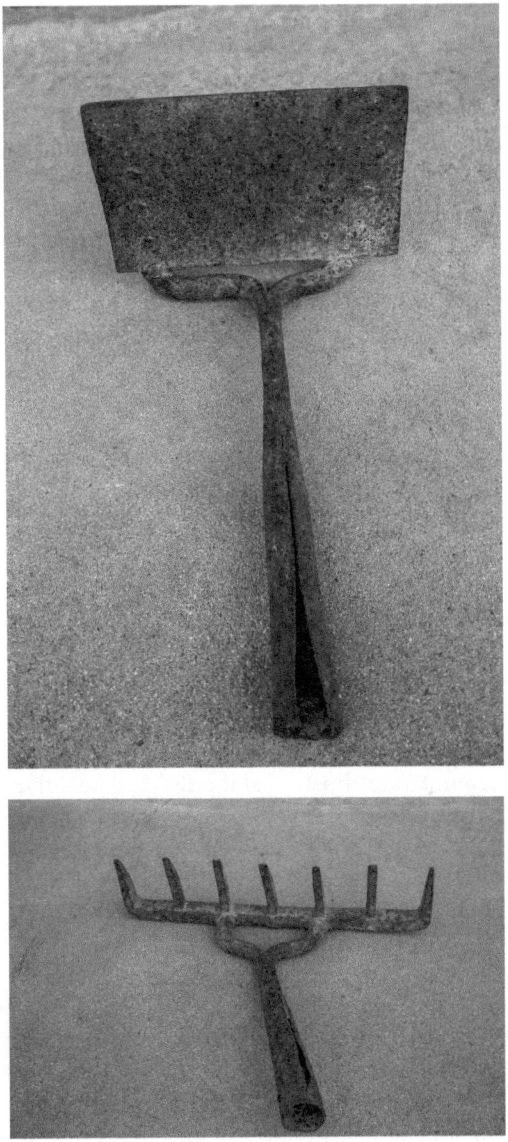

Fig. 6.1 Chinese hoe and rake, collection of Hay Goal Museum, Hay, New South Wales. Photograph by Margie McLelland, Wildprints.

heavy beam which levelled the soil and crushed the clods into a fine tilth.[53] The hoes they used to weed between the rows of growing plants and aerate the soil became more sophisticated over time, making the heavy physical work of gardening easier. Chinese gardeners replaced hand hoes with commercially manufactured hand-operated single or double wheel hoes such as the 'Planet Junior'. Later they used small mechanised rotary hoes and on larger acreages, multi-row cultivators.[54] Similarly in Australia, the range of gardening implements used varied according to the size of the garden. In southern and western New South Wales, for example, horse-drawn harrows and ploughs were used in larger gardens.[55]

Individual gardeners adapted agricultural implements from the materials they had available, particularly in remote areas of Australia. One example is a four-tined garden rake recovered from the site of Ah Toy's garden on the Palmer River goldfields in far north Queensland. The rake was ingeniously made by modifying a broken ten-tined sluicing rake originally used in mining operations to rake alluvial gravel. Three of the outer tines of the rake had broken off. Three tines on the opposite end were later sawn off to balance the tool, so it became a useful garden rake.[56]

Thus, rather than being discarded, the rake originally used in gold mining operations took on new life as a garden rake. It illustrates how artefacts can be modified through their life cycle and take on new roles, being used for tasks never intended or imagined by their original makers. In this process, artefacts accumulate additional layers of context and meaning. The rake demonstrates bush 'making do' and adaptive reuse – improvisation common to both Chinese and Europeans living in the Australian outback, where metal implements were valued and not lightly discarded. It also represents the economic transition between mining and market gardening as the main source of subsistence for Chinese immigrants, the continuity of Chinese horticultural practices and the hard manual labour involved.

Chinese market gardeners in regional New South Wales adapted and improvised their agricultural implements. Surveying the site of the market garden in Goulburn operated by the Nomchong family until the 1950s, McGowan recorded a large clod crusher, the roller studded with iron pegs from the nearby railway line.[57] On the site of a Chinese market garden on Bedooba Station, north of Gilgunnia in western New South Wales, he found numerous improvised agricultural implements, including a single blade plough and seeder.[58] The collection of

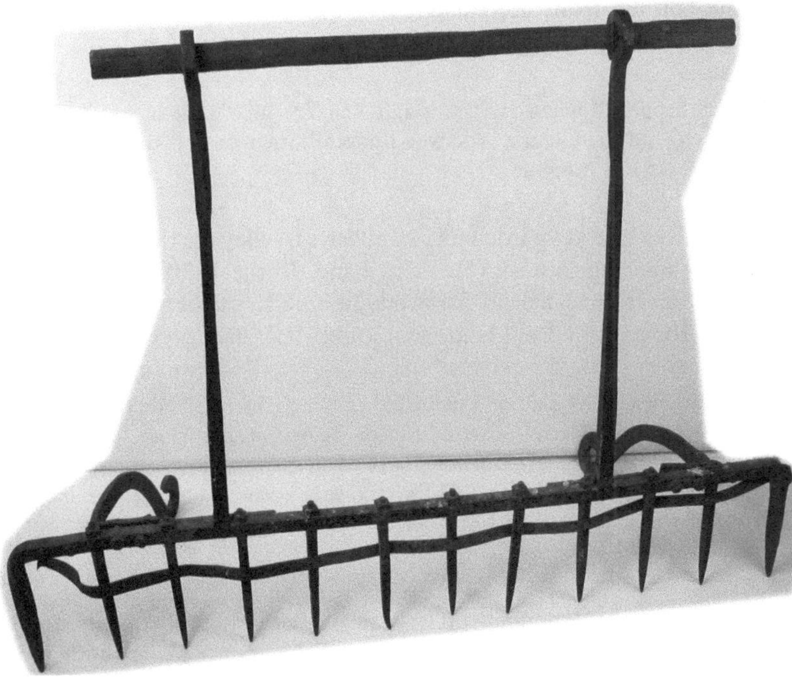

Fig. 6.2 Chinese harrow, collection of Museum of the Riverina, Wagga Wagga, New South Wales. Courtesy Museum of the Riverina.

the Museum of the Riverina includes agricultural implements improvised by Chinese market gardeners, such as a plough and small harrows or seeders (Fig. 6.2). The wheels of these attachments were often recycled from children's tricycles or prams.[59]

There are many other cases of Chinese market gardeners improvising agricultural implements using whatever materials they had available, to make the tasks of harvesting and preparing produce for market easier. In New Zealand, Lee Bak Yew, who started a market garden in Oamaru in 1952, was keen to make his operations more efficient. He and his partners changed from watering by hand to using hoses and experimented with making machines to harvest Brussels sprouts, which were labour intensive to pick. In the 1960s they made a machine to strip the sprouts from the stalks using bits of scrap metal.[60] Lee Bak Yew's son Peter Lee recalls their trial and error approach:

> This all happened when my father and Ken Yee used a bit of scrap metal from here and there and a little wee Bunsen burner to heat this and bend that to make this machine.[61]

Some inventive market gardeners found ways to alleviate the labour intensive task of washing root vegetables by hand. In the 1950s Jack Chong, a market gardener in Manaia in Taranaki, designed a carrot washer which he built with the aid of a local engineering firm. His daughter Sue recalls:

> It was the first of its kind and consisted of a long barrel, made of wooden slats, rotated by a petrol-powered motor. It was located close to the river…water was pumped up from the river and sprayed onto the rotating carrots from sprinklers fixed above the revolving washer. Covered in soil, the carrots were fed in from the top and fell into a box at the bottom all wet but nice and clean.[62]

Chinese market gardeners also ingeniously adapted commercially manufactured labour saving devices such as wheel hoes to make them more efficient. Peter Lee, the son of Lee Bak Yew, recalls how his cousins made a mechanical planter using pieces of tin and bicycle wheels, and modified machines for sowing seeds and spreading fertiliser. He also describes how his father adapted the hand operated Planet Junior wheel hoe for use with a tractor:

> We used to sow seeds by pushing a Planet Junior row after row. So he [my father] sent me to Christchurch to go to this company to see if I can find this model of a Planet Junior. So I went and I had to buy four of them and he made a rack and hung the handles of the machines on the rack and towed it behind the tractor rather than having to push them over one row at a time.[63]

A striking contemporary example of how little some aspects of market gardening have changed is the garden of Gordon Ha in Botany, one of the longest surviving Chinese market gardens in Sydney (Fig. 6.3). Today Ha and his cousin cultivate Asian herbs and Chinese vegetables such as bok choy and cho sum for a niche market instead of the European vegetables their parents grew. They use rotary hoes and a tractor to cultivate their 5.5

Fig. 6.3 The market garden worked by Gordon Ha and his cousin in Botany, Sydney, 2011. Photograph Joanna Boileau.

acre plot. However, they still use traditional labour intensive methods, harvesting their crops by hand.[64]

Motor Vehicles and Tractors

The development of the internal combustion engine in the nineteenth century revolutionised road and later rail transport. From the 1930s market gardeners in Australia and New Zealand began to purchase motor vehicles to transport their produce, and later tractors to cultivate the soil. The rate at which market gardeners generally, and Chinese market gardeners in particular, mechanised their operations varied. There is evidence that some Chinese market gardeners were slow to take up the use of motor vehicles and tractors. In her recollections of her family's market garden in the Sydney suburb of Granville, Marjorie Long Dodd recalls that her father bought a truck in the 1930s, a 1929 Chevrolet, but did not learn to drive. For many years his neighbour drove the truck to take his vegetables to the city markets each week.[65] Owen Bennett, who grew up in Ryde, another Sydney suburb, in the 1930s, recalls that Charlie, a local Chinese market gardener, never had his own truck, relying on a European carrier to take his produce to the markets.[66]

However, in some areas gardeners were quick to adopt machinery. In Ohakune in New Zealand, for example, the laborious methods of removing stumps by hand were quickly abandoned with the appearance of the first bulldozer in the district in 1937. Local farmers, both European and Chinese, were initially sceptical of the ability of the machine to remove the heavier roots and stumps, but their doubts were dispelled when the Ministry of Works demonstrated it. Kong Shem came to Ohakune from Auckland in around 1935. He remembers the arrival of the first bulldozer in the district and how it could clear an acre of land in a day.[67] Mechanisation opened up new prospects for market gardening in Ohakune, enabling larger acreages to be brought into cultivation and enough vegetables to be produced to meet the needs of a large proportion of the North Island population.[68] This also illustrates the role of government agencies in the diffusion of innovation and the benefits of farmers being able to see new technology in operation in the field.

Other market gardeners bought motor vehicles and learned to drive as early as the 1920s. Quong Sing, who ran a large market garden in the Auckland suburb of Mount Roskill in the 1920s and 1930s and employed over 70 men, had several trucks. Neighbour Tom Grinter was often called over on cold winter mornings to help start them up.[69] Some men became skilled mechanics and specialised in servicing the vehicles of their compatriots, such as Jim Young and Mun Chew Young in Hawke's Bay in the 1940s.[70]

The main reason Chinese market gardeners did not buy motor vehicles in the 1920s and 1930s was probably their prohibitive cost rather than conservatism. It is likely that Chinese market gardeners, and many European gardeners as well, only bought motor vehicles and tractors when they could afford them. The intensive cultivation techniques used on small holdings and the typical layout of long, raised beds with narrow ditches between them meant that tractors had limited application. They were mainly used for preparing the ground before planting and after World War II, when more sophisticated attachments became available, for weeding, seeding and spreading fertiliser. Thus it was more economic to hire equipment or use contractors when necessary rather than make a major capital outlay on equipment that was not used regularly.

Some enterprising market gardeners such as Harry Gow and his son Jim in Otaki invested in tractors and did contract work for other growers. In 1942 Harry bought one of the first tractors imported to New Zealand from America, and Jim used it to do contract work for local growers.[71] During World War II there was a surge in farm mechanisation, particularly after the United States entered the Pacific war in 1942 and New Zealand became a major supplier of vegetables to the armed forces. Market gardening was now an essential industry, and despite being officially classified as aliens, Chinese gardeners were at the forefront of efforts to increase production. Large quantities of tractors, sprayers, pumps and vegetable planters were imported from the United States under a lend lease agreement.[72] One market gardener who obtained a licence to import tractors during the war was G.Y. Toy, who supplied a dehydration factory in Pukekohe. Toy was one of the largest growers in Hamilton from the 1940s; he had 180 acres in the Rukuhuia area which he and his family cultivated until the 1980s.[73]

Despite these wartime developments, in the early 1950s there was still relatively little mechanisation on market gardens in New Zealand. From the early 1960s mechanisation intensified as city market gardeners moved their operations to the urban fringes and average acreages increased. With the increased distance involved it became economic to own a truck to transport vegetables, and market gardeners purchased tractors and did their own ploughing and disking rather than contracting it out. Rotary hoes and specialised equipment to plant, fertilise, weed, spray and harvest crops were widely available.[74]

Motor Vehicle and Agricultural Equipment Manufacturers

To introduce market gardeners to the latest agricultural tools, some agricultural equipment manufacturers employed Chinese salespeople and advertised in Chinese language media. They also arranged field demonstrations of their products, an excellent way for market gardeners to evaluate their performance under local conditions. Allen and Company, the American manufacturers of the Planet Junior range of farming implements, exported their products to New Zealand from the early 1890s and held field demonstrations around the country in 1894.[75] In 1898 their sales representative, a Mr Sandford, arranged an exhibition of Planet

Junior products at Ming Quong's market gardens in Royal Oak, Auckland.[76]

As Chinese market gardeners began to purchase motor vehicles in the 1920s and 1930s, and later tractors and other agricultural equipment, these industries began to look to the Chinese market. In New Zealand, the International Harvester Company employed Chinese staff. Mun Chew Young, from Hawke's Bay, served an apprenticeship with International Harvester in Palmerston North in the 1940s.[77] A member of the Chinese community in Dunedin interviewed by Pawakapan gained a qualification as a skilled mechanic and from 1936 to 1953 he was employed by International Harvester selling tractors and trucks, especially to Chinese market gardeners.[78]

Advertisements in the *New Zealand Chinese Growers' Monthly Journal* in the 1950s provide evidence for a growing market for farm equipment among Chinese market gardeners. The journal began publication in August 1949 with a circulation of around 2,000. From the first issue the editors actively sought advertising from firms with business connections with the Chinese community.[79] Between 1949 and 1954, regular advertisements appeared for tractors and specialised attachments (for example, the advertisement for the British Ransomes Crawler tractor illustrated in Fig. 6.4).[80] There were also advertisements for pumps and irrigation systems such as the Buckner Giant Overhead Sprinkler System, placed by Brown Brothers Engineering. Their motto was 'profit flows when water flows'.[81]

In Australia, advertisements placed by car dealers, automobile maintenance companies, and driving schools began to appear in the *Tung Wah Times* in the late 1920s. They included Reo Motors Ltd; Manly Ford; the Sydney Automobile Maintenance Association; the Australian Roma Oil Company and Fowlers Motor College.[82]

Transport

During the nineteenth century roads, railways, ships and riverboats connected the isolated centres of population in Britain's colonies in the Antipodes and linked primary producers with their markets. Before the great railway building projects undertaken by colonial governments from the 1880s, roads were poor and shipping was the major form of transport for freight and passengers. The first railways built in Australia did not only originate from the capital cities, but fanned out from other major coastal

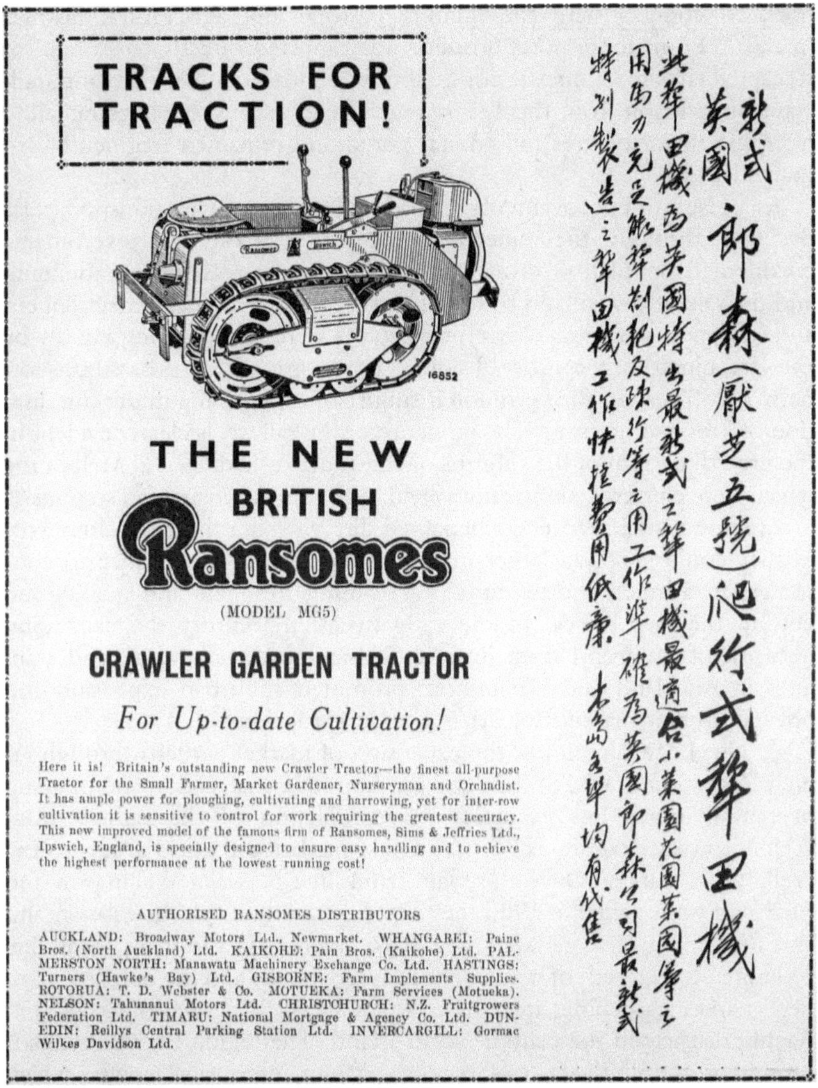

Fig. 6.4 Bilingual advertisement for Ransomes Crawler garden tractor, *New Zealand Chinese Growers' Monthly Journal*, 1 March 1950, p. 3.

ports to connect with the mining, pastoral and agricultural districts inland.[83] From these ports produce was exported directly to Britain, or to capital cites for domestic consumption. Similarly in New Zealand, railway construction from the 1870s began as separate short lines radiating from the major centres (all coastal ports) and remained fragmented for many decades.[84]

As a result of improvements in road, rail and sea travel, transport costs declined through the nineteenth century.[85] Colonial governments encouraged population growth in rural areas through railway building and the sale of crown land to settlers on credit as well as closer settlement and irrigation schemes. Urban populations were also growing rapidly by the late nineteenth century, leading to the growth of urban transport networks. These in turn promoted suburban expansion and decentralisation of jobs and housing. The influence of the railway is clearly evident in the growth pattern of the suburbs of cities such as Sydney and Melbourne which were clustered along railway and tram routes and around stations.[86]

Chinese market gardeners benefited directly from these developments as they could supply a larger market and deliver their produce to consumers in a much shorter time, maintaining freshness and quality and getting the best prices. In the early twentieth century the expanding network of train and tram lines in Sydney and Melbourne, and tram lines in Auckland and Wellington, promoted suburban expansion and opened up new markets for urban market gardeners.[87]

As Nigel Murphy notes, the expansion of market gardens throughout New Zealand followed in time and location the construction and opening of railway lines. For example, the completion of the railway from Wellington to Otaki in 1886 gave Otaki market gardeners access to the Wellington market. Once the main trunk line between Wellington and Auckland was opened in 1908 they supplied the larger towns *en route* and also the dining cars on trains.[88] The completion of the main trunk line promoted the growth of new towns and settlements and the opening up of new market gardening areas to serve them, such as the Ohakune and Raetihi districts in the central North Island. The region became a major centre for the cultivation of root vegetables, especially carrots, which became the iconic symbol of Ohakune.[89]

The expansion of rail networks in Australia in the late nineteenth century also supported the growth of market gardens in rural areas. In western New South Wales, market gardening centres developed near country towns with good transport connections to Sydney such as

Kelso, Wellington and Bathurst.[90] In western Queensland and the gulf region, Chinese market gardeners established gardens to serve the settlements which grew up along established stock routes and the railway line from Townsville to Hughenden.[91] Rail transport was not suitable for delicate, highly perishable produce such as salad vegetables and herbs – these continued to be produced close to the point of sale and transported by road.

The advent of motorised trucks in Australia and New Zealand from the 1920s challenged rail transport. In Otaki, for example, Pop Edwards began a small transport business in 1923 with one truck transporting butter from the Otaki dairy factory to Wellington. In 1929 he extended the service to vegetable growers, two days a week, to meet the Monday and Thursday produce auctions. As Alex McLellan recalls, at that time most Chinese and European market gardeners carried their produce to Wellington by horse and cart, leaving Otaki at 4 am. So popular was Edwards' new service he was soon working six days a week, and he had to buy extra trucks and recruit more drivers.[92] In an attempt to preserve the State's rail monopoly from the 1930s the New Zealand government passed legislation restricting the transport of goods by road in areas where rail operated. However, from the 1950s trucks were increasingly used to transport vegetables.[93] They proved to be faster and more efficient than rail.

Water Storage and Irrigation

Chinese market gardeners in arid areas of Australia made inventive use of storm water flows and soakages to harness irregular rainfall. The dams and water races they constructed are prime examples of technological and environmental adaptation. There are many examples of practical 'making do' in the variety of methods Chinese market gardeners used to raise and store water and store liquid manure on their gardens. On a Chinese market garden site in Cobar recorded by McGowan, storm water was channelled via sluice gates into two large dams on the gardens. Water was pumped from the dams into a water tower, then reticulated into large 440 gallon (2,000 litre) ship's tanks placed at strategic points around the garden. The tanks were partly sunk into the ground, so that hand-carried watering cans could be easily filled, in an adaptation of the in-ground water reservoirs typical of gardens in southern China.[94]

In Pine Creek in the Northern Territory, Nellie Fong recalls that her husband watered the plants in their market garden using a kerosene tin and stored liquid manure in 44 gallon drums, in place of the traditional large ceramic jars used in China. She vividly recalls the stench of the fertiliser, fortified with oxblood:

> ... and when they kill, the ox blood, we always take the tin out there for them to drain the ox blood into the tin. And that's our fertiliser, the liquid one – and can they smell! The whole garden smell ... We put them in a drum first, a big 44 gallon drum. And when we needed to fertilise them we just put them in the bucket and then dip it in water and then water it. For the seedlings it would be a very fine one, very thin one, and for the mature one it would be a stronger one.[95]

Kerosene tins were favoured by Chinese market gardeners for making watering cans, with the addition of spouts to produce a thin stream of water. In 1878 a correspondent to the *Queenslander* newspaper wrote:

> Kerosene tins, each holding four gallons, are held in favor by the Chinese for making sprinkling pots. To these tins, and also to round tins of about the same capacity, spouts of tin are soldered. The spouts are about two feet in length. A man takes two of these tins upon a stick of bamboo across his shoulders; the stick usually is about five feet long. John trots off ... to his waterhole or well. Dipping down one shoulder he fills a tin; then he dips down the other without taking the load from his shoulders, and trots back to where the water is to be applied. He has a hand upon each tin, and as he trots along the alleys the water is poured out in a flat, full stream with a shaking motion that thoroughly wets two feet on each side. This goes on until the whole of the cultivation is soaked.[96]

Kerosene tins fulfilled many other useful purposes. In New Zealand, for example, with the addition of a handle, they were used as containers for blood and bone or bone dust, or for harvesting peas in the Wairarapa.[97] Some market gardeners such as Kim Young in Ohakune adapted smaller tins for sowing fine seeds such as carrots, by punching holes in the bottom.[98]

In the Sydney suburb of Manly in the early twentieth century, Chinese market gardeners cultivated crops along the streams running into the Manly and Curl Curl lagoons. Local resident Charles Goodman recalls that they used baths to store liquid fertiliser, a smelly mix of pig manure and nightsoil.[99] In Granville, Sydney, in the 1930s the Leung (anglicised to Long) family had the luxury of galvanised iron pipes running through their market garden. At every second tap was an old bath which was filled with water to make filling the watering cans easier. At other taps there were rectangular water storage tanks dug into the ground, with cleated planks for walking down into the water to fill the watering cans, balanced in typical Chinese style on each end of a bamboo shoulder pole. According to Ma Leung's daughter Marjorie, the watering cans were manufactured specifically for Chinese gardeners. The back-breaking work of hand watering came to an end in 1940 when Marjorie's brother Norman installed a Skinner overhead irrigation system with sprinklers fixed on hardwood posts 6 feet above the ground.[100]

Traditional Chinese methods of raising irrigation water from watercourses and the characteristic network of irrigation channels and water reservoirs traversing the gardens, endured in Australia and New Zealand. There are numerous descriptions of the traditional foot-powered chain pump in use. In 1886 the Melbourne *Argus*, for example, published an account of the 6-acre garden of Ah Sam on the Barwon River in Geelong. The report describes the age-old methods used for raising water and the water reservoirs scattered throughout the garden.

> The land, formed of two slopes, north and south, runs parallel with the river, and in the centre, on top of the rise, and at about fifty feet from the Barwon, there is a Californian pump, worked by three men by treadle motion. In a comparatively short time, sufficient water can be pumped up to flood the whole six acres; but with an eye to economising labour in seasons when so much moisture is unnecessary, Ah Sam has constructed slabbed pits at every few yards along the various paths, and the water, as it flows from the shoot, passes along the different channels, and fills these pits before it can overflow the ground generally.[101]

In the early 1900s Chinese market gardeners in Forbury, South Dunedin used foot-powered chain pumps to drain water from their low-lying land, emptying it into drainage channels along the adjoining road. In 1902 the Reverend Alexander Don recorded in his diary:

> The Chinese on the low-lying Forbury market-gardens are applying the machine that in China they use to raise the water from streams up to their rice fields, here to raise the water off their gardens into the street open drains to run off...[102]

Figure 6.5 shows a chain pump in use in South Dunedin. It is very similar in design to the chain pump photographed by King in China in the early 1900s, illustrated in Chapter 2 (Fig. 2.1).[103]

Fig. 6.5 A foot-powered chain pump in use in South Dunedin, early 1900s. Courtesy Kings High School, Dunedin.

In many cases water was raised by horse power, using a whim or horse capstan.[104] In western New South Wales in the 1890s, whims were used in Chong Lee's garden adjacent to the Chinese camp on banks of the Lachlan River at Hillston and in the large garden (known as the Garden of Eden) in the Chinese camp on the banks of the Darling River at Bourke.[105] They were also used to raise water from wells, for example in Gilgandra and Tenterfield.[106] The horse was not generally used as a draught animal in southern China, but in Australia and New Zealand Chinese market gardeners quickly made use of it. Many observers commented on the care and attention Chinese market gardeners gave their horses, as they were so important to their gardening operations.[107] Tom Parr, who worked in the Hillston gardens in his school holidays for 5 shillings a week in the early 1900s, recalls that the Chinese gardeners raised water from the river using small buckets holding about 2 litres, fastened to an endless chain driven by a horse.

> There were no engines in those days, none for pumping water for market gardens. Their method of getting the water up to the surface from the river was by means of small buckets, say perhaps holding half a gallon... These were fastened to an endless chain and the buckets completed the circle, thus circulating a continuous stream of water. This setup was driven by one horse going round and round continuously. He had to be driven or he would stop. The horse was blindfolded, mostly because he got too cunning and it necessitated someone to keep him going.[108]

In other cases the whim was powered by a windmill. McGowan records that the two market gardens established by 1896 in Booligal in western New South Wales were irrigated by windmills which pumped water from the Lachlan River.[109] In 1902 the *Brisbane Courier* described a home-made windmill with canvas sails used by Chinese market gardeners on a plot of 'several acres' in the Brisbane suburb of Enoggera. Windmills were still in use in New Zealand in the 1930s. In 1933 Chun Ding and his son Chew Chung Ding moved their market gardening operations from the Kaikorai valley to Outram where they used a windmill to pump water from an underground bore into 44 gallon drums. A few years later they bought a Fordson tractor and attached two of the drums to it to carry the water for irrigating their crops.[110]

Chinese market gardeners turned to new power sources, key developments of the British industrial revolution which were transferred to Australia and New Zealand from the late nineteenth century. Inanimate power sources such as coal, oil, petroleum and steam replaced human and animal sources in many industries.[111] Horse-powered whims were succeeded by steam pumps in some areas of Australia by the late nineteenth century. In 1902, for example, the *Brisbane Courier* reported on the successes in irrigation achieved by an unnamed Chinese market gardener, in Warwick, west of Brisbane.

> He has thirty acres of land under cultivation, and on the bank of the river has erected a four horse power steam pump. By this means the water is raised from the river, and it is then conveyed through long shallow trenches to different parts of the ground, and lodged in small dams, which are located at the side of the trenches, and from these dams it is spread over the plots in the usual way.[112]

Steam pumps were in common use until the 1920s, but declined in the 1930s with the introduction of electric pumps. In the 1920s and 1930s Chinese market gardeners in Camden, New South Wales pumped water from the Nepean River using a four-inch centrifugal pump, belt-driven and powered by a portable 15 horsepower Marshall or Ruston steam engine on four wheels. Shafts could be fitted to the movable turntable at the front of the engine mount so that draught horses could move the pump to wherever it was required. Local historian R. Nixon notes the important role played by the 'engine man' in the syndicates which operated the market gardens. Ranking third in importance after the 'boss' of the garden, who had the largest financial stake in the enterprise, and the man responsible for marketing produce, the engine man maintained the irrigation pumps along the river.[113] Sometimes he sought the assistance of local European tradespeople such as Joe Kearns, who ran a motor garage in Camden, or Stan McKnight, the engineer at the local milk factory. By the late 1930s steam engines were wearing out and parts were no longer available. Electric pumps required far less maintenance than steam pumps, releasing another man to work on other tasks in the gardens.[114]

Mechanised irrigation systems would have revolutionised gardeners' lives, releasing them from the daily toil of hand watering and saving much time and labour. In New South Wales, petrol pumps were used in Cobar, Bourke and Goulburn from the 1920s.[115] In the Sydney suburb of Willoughby, a centrifugal pump powered by an electric motor was installed in the King family garden in the 1920s or 1930s, to raise water from Sailors Bay Creek.[116] The water was distributed through the garden via galvanised iron pipes to several taps. Sprinklers attached to rubber hoses could be moved around the garden as required. In 1920 a Queensland newspaper reported the handsome returns achieved by a Chinese market gardener in Mitchelton, north-west of Brisbane, as a result of intensive irrigation using an oil engine and pump. His tomato crop that season brought him a gross return of £1,000 on the Melbourne market.[117]

Mechanical pumps were costly in the late nineteenth century. In 1892 the *Riverine Grazier* reported that in Hay 'the company of Chinese gardeners who lease from the Council acres of land nearly opposite the slaughter yards have just had erected an engine and pump at a cost of £280'.[118] This represents a major capital investment at a time when workers on Chinese market gardens earned around £52 per annum.[119] Thus it is likely that only the most successful market gardening syndicates could afford this technology, while many others continued traditional animal or human-powered methods of raising water. As this case indicates it is unwise to make general statements about the pace and timing of technological change, for the location, size and financial success of Chinese market gardens were important factors in how readily gardeners adopted technological advances. Thus the adoption of mechanisation was unevenly spread, both geographically and temporally.

Horse-powered pumps and horse-drawn vehicles were in use in many areas long after steam pumps and, later, petrol and electric pumps and motor vehicles became widely available. For example, the Chinese market gardener Georgie Ah Ling, who first came to the Victorian town of Donald to assist his cousin in his market garden in the late 1920s, continued to use traditional Chinese agricultural methods throughout his long life. After his partner died in the late 1940s he worked alone on his small garden. He hand watered his crops, never sprayed them with pesticides and delivered his produce by horse-drawn vehicle until the early 1980s. However, at some time he purchased a Clutterbuck oil engine to pump water from the nearby reservoir, easing the heavy work of lifting water by hand.[120] Preserved in the Agricultural Museum in Donald, this engine is an Australian-made copy of the British Blackstone oil engine and was manufactured from around 1916 by James Martin and Co. in Gawler, South Australia for Clutterbuck Brothers.[121]

Fertilisers

The two essentials of market gardening were water and manure to maintain soil fertility. Chinese market gardeners adapted their methods of fertilising their fields and controlling pests in Australia and New Zealand by taking advantage of the availability of new products. In the late nineteenth and early twentieth centuries they had a ready supply of animal manure to maintain the fertility of their soil, provided by horses, cows, pigs, sheep and chickens, even in cities. The suburbs of cities in Australia and New Zealand were still teeming with animal life. In her study of food production in Australian cities, Andrea Gaynor vividly describes the crowing of cocks at daybreak, the lowing and stamping of dairy herds, rats scuttling in laneways and horses hauling delivery carts.[122] In Sydney, Melbourne, Brisbane and Perth, the main source of fertiliser was horse manure and straw, which came from racing, livery and tramway stables.[123] Sheep were another source of fertiliser. Gardeners in Camden, west of Sydney, for example, purchased large quantities of sheep manure from the Flemington sale yards.[124]

Urban market gardens were generally located close to water sources, such as lakes and creeks, and close to other industries requiring water, such as slaughterhouses, tanneries, dairies, poultry farms, piggeries and stables. These industries all supported market gardening as they provided ready sources of fertiliser. The Sydney suburb of Botany was known as Struggletown in the late 1880s and 1890s. It had an unsavoury reputation due to the sights, sounds and smells of the glue works, tanneries and fellmongeries which were interspersed between Chinese market gardens and the houses of factory workers and fishermen.[125] In Perth, Chinese market gardens were located around the fringes of Smith's Lake north of the city, side by side with piggeries, dairies and dumps for nightsoil.[126]

In country towns the surrounding primary industries provided plentiful manure from sheep, cattle, horses and pigs, for example in Deniliquin, New South Wales and Otaki in New Zealand.[127] Gardeners also collected waste from local slaughter yards for use as fertiliser, performing a useful recycling service, for example in the Perth suburb of Bayswater in the early 1900s and in Daylesford, Victoria, in the 1860s.[128]

Chinese market gardeners used prodigious quantities of manure, especially on sandy, peaty soils. In Perth, 200 tons of stable manure was ploughed into one 6-acre garden in Bayswater during the non-growing season. This was mainly animal manure from neighbouring stables and dairy farms and waste from the sewerage treatment works at Welshpool.[129] Thus Chinese market gardeners made a significant contribution to recycling the huge quantities of manure produced by animals in cities.[130]

In many cases Chinese market gardeners in Australia and New Zealand continued traditional methods, collecting their own urine and nightsoil to produce liquid manure. According to David Horsfall, in the late nineteenth century Chinese market gardeners in Victoria were the collectors of nightsoil 'in scores of country towns'.[131] For example, there are reports of gardeners collecting nightsoil in Geelong in 1870.[132] Human waste was often collected under cover of darkness, hence the euphemism nightsoil. The use of nightsoil was controversial and by the late nineteenth century, as local governments became more aware of public health and sanitation issues, came under closer scrutiny and regulation. In Western Australia in the early 1900s, the Fremantle municipality sold nightsoil to gardeners. After the Engineer for Roads and Bridges submitted a report complaining about the state of the gardens along the Jandakot Road in 1908, all sales of nightsoil were banned and the Fremantle Board of Health's health inspector was removed from office.[133]

By the early twentieth century local government authorities increasingly regulated the keeping of animals within town and city boundaries, in the interests of public hygiene. However, dairies, stables, market gardens and poultry farms remained scattered through the expanding suburbs of urban areas well into the 1920s and 1930s. The oral histories collected by the City of Ryde Council in Sydney paint a picture of a semi-rural landscape, dotted with Chinese market gardens, poultry farms, orchards, stables and dairies.[134] Horses in particular were a familiar presence in suburban streets. Well into the twentieth century they were used to deliver bread, milk, fruit and vegetables and many other goods and they were an integral part of the localised structure of food production and retailing in the days before refrigeration.[135]

By the interwar period supplies of manure in cities were declining as the number of horse-drawn vehicles decreased, suburban dairies, stables and poultry farms were displaced by housing and industry and sewerage systems were constructed. However, large areas of cities and many country towns were still serviced by night carts.[136] Like other market gardeners, Chinese market gardeners turned to other sources of fertiliser and increasingly purchased commercial products such as blood and bone and superphosphate. Articles published in the *Tung Wah Times* between 1902 and 1930 indicate the interest with which Chinese market gardeners followed the latest developments in agriculture, both in China and Australia. For example, a report was published of a Sydney company winning an award for high quality fertilisers in 1902, and in 1908 there was a report of a new chemical fertiliser being manufactured in Foshan, in Guangdong.[137] The newspaper also provided advice on the best fertilisers for specific crops.[138] It is difficult to quantify what proportion of Chinese market gardeners used commercial fertilisers and to what extent they continued to use animal manures. The relative cost was obviously a factor, and the size and financial success of the market gardening enterprise. As the *Tung Wah Times* reported in 1930, high customs duties increased the cost of imported fertilisers.[139]

The availability of commercially manufactured fertilisers from the 1840s, as advances in knowledge of the role of trace elements in plant nutrition were taken up by industry, was a major advance in agriculture in Britain.[140] The German chemist Justus Von Liebig's research paved the way for the development of chemical fertilisers. Von Liebig was well aware of the prior example set by Chinese farmers, describing them as 'the most admirable gardeners and trainers of plants, for each of which they understand how to prepare and apply the best adapted manure'.[141] Superphosphates, made by treating phosphate rock with sulphuric acid, made trace elements more readily available to plants and were easier to apply than traditional animal manures and composts.[142]

These developments reached Australia and New Zealand, where domestic fertiliser industries were established from the 1880s.[143] In New Zealand, for example, local production of superphosphate began in 1882 at the Kempthorne, Prosser & Company plant near Dunedin.[144] The rapid expansion of commercial fertiliser industries in Australia and New Zealand, mainly for use on pastures, was stimulated by the advent of refrigeration in 1882 and the rapid growth of the meat and dairy industries. There was also a developing export trade in blood and bone and superphosphate from Australia to New Zealand by the early 1900s.[145] The growth of fertiliser industries supported the development of the market gardening industry, and Chinese market gardeners were important domestic customers. In 1910, Mr H. Anderson, Undersecretary for Agriculture in New South Wales, observed:

> Chinese vegetable gardeners are the best customers for our highest quality of blood and bone manure, and I need hardly remind you that the Chinese and Japanese were intense cultivators thousands of years before New South Wales was discovered. These nations have nothing to learn from us with regard to petite culture and the utilisation of every possible source of fertility for the soil.[146]

Many Chinese market gardeners were willing to adapt their traditional gardening practices to the changing world around them and adopt more scientific approaches to horticulture. One was James Chung Gon, in Launceston. By the late 1940s, if not earlier, he was experimenting with artificial fertilisers and adopting scientific methods of soil analysis and the addition of trace elements. In 1953 the *Mercury* newspaper in Hobart reported:

> In addition to the high cost of labour, he is finding it more difficult to grow good crops on the garden because it is impossible to obtain adequate supplies of stable manure. Even artificial fertilisers, like blood and bone, are hard to obtain, he says, and they have become extremely expensive . . . In recent seasons Mr Chung Gon has been experimenting with various types of artificial manures to replace the hundreds of cart-loads of stable manure that once were carted into the garden every month. He uses six tons of blood and bone, half a ton of ammonia and a quarter of a ton of potash every year. Latest tests have been with a special mixture of a dozen different trace elements.[147]

Pest Control

Chinese market gardeners in Australia and New Zealand continued their traditional methods of pest control, well into the twentieth century. For example, Esma Smith, who grew up in Forbes, New South Wales in the 1940s, remembers that her father walked up and down his rows of vegetables painstakingly collecting beetles in a small funnel-shaped wooden container. He would then pour hot water through it.[148] This age-old Chinese method of hand picking was recommended in British pest control manuals distributed in Australia in the 1890s as the surest and most cost-effective method of controlling pests such as the caterpillars of cabbage moth.[149]

Arsenic-based pesticides were developed in Europe and North America in the nineteenth century; they included Paris Green (a copper-based arsenate), London Purple and arsenate of lead.[150] In fact arsenic-based pesticides had been known in China for centuries.[151] They were used by Chinese market gardeners in Australia and New Zealand, for example by James Chung Gon in Launceston, and were highly toxic to animals and humans.[152] Other methods of pest control used by Chinese and European gardeners included kerosene emulsion, tobacco dust and derris dust. Kerosene emulsion (a mix of kerosene, soap and water) was particularly effective for treating cabbage moth. In 1910 a correspondent to a Queensland newspaper reported:

> A large grower, an up-to-date Chinese gardener, told me that he used nothing else on his cabbages, and that where the moth was plentiful he gave them regular sprayings until the plants got ahead of the pest.[153]

Tobacco dust, a traditional remedy familiar to both European and Chinese gardeners, was widely used in the nineteenth century to combat aphis and thrips, although repeated applications were required for the treatment to be effective.[154] Derris dust was developed in the 1930s in the United States as a less toxic alternative to arsenic-based pesticides, and after local testing, was in use in Australia by 1935 and in New Zealand by 1936.[155]

In their recollections of the 1930s some European observers comment that Chinese market gardeners were conservative and reluctant to use any modern sprays, so the quality of their produce was not as good as that of European or Italian growers.[156] However, there is evidence that Chinese market gardeners adopted modern artificial pesticides while continuing to use more traditional methods. Chung Gon, for example, continued to use traditional methods to control white cabbage butterfly through to the 1940s, interplanting cabbages with lettuce. He sprayed young plants with arsenate of lead and treated more mature plants by dusting lightly with derris dust. He also used tobacco dust to control aphis. To control the cabbage butterfly he placed a lamp with a naked light in the garden at night; the pests would fly into the flame and die.[157] Drawing on his Chinese experience, to protect the fruit in his orchard from birds he suspended large, slow-burning coils of incense from the trees with dozens of small fire crackers attached to them. As the spiral burned the crackers ignited and exploded, dropping into a tin drum on the ground below.[158]

The pesticide and herbicide industries expanded after World War II. These industries supported the development of the market gardening industry, as market gardeners turned from traditional pest control methods to commercial products.[159] Advertisements in the *Tung Wah Times* for arsenate of lead pesticides (manufactured by Coopers) and derris dust (marketed by McDougall Brothers under the brand name Katakilla) suggest that both products were used by Chinese market gardeners in the 1930s.[160] By the 1960s artificial pesticides and herbicides were widely used. Between 1962 and 1964 the Chinese language *New Zealand Chinese Growers' Monthly Journal* included advertisements placed by Henry H. York and Co. (for Folidol M50, a new organic phosphorous compound recently developed by Bayer in Germany); Ivory Spray Chemicals; Shell Agricultural Chemicals; and Dow Chemicals International.[161]

New Growing Methods

Particularly innovative Chinese market gardeners kept abreast of the latest technological developments in Australia, New Zealand and overseas, as well as changing consumer tastes and the changing demands of food manufacturers. They experimented with new growing methods and crop varieties. One such leader in his own community was Thomas (Tom) Wing of Hawke's Bay in New Zealand. He worked in partnership with his younger brother Alan Wing and Richard (Dick) Smith. They ran market gardens and a fruit orchard. Tom and his wife also ran a fruit and vegetable shop nearby in Hastings for many years. In the 1960s Tom imported lettuce and cauliflower seeds from California after reading about them in the US magazine *Saturday Evening Post.* He grew lettuces on polythene, a new method developed in America which had only just been introduced to New Zealand. Iceberg lettuces thrived in New Zealand; they grew so big that sometimes only four would fit in the old wooden banana cases. Tom marketed them as 'Krispy Cut'. The cauliflowers also did well and he invented 'Tight and White' as the market brand. Tom's son Robert Wing recalls:

> 'Tom was the thinker of the brothers and from what I remember, the worker of the two.'[162]

Joe and Fay Gock, market gardeners in Mangere, were also innovators, entrepreneurs and community opinion leaders. During the 1960s they developed modern methods of growing kumara in hotbeds. With the assistance of the Department of Industrial and Scientific Research they developed a temperature controlled kumara storage and curing shed which extended the length of time the crop could be stored and dramatically reduced losses.[163] Innovation and developing new product to meet changing consumer demands has become even more important with the changes in the market gardening industry over the last two decades and the rise of large supermarket chains.[164]

Seeds and New Crop Varieties

In seeds lay the germ of market gardeners' hopes and their future prosperity. Chinese market gardeners introduced Chinese plant species to Australia and New Zealand, applied their intensive horticultural methods to European crops and developed new crop varieties. Many new immigrants, both European and Chinese, brought the seeds of trees, flowers and vegetables with them to New Zealand and Australia. Paul Fox documents the key role which the burgeoning plant trade between Britain and Australia, conducted by nurserymen on both sides of the globe, played in refashioning the Australian landscape as it introduced exotic flora from around the world.[165]

The Chinese entrepreneurs who first established market gardens on the goldfields in Australia and New Zealand in the 1860s brought seeds with them from China or obtained them through local Chinese merchants. In 1864 a correspondent to the *South Australian Advertiser* in Adelaide wrote an account of the vegetables imported from China by Chinese who settled in Victoria, which included varieties of Chinese cabbage, melon seeds and fresh ginger. The most popular vegetables were varieties of the *Thornea* genus, which he noted were available in Melbourne and could be grown successfully from seed in the Victorian climate.

> There are several varieties resembling turnips, but ribbed like a sweet melon *(Thornea Battatas)*, and weighing from half a pound to three pounds, sweet taste, and good either roast or boiled, and highly nutritious. Another variety resembles in form beetroot or wurtsel, weighs two to three pounds, and being very palatable is often used raw; the flavor is that of the sweet chestnut. The mode of propagation is from seed.[166]

Chinese gardeners in New Zealand also imported the seeds of Chinese vegetables direct from China, as evidenced by an incident reported by Alexander Don. In 1883 he met a Chinese man *en route* to the Round Hill goldfields from Riverton who was carrying a sack of turnip seed. He told Don that he had imported the seed from China because he considered it was better than the local variety.[167]

Chinese market gardeners quickly turned to growing a wide variety of European vegetables for the European market and commercial nurseries supplied them with seed. Seed and plant nurseries became a burgeoning industry in Australia and New Zealand from the mid nineteenth century, as plants and seeds from around the globe were propagated in the British colonies.[168] James Beattie and John Adam document the wide range of plants from Asia that came into New Zealand, highlighting the polyglot, multicultural nature of the British Empire and the complex circulation of ideas, people, goods and plants between continents. As Beattie notes, little has been written about the important role of Chinese immigrants in introducing gardening practices and plants from their homeland.[169]

Colonial nurserymen were at the forefront of experimentation in what plants could be grown in different climatic zones in Australia and New Zealand.[170] They also saw the opportunity for marketing their products to Chinese gardeners. One was the entrepreneurial nurseryman Thomas Lang of Ballarat, who imported a million plants into Victoria between 1858 and 1870. Quick to capitalise on the commercial possibilities of the expanding goldfields towns, Lang imported seeds from as far afield as the Himalayas and Shanghai. To cater for the large Chinese population in the Ballarat region, he prominently displayed the seeds he imported from China in the window of his Ballarat store, with Chinese labels.[171]

Chinese merchants also played their part in introducing Chinese plant species. In northern Australia many of the subtropical fruit and vegetables familiar to immigrants from southern China could easily be cultivated and there was a significant trade in seeds and preserved foodstuffs through the busy port of Darwin, Australia's gateway to Asia. Eddie Quong grew up on his parents' market garden in Pine Creek in the Northern Territory which his grandparents, newly arrived from China, had established in the late 1880s or early 1890s. They grew rice, fruit and vegetables, including Chinese vegetables to supply the large Chinese population in the Territory. Quong recalls that his grandmother did not bring vegetable seeds with her, but they could easily be obtained from China. At that time there were frequent shipping services between Hong Kong and Darwin, and regular contacts between Chinese people living on the coast and those living inland at Pine Creek.[172]

Farmers have been modifying the genetic makeup of the crops they grow for thousands of years by selecting for traits such as higher yield, faster growth and resistance to disease and pests. To economise many Chinese market gardeners continued to collect seeds from their own plants in addition to purchasing them from commercial seed suppliers. In the early twentieth century in Western Australia, rural and metropolitan gardeners exchanged seeds from their own plants; the change of soil and climate was thought to benefit seed strains.[173] In New Zealand in the 1930s, Alice and Percy Lee enlisted their children to collect seed from their 11-acre market garden in Christchurch.[174] George Sue, who grew up in a market gardening family in Otaki in the 1940s, recalls that Chinese market gardeners saved the seeds from their plants, particularly the seeds of the Chinese vegetables which they grew for their own consumption. Only a few varieties of Chinese vegetables could be grown in the New Zealand climate, and with the political upheavals taking place in China in the 1930s and 1940s imported seeds were difficult to obtain.[175] In the 1950s, before modern hybridised seeds were widely available, many gardeners were secretive about their own seed varieties. Percy Luen of Luen Brothers market garden in Mangere recalls:

> It would not be protocol to ask a fellow grower where they got a certain variety of seed from... To keep the strain of seed exclusive to themselves, most gardeners saved the seeds of crops that had performed well.[176]

By the late nineteenth century seed merchants were large business enterprises. Carter and Sutton in London, for example, targeted intending migrants to the colonies. In Australia and New Zealand, Arthur Yates and Co. marketed the latest varieties of plants suitable for local conditions and published artistically designed mail order catalogues and gardening guides.[177] Entrepreneurial Chinese businessmen also saw the commercial opportunities for supplying their market gardening compatriots with seed. Chan Moon Ting (also known as James Chin Ting), who emigrated to New Zealand with his two brothers in 1886 and ran a general store in Wellington, established the Te Aro Seed Company in 1913 in partnership with Yee Chong Wing. They later established a nursery and large glasshouse in nearby Hataitai where they propagated a wide variety of seedlings. As the company grew sales representatives travelled around the country advising Chinese market gardeners on the latest seed varieties and taking orders.[178]

Commercial advertising provides an indication of the potential market for seeds among Chinese gardeners. Seed merchants were aware of the predominance of Chinese in the market gardening industry and placed advertisements in Chinese language newspapers, despite their relatively small circulation. In the early 1900s in Sydney, seed merchants Anderson & Co. and George Austen & Co. employed translators to write to Chinese gardeners and advertised in the *Chinese Australian Herald*.[179] Seed merchants also advertised regularly in the *Tung Wah Times* from 1899, particularly between 1917 and 1927.[180] In New Zealand, the seed merchant R. A. Nicol advertised regularly in the *Man Sing Times*, published only from 1921 to 1922.[181] A number of seed merchants advertised in the *New Zealand Chinese Growers' Monthly Journal* from its inception in 1949, including Arthur Yates and Co. and F. M. Winstone Seeds (Fig. 6.6). Te Aro Seeds Ltd, trading as Yee Chong Wing & Co., was the most frequent advertiser.[182] From the 1970s, advances in plant breeding and hybridisation provided growers with a wide range of crop varieties selected for frost and disease tolerance.[183]

Individual Chinese market gardeners also played an important role in improving plant varieties through breeding. In New Zealand in the 1950s, for example, Peter Lee's father Lee Bak Yew, a market gardener in Oamaru, developed his own seed stock from a single pod of Brussels sprouts. The seeds were originally from the Suttons variety 'Full Basket', imported from England, but Peter believes his father's variety was better adapted to local conditions and superior in flavour.[184] In the late 1950s Joe and Fay Gock in Mangere experimented with growing seedless watermelons for the summer market. They obtained seed from America and successfully developed their own variety suited to New Zealand growing conditions.[185]

Thomas Wing was at the forefront of the citrus industry in Hawkes Bay, growing oranges, grapefruit and mandarins at Eskdale, north of Napier. Although locally grown mandarins are commonplace in New Zealand these days, this was not the case in the late 1950s when he planted the orchard. Tom Wing's children recall:

> Growing mandarins at Eskdale was a bit of a gamble but Thomas reckoned there was a micro climate there and that proved correct. He also experimented in growing grapes and supplied root stock to the first Eskdale vineyards. We remember him rigging up an alarm system at home and

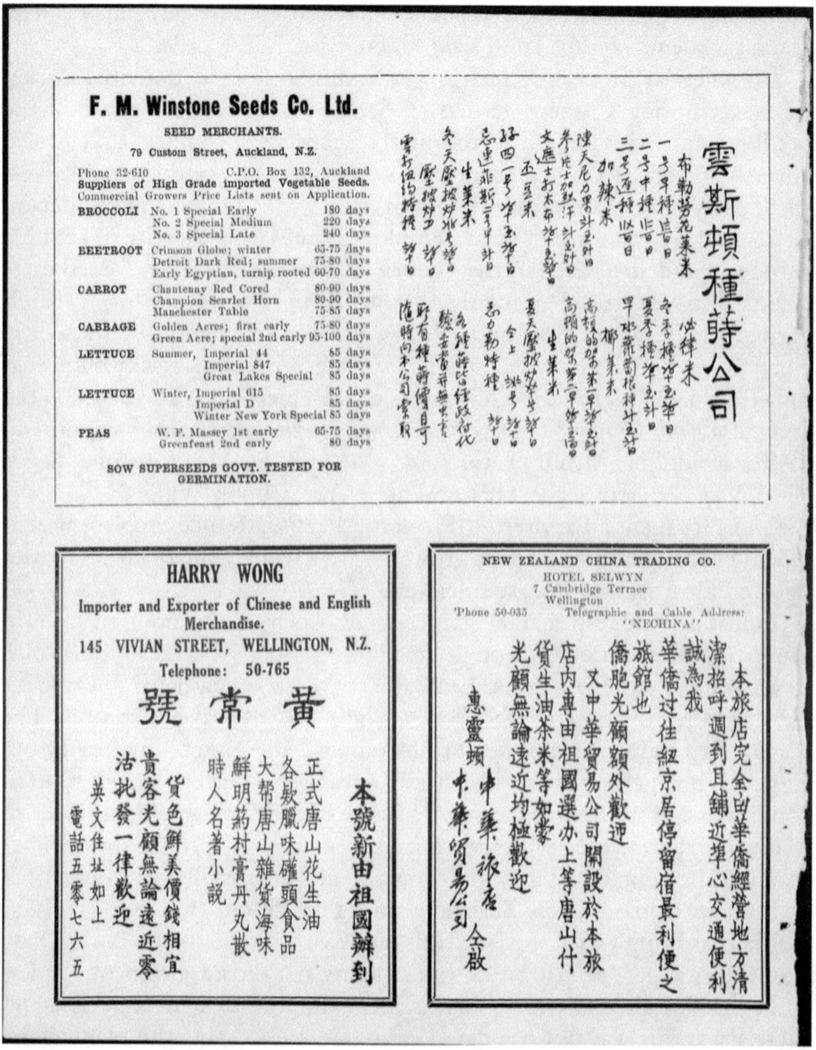

Fig. 6.6 Bilingual advertisement for F. M. Winstone Seeds, *New Zealand Chinese Growers' Monthly Journal*, 1 March 1951, p. 2.

> roaring off in the middle of the night in his truck to go and light the smoke pots in the orchard when there was one of the (frequent) heavy Hawkes Bay frosts. He'd come home smelling of tar from the smoke pots. He used to drive from Hastings to the Eskdale orchard early in the morning every day before the rest of the family was awake, and return late at night.[186]

Two other innovators and community leaders were Albert Young and Poy Young, who were initially partners in a market garden and store in Napier in the 1930s, Loo Kee and Co. In 1944 they established separate enterprises. Albert Young established a thriving market garden of 30 to 40 acres at nearby Bay View with other partners, supplying the Napier store and the Wellington market with vegetables and Watties with tomatoes for sauce. Albert had a commanding personality and was always ready to offer advice and encouragement to his compatriots.[187] Poy Young established a 30-acre garden at Bay View with his wife. He gained a reputation for the quality of his produce and set the standard for other growers. Brian Young, the son of Albert Young, recalls:

> Poy Young was number one by a mile . . . They were known around here as a top grower. Poy Young was a visionary. He was one of the first to purchase and use a tractor-driven rotary hoe. He was the first to experiment with Brussels sprouts in the Hawke's Bay. They were early in the season and grown from 1958.[188]

Built Structures

Built structures are another aspect of Chinese market gardening which developed over time, mirroring technological and economic changes. As Chinese market gardeners prospered and adopted technological developments, they made improvements to their accommodation and built additional structures to house horses and store carts, fertilisers, pumps and other farm equipment. From the earliest years of Chinese market gardening in Australia, there was a range of built structures on gardens, including underground cellars for storing produce, the precursors of twentieth-century cool rooms. An 1863 account of a Chinese market garden on the Loddon River in Victoria describes the accompanying structures.

> On entering the garden the visitor's attention is first attracted by the odd look of the quadrangle in which he finds himself. One side is formed by the most comfortable looking stable, hay store and pigsty; another by the

> dwelling house and bedrooms of the Chinese; the other two by the fencing of the garden. In the yard is a cellar for the preservation of cut vegetables, and a large piece of land is covered in with branches laid on poles which serve as a sort of arbor under which the owners appear to dine.[189]

Gardeners made use of whatever building materials were available. Eardley and Eileen Gifford describe the functional cottages erected for the workers on Chinese market gardens in the St George district in Sydney, 'built without levels or plumb lines, always in a glorious higgledly piggledy fashion'. Tree trunks were used for uprights, galvanised iron sheets for walls and roof, and the small windows were covered with chaff bags nailed along the lintel, keeping out flies and mosquitoes.[190] In her history of market gardening in Melbourne, Joanne Monk notes the growing number of built structures on gardens and the use of corrugated iron on buildings and sheds from the 1920s.[191] The photograph in Fig. 6.7, taken in around 1925, shows a house and sheds on a Chinese market garden in Kalinga, Queensland. The roofing is corrugated iron.

Larger, more prosperous market gardening operations had more capital to develop infrastructure, including accommodation for workers. Ming Quong, a successful entrepreneur in Auckland in the late 1880s and 1890s, leased a 22-acre portion of the Epsom farm of prominent businessman John Logan Campbell as a market garden. He also had a house and 23 acres of land in Surrey Crescent closer to the city where a number of his kinsmen lived and worked. The latter property included 18 acres of vegetable crops and 5 acres of orchard. In 1894 he advertised his market garden in Surrey Crescent for lease. It included: 'a good 5 roomed house, with cow shed, stables, piggeries, outbuildings etc' and he even had town water and a telephone laid on.[192] From 1895 Quong operated a grocery and fruiterer business in the city.[193]

In New Zealand accommodation needs on market gardens changed from the late 1930s as wives and children were allowed to enter the country and the structure of market gardening enterprises began to change from all-male partnerships to family-run businesses. For example, Norman Yee and his cousin George Tong established a market garden near Oamaru in the 1930s. After their wives and children arrived from China in the late 1940s and early 1950s the cousins established separate family businesses. By the 1960s the buildings on Norman Yee's garden included the family house, several storage sheds, bunkhouses for the workers and a separate kitchen-dining room.[194] James Ng describes the

Fig. 6.7 Chinese market gardeners at work in their garden at Kalinga, Brisbane, c. 1925. A house and sheds are visible in the background. Courtesy State Library of Queensland, Record no. 112724.

buildings on the King Brothers' market garden in Ashburton which was established in 1892 and operated until around 1966. After the families of the nine partners in the business, most of whom were related, arrived in New Zealand from the late 1930s the cooperative business structure of the firm was maintained. By the 1950s there was a small village on the 150-acre garden, including several cottages, numerous sheds housing a communal kitchen, dining room, wash room and additional bedrooms, as well as a workshop and a garage to house a truck and a car.[195]

In the early 1900s Chinese market gardeners in the Melbourne suburb of Hawthorn and in Otaki in New Zealand raised seedlings under glass.[196] Glasshouses and cool rooms came into use between the wars. Among the earliest market gardeners in New Zealand to establish a full-scale glasshouse was Joe Ah Chan, who established a market garden in Thames in 1923 with his wife and family, and grew glasshouse tomatoes.[197] Glasshouses were a particular asset in the colder climate of the South Island, extending the growing season of many crops. In the 1930s Chinese growers in Greymouth (William Young), Oamaru (Young Sai Kit) and Christchurch (Alice and Percy Lee) were all growing tomatoes, lettuce and spinach in glasshouses in addition to outdoor crops.[198] In the North Island, Lee Oy Ling was one of the first Chinese growers in Mangere to establish a glasshouse where he grew large quantities of tomatoes and beans from the late 1930s.[199]

From the mid 1950s commercial glasshouse production of tomatoes and to a lesser extent cucumbers, beans and lettuce became an important feature of the horticultural industry in New Zealand. This had traditionally been the preserve of Europeans, but an increasing number of Chinese growers took up glasshouse growing, for example in Oamaru, Christchurch, Lower Hutt and Otaki.[200] In Auckland glasshouses with automatic heating and watering systems became popular as Chinese growers came under increasing pressure from residential and industrial development and intensified their operations.[201] This mode of production promised higher returns per acre and was suited to the small size of landholdings, for example in Mangere, Avondale, Onehunga and Otahuhu.[202] Lee and Lam note that many of the Chinese growers who took up glasshouse growing from the early 1960s such as Harry Luen, George and Percy Luen, Jack Chong, Poy Lee and Sherman Lowe were young men, a second generation of growers who were educated in New Zealand and quickly gained the scientific and technical knowledge required to grow under glass.[203]

While glasshouses reduced labour costs they represented a considerable capital investment. They were also expensive to run and subject to rising fuel costs, whether coal in the earlier years or later diesel or electricity. Some growers, for example Peter Lee in Oamaru, gave up heating their glasshouses, planting later in the season to avoid early frosts.[204]

A significant adaptation of traditional Chinese horticultural practices in Australia and New Zealand was fencing. In southern China garden plots were generally unfenced, but in towns and cities in Australia and New Zealand animals were abundant and fences were needed to keep them out. Fences were constructed of whatever materials were available, including timber posts and wire lined with gorse or brush, timber posts and rails, or wooden palings and later barbed wire.[205] Often fences were ineffective barriers. Contemporary newspapers regularly reported cows, sheep or chickens breaking into the plots of Chinese market gardeners and destroying their crops. In 1897, for example, Ah Tip's 'splendid' 8-acre garden in Coburg, Victoria was destroyed by a flock of 400 starving sheep which broke though the wire fence. The drover was later charged with allowing the sheep to trespass and not allowing them to graze sufficiently while on the road.[206]

The first Chinese arrivals on the Victorian goldfields staked out their claims and surrounded their dwellings and gardens with protective fences. A rare insight into the world of early Chinese immigrants to Australia is the autobiography of Ah Siug Jong, a miner on the central Victorian goldfields from the mid 1850s. Born in Canton around 1837, Ah Siug Jong arrived in Victoria in 1855.[207] His autobiography includes a map and detailed description of his tent and garden on the diggings, which was surrounded by a bark fence (Fig. 6.8).[208]

Jong's plan shows a complex of five tents, with a narrow walkway between them and three separate cook houses. There are two small gardens adjacent to the tents in the upper part of the complex and a larger garden in the lower right-hand corner, probably communal. Jong marks the location of furniture (beds, tables, storage boxes for clothes and tools, and even the 'piss pot' in the yard) and the activities carried out in different spaces. The whole complex is surrounded by what appears to be a fence and a front gate is clearly marked. A notation along the right-hand side reads 'door lock up', an indication of the occupants' concern for security. Jong describes in detail his own dwelling and garden:

> Tent 3 piece calico, 6 yds long, 3 yds wide, walls 2 yds high. Front yard 1 perch [c. 5m] long, about 4 yds wide. Side garden 6 yds long and 2 yds wide [c. 6m by 2m], with a door. Here I built house for fowls. I liked living there and made

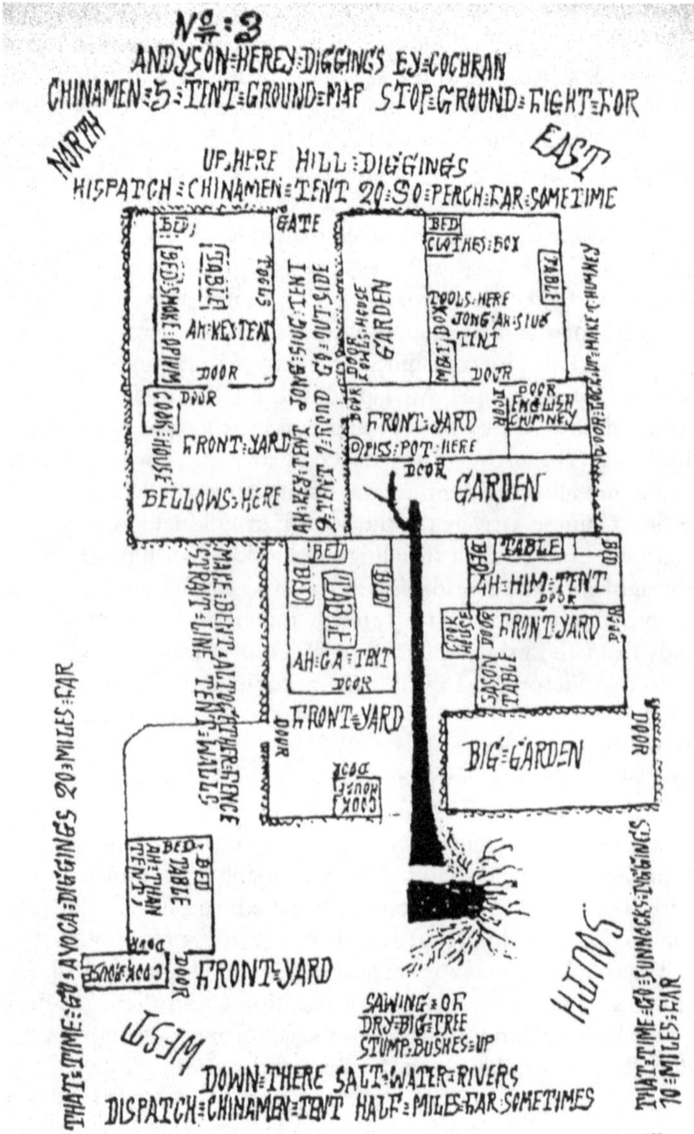

Fig. 6.8 Plan of tents and gardens on the Andersons Hill diggings, Victoria, late 1860s, drawn by Ah Siug Jong. Ah Siug Jong, Diary, State Library of Victoria, p. 6 [original ms. 1866–1872].

> a garden. I made the garden in August 1866 and by October 1866 I had melons and vegetables growing. The garden fence was made of bark.[209]

The division of spaces in this tent complex is designed to meet the needs of single men living in the often hostile environment of the goldfields, with its surrounding fence, open courtyard spaces for communal activities and individual tents for sleeping and storage of personal effects. It represents an adaptation of the traditional layout of houses in southern China which were designed to meet the needs of extended family units. Typically these comprised a series of rooms for sleeping, storage, and cooking opening onto a central courtyard used for communal activities. The narrow lane-ways between houses were also used for communal activities.[210]

Conclusion

As Frost has observed in Australia, Chinese agriculture was complex and dynamic. The skills of Chinese immigrants lay in their entrepreneurship and innovation, adapting their traditional horticultural techniques to the Australian environment.[211] Not only were Chinese immigrants themselves highly mobile, the technologies, skills and ideas they brought with them were also highly mobile. The movement of these technologies, skills and ideas accompanying networks of migration was central to processes of technological change. This chapter has shown how Chinese market gardening underwent technological change and adaptation from the time the first Chinese gold miners in Australia and New Zealand began to grow crops commercially to supplement their incomes. It has explored the complex interplay between the continuity of traditional Chinese horticultural practices and the adaptation of these technologies to new social and environmental conditions.

There was considerable continuity in Chinese horticultural practices in Australia and New Zealand, which involved a high degree of manual labour and limited application of mechanisation due to the small size of plots, intensive cultivation methods and the wide variety of crops grown. However, Chinese market gardeners adapted their traditional horticultural techniques to new environments and markets and adopted European technological developments which were appropriate to the scale and work organisation of their enterprises. These were both process and product innovations, including steam and petrol powered irrigation pumps, motor vehicles and tractors as well as commercially produced fertilisers, pesticides and improved seed varieties.

The factors influencing the take-up of technology are complex and interrelated. Firstly, they must be considered within the context of complex movements and circulation of people around the globe. Thus, for Chinese immigrants key social and economic factors were their desire to maintain links with kin and home village in China and their obligation to support family members. Other factors were: the high capital cost of equipment such as pumps, irrigation systems, motor vehicles, tractors and greenhouses; the scale and financial success of market gardening enterprises; and fluctuations in the supply of labour on gardens, mainly in response to changing government immigration policies.

The evidence presented in this chapter challenges the stereotype of Chinese market gardeners as conservative peasants resistant to change. It highlights the role of particularly innovative and entrepreneurial market gardeners as change agents within the Chinese community as they modified existing technology, experimented with new crop varieties and growing techniques and passed on their practical experience to others. Commercial firms such as agricultural equipment manufacturers and seed merchants played an important part in the diffusion of ideas and innovation, for example by employing Chinese salespeople, advertising in Chinese language media and holding field demonstrations. The food processing and produce marketing industries also played a key role, acting as intermediaries between growers and changing consumer demands.

Thus, a detailed study of Chinese market gardening demonstrates that technological change involves complex interactions between migrants and the technologies of their adopted country. Complementary industries were crucial in supplying the raw materials or production inputs required for market gardening, including manure, fertilisers and pesticides, seeds and mechanical equipment such as pumps, irrigation plant and cultivation tools. Technological developments in transport facilities, in particular the expansion of railways and tramways, were also crucial in expanding the market reach of market gardeners. Technological change is also shaped by social forces. The role of the social environment in the history of Chinese market gardening in Australia and New Zealand is exained in the following chapter.

Notes

1. Jan Todd, *Colonial Technology: Science and the Transfer of Innovation to Australia*, Cambridge, 1995; Warwick Pearson, 'Water power in a dry continent: the transfer of watermill technology from Britain to Australia in

the 19th century', PhD Thesis, University of New England, Armidale, 1995, pp. 27–28; and Joe Powell, *Mirrors of the New World: Images and Image-makers in the Settlement Process*, Folkestone/Hamden Connecticut, 1977, pp. 75–77.

2. Darwin Stapleton, *The Transfer of Early Industrial Technologies to America*, Kansas, 1987, p. 25; Pearson, 'Water power in a dry continent', pp. 1, 16, 23; Nathan Rosenberg, *Exploring the Black Box: Technology, Economics and History*, Cambridge, 1994, pp. 1–2.
3. Recent scholars interpret the technological changes that occurred in agriculture in Britain from the late eighteenth century as a patchy, uneven process rather than a revolution comparable to the industrial revolution (Norman McCarol, *British History 1815–1906*, Oxford, 1991, p. 81).
4. Robert Allen, 'Agriculture during the industrial revolution' in Roderick Floud and Donald McCloskey (eds.), *The Economic History of Britain Since 1700*, Vol. 1, 1700–1860, Cambridge, New York, 1994, p. 115; and Patrick O'Brien, 'Modern conceptions of the industrial revolution', in Patrick K. O'Brien and Roland Quinault (eds.), *The Industrial Revolution and British Society*, Cambridge, 1993, p. 19. The economic data available is open to further reinterpretation as more sophisticated economic modelling becomes possible (Mark Crafts, 'The industrial revolution' in Floud and McCloskey (eds.), *Economic History of Britain*, p. 59).
5. These factors included the decline of the family farm, the employment of more adult males and an increase in seasonal and day labour, with resulting underemployment and unemployment in rural areas (Allen, 'Agriculture during the industrial revolution', pp. 105–106; McCarol, *British History 1815–1906*, p. 222).
6. The indices used include volume of farm output, the sum of farm wages, profits and rents; and a measure of food supply versus demand (Allen, 'Agriculture during the industrial revolution', pp. 101, 110; and O'Brien, 'Modern conceptions of the industrial revolution', p. 22).
7. Allen, 'Agriculture during the industrial revolution', pp. 111–113; and McCarol, *British History 1815–1906*, p. 82.
8. N. Von Tunzelmann, 'Technological and organizational change in industry in the early industrial revolution', in O'Brien and Quinault (eds.), *The Industrial Revolution*, p. 259. See also Joel Mokyr, 'Technological change 1700–1830', in Floud and McCloskey (eds.), *Economic History of Britain Since 1700*, pp. 14–15.
9. Colin White, *Mastering Risk: Environment, Markets and Politics in Australian Economic History*, Melbourne, 1992, p. 155; Stapleton, *Transfer of Early Industrial Technologies*, pp. 26–27.

10. N. Von Tunzelmann, 'Technology in the early nineteenth century', in Floud and McCloskey (eds.), *Economic History of Britain Since 1700*, Vol. 1, p. 271.
11. White, *Mastering Risk*, pp. 157–158.
12. Powell, *Mirrors of the New World*, p. 78; and Stapleton, *Transfer of Early Industrial Technologies*, pp. 4–5.
13. Todd, *Colonial Technology*, pp. 11, 128, 146–149; and Ann Moyal, *A Bright and Savage Land: Scientists in Colonial Australia*, Sydney, 1986, pp. 178–179, 192.
14. Barry McGowan, 'Chinese market gardens in Southern and Western NSW', *Australian Humanities Review*, Issue 36, 2005; Warwick Frost, 'Migrants and technological transfer: Chinese farming in Australia, 1850–1920', *Australian Economic History Review* Vol. 42, No. 2, 2002, pp. 113–131; Jock Collins, 'Chinese Entrepreneurs: The Chinese diaspora in Australia', *International Journal of Entrepreneurial Behaviour and Research*, Vol. 8, No. 1/2, 2002, pp. 113–133; Lily Lee and Ruth Lam, *Sons of the Soil: Chinese Market Gardeners in New Zealand*, Pukekohe 2012; and Nigel Murphy, *Success Through Adversity: A History of the Dominion Federation of New Zealand Chinese Growers*, Pukekohe, 2012.
15. Frost, 'Migrants and technological transfer', p. 113.
16. The following works, for example, make no reference to Chinese farmers: Garry Hawke and Ralph Lattimore, *Visionaries, Farmers and Markets: An Economic History of New Zealand Agriculture*, Wellington, 1999; NZ Rural Press, *New Zealand Agriculture: A Story of the Past 150 Years*, Auckland, 1990; and Benjamin Evans, *A History of Agriculture, Production and Marketing in New Zealand*, Palmerston North, 1969.
17. Lee and Lam, *Sons of the Soil.*
18. Frost, 'Migrants and technological transfer', p. 116.
19. Ibid., pp. 116–122.
20. Warwick Frost, 'Australia Unlimited?: environmental debate in the age of catastrophe, 1910–1939', *Environment and History*, Vol. 10, No. 3, August 2004, pp. 286–287; and Geoff Park, *Nga Ururora: Ecology and History in a New Zealand Landscape*, Wellington, 1995, pp. 13–15.
21. Moyal, *Bright and Savage Land*, pp. 170–171.
22. Liliane Hilaire-Perez and Catherine Verna, 'Dissemination of technological knowledge in the Middle Ages and early modern era: new approaches and methodological issues', *Technology and Culture*, Vol. 47, No. 3, July 2006, p. 563; Nathan Rosenberg, *Inside the Black Box: Technology and Economics*, Cambridge, 1982, p. 19; Stapleton, *Transfer of Early Industrial Technologies*, p. 565.
23. *Evening Post*, Wellington, 21 June 1916, p. 12.

24. Geoffrey Stedman, 'The South Dunedin Flat: a study in urbanisation 1849–1965', MA Thesis, University of Otago, Dunedin, 1966, pp. 133–134.
25. See, for example, Weston Bate's discussion of European and Chinese market gardeners in Brighton, Victoria in the early 1900s (Weston Bate, *A History of Brighton*, Melbourne, 1983, pp. 363–364).
26. For example, Tom Wing of Hawke's Bay and Joe and Fay Gock of Mangere (Lee and Lam, *Sons of the Soil*, pp. 255, 258–259, 411–418).
27. Everett Rogers, *Diffusion of Innovations,* New York, 2003, pp. 256–257, 400.
28. See, for example, Robert Chambers, Arnold Pacey and Lori Ann Thrup (eds.), *Farmer First: Farmer Innovation and Agricultural Research*, London, 1989, p. 3; Ian Scoones and John Thompson (eds.), *Beyond Farmer First: Rural People's Knowledge, Agricultural Research and Extension Practice*, London, 1994, pp. xiii, xv; and Paul Richards, *Indigenous Agricultural Revolution: Ecology and Food Production in West Africa*, London, Melbourne, 1985, pp. 159–161.
29. Gordon Clark, *Innovation Diffusion: Contemporary Geographical Approaches*, Norwich, 1984, pp. 14.
30. Niti Pawakapan, 'The Chinese in Dunedin between the 1920s and the 1930s', MA Thesis, University of Otago, Dunedin, 1987, p. 99.
31. This harks back to the traditions of imperial China. Just as families selected and supported their brightest and best to become government officials, the extended families of migrants to Australia and New Zealand collected and paid the fares of their most enterprising sons to go overseas to make their fortunes. Andrew Piper, pers. comm., July 2012.
32. Donald History and Natural History Group, *Georgie Ah Ling: Donald's Friend*, Donald, 2008, pp. 11, 14, 19, 27.
33. A disc scarifier is a machine with spikes or prongs used for loosening soil and weeds prior to planting. A moulding machine is a heavier device used to break up and turn over large lumps of soil, at a greater depth (*Oxford Dictionaries*, http://oxforddictionaries.com/definition/scarifier?; and Global Farm Machinery Network, http://agmachine.ning.com/photo/molding-machine?, accessed 10 May 2012).
34. Donald Hunt, 'Market gardening in metropolitan Auckland', MA Thesis, University of New Zealand, Auckland, 1956, p. 19. This difference was maintained over the next two decades, although average acreages increased. In her 1977 study Lee found that the average size of Chinese market gardens in the Auckland region was 10 ha (24 acres), compared to 14 ha (34.5 acres) for the gardens of other ethnicities (Thong Ling Lee, 'Chinese market gardening in the Auckland region', MA Thesis, University of Auckland, 1974, p. 25).

35. See, for example, in New Zealand, Pamela Wai Shing, 'Locational and structural changes of market gardening in Pukekohe–Bombay–Patumahoe', MA Thesis, University of Auckland, 1977, p. 93; and Thong Ling Lee, 'Chinese market gardening in the Auckland Region', MA Thesis, University of Auckland, 1974, pp. 34, 36.
36. Board of Studies, New South Wales, 'Elders of the Yiu Ming Society speak about their experiences as Chinese in Australia', http://www.teachingheritage.nsw.edu.au/, accessed 13 May 2012.
37. Lee and Lam, *Sons of the Soil*, p. 281.
38. Ibid., p. 216.
39. Bill Wong interviewed by Lily Lee, 22 October 2006, quoted in ibid., p. 265.
40. Two prominent intellectuals were Liang Quichao, a fierce critic of Imperial China, who spent at least six months in Australia in 1900 giving lectures and writing essays, and the journalist Tang Caixhi, who came to Sydney in 1901 as editor of the *Tung Wah News* (John Fitzgerald, *Big White Lie: Chinese Australians in White Australia*, Sydney, 2007, pp. 11, 116–117.)
41. *Grey River Argus*, 21 December 1910, p. 2.
42. Murphy, *Success through Adversity*, p. 16.
43. Lee and Lam, *Sons of the Soil*, pp. 337, 370, 378.
44. Ibid., pp. 432–433.
45. *Tung Wah Times*, 30 November 1907, p. 7 c. Tangyes Ltd was a British engineering firm, founded in Birmingham by Richard Tangye and his two brothers in 1856. It operated until 1957. The firm specialised in manufacturing machine tools, including hydraulic and centrifugal pumps and a wide range of other power machines. Tangyes Ltd exported their products worldwide, including to Australia and other British colonies (Grace's Guide: British Industrial History, http://www.gracesguide.co.uk/Richard_Tangye, accessed 11 May 2012).
46. Liu Weiping, 'Chinese newspapers in Australia from the turn of the century', Chinese Heritage of Australian Federation, http://www.chaf.lib.latrobe.edu.au/chinese_newspapers.htm, accessed 11 May 2012. The first Chinese language newspaper published in Australia was the *Chinese Australian Herald,* launched in Sydney in 1894, followed by the *Tung Wah News,* launched in 1898.
47. *Tung Wah Times*, 28 March 1900, p. 3 b.
48. *Tung Wah Times*, 3 January 1900, p. 2, b-c; 15 January 1910, p. 7 a; and 22 May 1926, p. 7 b-d.
49. Bing Wong interviewed by Lorna Wong, Wellington, 30 July 2007, 21 Voices Project, New Zealand Chinese Oral History Foundation, 11-OH-15, Auckland City Library.
50. *Argus,* Melbourne, 14 October 1937, p. 1.

51. *Bruce Herald*, 24 August 1870, p. 7.
52. George Lee Kim interviewed by Paul McGregor, 17 May 1994, National Library of Australia, Oral TRC 3522/4/4, Tape 1.
53. Anne Thorpe *et al.*, 'Otaki's market gardens', *Otaki Historical Society Historical Journal*, Vol. 30, 2008, p. 26.
54. Ibid., p. 26. The Planet Junior line of small farming tools was produced by the Philadelphia manufacturer S. L. Allen and Company from 1870. By 1890 the firm had carved a significant niche in American agricultural manufacturing (*Taranaki Herald*, 5 October 1894, p. 2; and *Wanganui Herald*, 5 October 1894, p. 2.).
55. Barry McGowan, *Tracking the Dragon: A History of the Chinese in the Riverina*, Wagga Wagga, 2010, p. 20.
56. Ian Jack and Kate Holmes, 'Ah Toy's Garden: a Chinese market garden on the Palmer River Goldfield, North Queensland', *Australian Journal of Historical Archaeology*, Vol. 2, 1984, p. 56.
57. Barry McGowan, 'Ringbarkers and market gardeners: a comparison of the rural Chinese of New South Wales and California', *Chinese America: History and Perspectives*, San Francisco, 2006, p. 36.
58. Barry McGowan, 'Chinese market gardens in Southern and Western NSW', *Australian Humanities Review*, Issue 36, 2005, n.p.
59. McGowan, *Tracking the Dragon*, p. 22.
60. Lee and Lam, *Sons of the Soil*, p. 72.
61. Peter Lee interviewed by Lily Lee, 12 March 2008, Oamaru, quoted in Lee and Lam, *Sons of the Soil*, p. 75.
62. Helen Wong, *In the Mountain's Shadow: A Century of Chinese in Taranaki*, Auckland, 2010, pp. 52–53.
63. Lee and Lam, *Sons of the Soil*, p. 72.
64. Gordon Ha interviewed by Joanna Boileau, Sydney, 27 July 2010. Contemporary Chinese market gardeners in New Zealand also comment on the labour intensive nature of market gardening, despite high levels of mechanisation in cultivating and seeding (Allan Fong interviewed by Ruth Lam, 27 September 2007, quoted in Lee and Lam p. 493).
65. Marjorie Long Dodd, *It's a Long Story*, Sydney, 1997, p. 37.
66. Owen Bennett interviewed by Lesley Goldberg, 3 November 1998, City of Ryde Library Services.
67. Warwick Roger, 'True builders of the Waimarino', *Weekly News*, 12 July 1971, p. 12.
68. Merrilyn George, *Ohakune: Opening to a New World*, Ohakune, 1990, pp. 222–223.
69. Jade Reidy, *Not Just Passing Through: The Making of Mount Roskill*, Auckland, 2007, p. 60.
70. Lee and Lam, *Sons of the Soil*, pp. 256, 261.

71. Thorpe et al., 'Otaki's market gardens', p. 19.
72. Bee Dawson, *A History of Gardening in New Zealand*, Auckland, 2010, pp. 245–246.
73. Lee and Lam, *Sons of the Soil*, p. 306.
74. Stephen Young, '"To be happy for the rest of your life": Chinese market gardeners in New Zealand', Address in honour of 'A Barbarous Measure; The Poll Tax and Chinese New Zealanders', National Library of New Zealand, 2 October 2003.
75. *Taranaki Herald*, 5 October 1894, p. 2; and *Wanganui Herald*, 5 October 1894, p. 2.
76. *Auckland Star*, 26 October 1898, p. 6.
77. Lee and Lam, *Sons of the Soil*, p. 261.
78. Pawapakan, 'Chinese in Dunedin', pp. 90–91. For privacy reasons, Pawakapan did not use the full names of his informants.
79. *New Zealand Chinese Growers' Monthly Journal*, 1 September 1949, p. 7.
80. They included advertisements for the Newman British Lightweight tractor with four-speed gearbox, the Bristol 20 tractor and the Fordson Major tractor manufactured by International Harvester.
81. For example, *New Zealand Chinese Growers' Monthly Journal*, 1 November 1949, p. 3; 1 January 1950, pp. 3, 4; 1 May 1950, pp. 2, 5, 6; and 1 July 1951, pp. 3, 11.
82. *Tung Wah Times*, 11 February 1928, p. 5 d; 30 March 1929, p. 6 a; 21 May 1932, p. 8d; 10 November 1928, p. 6 a. and 29 June 1929, p. 8 b.
83. For example, in the 1880s there were seven separate rail systems in Queensland, and eight in South Australia (Graeme Dawson, J.W. McCarty and Ailsa McCleary (eds.), *Australians, A Historical Library: Australians 1888*, Broadway NSW, 1987, pp. 94–96).
84. G.T. Bloomfield, *New Zealand: A Handbook of Historical Statistics*, Boston, 1984, p. 227; and Neill Atkinson, *Trainland: How Railways Made New Zealand*, Auckland, 2007, pp. 32, 56.
85. In Australia, for example, the cost of transporting fodder by rail or river steamer decreased from an average of 2s per ton per mile in the summer of 1830 to 6d per ton per mile in the summer of 1880 (D. B. Williams (ed.), *Agriculture in the Australian Economy*, Sydney, 1967, pp. 167–168).
86. Lionel Frost and Tony Dingle, *Sustaining Suburbia: A Historical Perspective on Australia's Urban Growth*, Economics and Commerce discussion papers, Discussion Paper No. 23.93, November 1993, pp. 6–7.
87. Lee and Lam, *Sons of the Soil*, p. 119.
88. Thorpe et al., 'Otaki's market gardens', p. 15; and Alex McLellan, 'Market gardening in Otaki', *Otaki Historical Society Historical Journal*, Issue 5, 1982, p. 61.

89. Nigel Murphy, 'A history of the horticultural industry', unpublished ms., 2010, pp. 11, 15–16. See also Lee and Lam, *Sons of the Soil*, pp. 119, 229.
90. Terry Kass, *A Thematic History of the Central West*, Sydney, 2003, p. 18.
91. Sandi Robb, 'Beyond the coast: Chinese settler patterns of the western gulf region, north Queensland', in Kevin Wong Hoy and Kevin Rains (eds.), *Rediscovered Past: China in Northern Australia*, North Melbourne, 2009, pp. 13, 15.
92. McLellan, 'Market gardening in Otaki', p. 62.
93. The first piece of legislation was the Transport Licensing Act 1931 (Murphy, 'History of the horticultural industry', p. 16).
94. Barry McGowan, 'Adaptation and organization: the history and heritage of the Chinese in the Riverina and Western New South Wales, Australia', *Chinese America: History and Perspectives*, San Francisco, 2007, p. 238.
95. Nellie Fong interviewed by Diana Giese, 28 December 1996, National Library of Australia, Post-war Chinese Australians oral history project, ORAL TRC 3543, transcript pp. 47–48.
96. *Queenslander*, 19 January 1878, p. 27.
97. Lee and Lam, *Sons of the Soil*, pp. 167, 141.
98. Ibid., p. 237.
99. Terry Metherill, 'Faster: Manly in the 1920s', unpublished ms., 2006, City of Ryde Library Service, p. 3.
100. Long Dodd, *It's a Long Story*, pp. 34–35. The Skinner overhead irrigation system, patented by Charles Skinner of Ohio in 1894, used pipes with nozzles supported on wooden poles with an oscillator at the supply end. The company he founded continued production until 1976 (Frank Roesch, 'Irrigating old style: the Skinner system', *Farm Collector*, July 2009, http://www.farmcollector.com/equipment/Irrigating-Old-Style-The-Skinner-System.aspx, accessed 29 January 2012).
101. *Argus*, 8 December 1886, p. 15.
102. Don Diaries 1899–1907, 1902, item 413, cited in Ng, *Windows on a Chinese Past* Vol. 1, fn. 77 p. 167. See also Pawakapan, 'Chinese in Dunedin', p. 38.
103. Franklin King, *Farmers of Forty Centuries*, London, 1949, p. 78.
104. A whim is a device similar to a windlass used in mining, consisting of a drum with a vertical or horizontal axle. A rope is wound around the drum, then through a system of pulleys and each end is suspended down the well or mine shaft. As the drum rotates one end of the rope, carrying an empty bucket, is lowered while the full bucket at the other end is raised. Later, whims were powered by steam engines (Henry Mosely, *The Mechanical Principles of Engineering and Architecture*, New York, 1869, pp. 203–204).
105. McGowan, 'Chinese market gardens', 2005, n.p.

106. Golden Threads: The Chinese in Regional New South Wales Database, 1999, Record S17, Notes, p. 1.
107. For example, see Donald History and Natural History Group, *Georgie Ah Ling*, pp. 6, 15, 26; and George Lee Kim interviewed by Paul McGregor, 17 May 1994, Tape 1, Australia-China Oral History Project, National Library of Australia, TRC 3522/4/4.
108. Tom Parr, *Reminiscences of a NSW South West Settler*, New York, 1977, p. 15.
109. Barry McGowan, *Dust and Dreams, Mining Communities in South-East New South Wales*, Kensington, 2010, p. 16.
110. Lee and Lam, *Sons of the Soil*, pp. 43–44.
111. See, for example, Crafts, 'The industrial revolution', p. 54; and Von Tunzelmann, 'Technological and organizational change in industry in the early Industrial revolution', in O'Brien and Quinault (eds.), *The Industrial Revolution and British Society*, pp. 256, 259.
112. *Brisbane Courier*, 6 August 1902, p. 2.
113. This was not a rigid hierarchy. The boss could also be the man responsible for marketing produce (R. Nixon, 'The Chinese community of Camden', unpublished ms., Camden Historical Society, 1976, pp. 3–5).
114. Ibid.
115. McGowan, 'Adaptation and organization', p. 238; and McGowan, 'Chinese market gardens', n.p.
116. The market garden was established by William Ah King in the 1880s and remained in the King family at least until the 1950s (Ian Rannard, *The Forgotten Gardens: The Story of the Last Market Gardens in Willoughby and Northbridge, NSW*, Douglas Park, 2005, pp. 14–16, 21–22).
117. *Northern Advocate*, Charters Towers, 8 January 1920, p. 1.
118. *Riverine Grazier*, 24 June 1892, quoted in McGowan, *Dust and Dreams*, p. 16.
119. New South Wales Legislative Assembly, *Report of the Royal Commission into Alleged Chinese Gambling and Immorality*, Votes and Proceedings of the Legislative Assembly of New South Wales, 1891–1892, Vol. 8, Sydney, 1892, p. 160.
120. Donald History and Natural History Group, *Georgie Ah Ling*, p. 11.
121. Don Beatty, 'Clutterbuck Oil Engine', *Gawler Machinery Restorers Club Inc. Newsletter*, 18, 1993, pp. 9–11. This is an interesting example of technological exchange between Britain and Australia, and the development of domestic Australian industry (*Advertiser*, Adelaide, 4 April 1917, p. 10).
122. Andrea Gaynor, *Harvest of the Suburbs: An Environmental History of Growing Food in Australian Cities*, Perth, 2006, p. 19.
123. *Brisbane Courier*, 16 January 1889, p. 3; and Gaynor, *Harvest of the Suburbs*, p. 23.

124. R. Nixon, 'The Chinese community of Camden', pp. 3–4.
125. Dawson et al., *Australians, A Historical Library*, p. 204.
126. In 1899 Chinese market gardeners around Smith's Lake were accused of polluting it. An investigation by an inspector from the Central Board of Heath revealed that the main sources of pollution were waste from the piggeries, dairy cattle drinking from the lake and nightsoil being dumped beside it (Gaynor, *Harvest of the Suburbs*, pp. 25–26).
127. *Sydney Morning Herald*, 13 January 1865, p. 8; Thorpe et al., 'Otaki's market gardens', pp. 26–27.
128. Gaynor, *Harvest of the Suburbs*, p. 23; and *Daylesford Mercury*, 13 July 1866, reprinted in the *Argus*, Melbourne, 14 July 1866, p. 6.
129. Anne Atkinson, 'Chinese market gardening in the Perth metropolitan region, 1900–1920', *Western Geographer*, Vol. 8, 1984, p. 47.
130. Andrea Gaynor estimates that the 2,632 horses in the City of Melbourne in 1891 would have produced 13,160 tons of manure, based on a figure of five tons per horse per annum (Gaynor, *Harvest of the Suburbs*, p. 23).
131. David Horsfall, *March to Big Gold Mountain*, Melbourne, 1985, p. 121.
132. *Bruce Herald*, Otago, 24 August 1870, p. 7.
133. Atkinson, 'Chinese market gardening in the Perth metropolitan region', p. 47.
134. Owen Bennett interviewed by Lesley Goldberg, 3 November 1998, City of Ryde Library Services, transcript pp. 17, 20.
135. In Sydney, for example, horses still remained in small numbers after World War II (City of Sydney Council, 'Sydney Stories: Sydney Horses' http://www.cityofsydney.nsw.gov.auAboutSydney/HistoryAndArchives/OralHistories HorsesInnerCity.asp, accessed 20 February 2012).
136. Gaynor, *Harvest of the Suburbs*, pp. 88–89.
137. *Tung Wah Times*, 19 April 1902 p. 3, e; and 24 October 1908, p. 3, b-d.
138. *Tung Wah Times,* 27 February 1926, p. 6, b-c.
139. *Tung Wah Times,* 6 December 1930, pp. 6–7, a-d.
140. G. J. Leigh, *The World's Greatest Fix: A History of Nitrogen and Agriculture*, Oxford, 2004, p. 75.
141. Von Liebig, quoted in Leigh, *World's Greatest Fix*, p. 109.
142. D. C. Edmeades, 'The agronomic effectiveness of lime reverted and dicalcic superphosphates: a review', *New Zealand Journal of Agricultural Research*, Vol. 43, 2000, p. 1.
143. Benjamin Evans, *A History of Agriculture, Production and Marketing in New Zealand*, Palmerston North, 1969, pp. 5–7; and Gary Hawke and Ralph Lattimore, *Visionaries, Farmers and Markets: An Economic History of New Zealand Agriculture*, Report to the Foundation for Research Science and Technology, Wellington, January 1999, pp. 1, 17.
144. *Bay of Plenty Times*, 13 December 1884, p. 2, 8 April 1885, p. 2; and *Taranaki Herald*, 29 November 1884, p. 2.

145. Evans, *History of Agriculture*, pp. 5–7; Hawke and Lattimore, *Visionaries, Farmers and Markets*, pp. 1, 17; and *Sydney Morning Herald*, 17 June 1912, p. 11.
146. *Advocate and the Emu Bay Times*, Tasmania, 5 November 1910, p. 7.
147. *Mercury*, Hobart, 28 July 1953, p. 12.
148. Esma Smith interviewed by Rob Willis, 4 September 2009, Forbes, NSW, National Library of Australia, Voices of the Bush oral history project, ORAL TRC 6125/11.
149. Gaynor, *Harvest of the Suburbs*, p. 29.
150. In 1894 the Massachusetts Experiment Station published the results of trials of a new pesticide, arsenate of lead, developed as an alternative to the even more toxic Paris Green. These results were quickly reported in newspapers in New Zealand and Australia, indicative of the ongoing intellectual and scientific exchange between Australasia and the United States in the late nineteenth century. See *Otago Witness*, 11 October 1894, p. 11; and *Traralgon Record,* 9 November 1894, p. 15.
151. Arsenic-based pesticides were recorded in China as early as 900 AD. Lead arsenate was first developed in the USA in 1892. It was widely used in Australasia, North America and England to control insect pests on fruit trees, garden crops and grasses. The synthetic organic insecticide DDT was not introduced until 1947 (Francis Peryea, 'Historical use of lead arsenate insecticides, resulting soil contamination and implications for soil remediation', Proceedings, 16th World Congress of Soil Science Montpellier, France, 20–26 August 1998, http://soils.tfrec.wsu.edu/leadhistory.htm, accessed 1 March 2012). See also E. Murphy and M. Aucott, 'An assessment of the amounts of arsenical pesticides used historically in a geographical area', *Science of the Total Environment*, Vol. 218, Issue 2–3, July 1998, pp. 89–101.
152. *Mercury*, Hobart, 28 August 1942, p. 3.
153. *Queenslander,* Brisbane, 19 November 1910, p. 30.
154. *Daily Telegraph*, Napier, 1 November 1882, p. 4; and *Clarence and Richmond Examiner*, 28 September 1895, p. 5.
155. *West Australian,* 3 April 1935, p. 11; and *Evening Post*, Wellington, 2 April 1936, p. 25. The active ingredient of derris dust is rotenone, a neurotoxin derived from the roots of several species of tropical plants. It is still popular today with organic gardeners for the control of caterpillars, aphids, thrips, and mites. It also has a long history of use as a fish poison (Pesticide Action Network, 'Fact sheet: rotenone', http://www.pan-uk.org/pestnews/Actives/rotenone.htm, accessed 5 March 2012).
156. For example, in Ryde, Sydney. Norman Boon interviewed by Pauline Curby, 1998, City of Ryde Library Service.

157. *Mercury*, Hobart, 28 August 1942, p. 3. During the war years the Australian government's Dig For Victory Campaign promoted the cultivation of fruit and vegetables by home gardeners and there was renewed interest in the techniques used by Chinese market gardeners.
158. *Mercury*, Hobart, 6 September 1950, p. 13.
159. It is interesting to note that debate over biological methods of pest control versus chemical methods arose in California in the 1880s, predating Rachel Carson's indictment of DDT in *Silent Spring* by 80 years (Ian Tyrell, *True Gardens of the Gods: California–Australia Environmental Reform 1860–1930,* Berkeley, 1999, p. 174).
160. *Tung Wah Times*, 19 July 1930 p. 8 d; and 5 September 1931, p. 8 c.
161. *New Zealand Chinese Growers' Monthly Journal,* 1 January 1962, p. 3; 1 February 1962, p. 6; 1 August 1963, p. 2; 1 July 1964, p. 4; and 1 August 1964, p. 3.
162. Adrienne Wing and Robert Wing, pers. comm.
163. Lee and Lam, *Sons of the Soil,* pp. 412–413. Kumara are difficult to store, requiring warm, well-ventilated conditions. Any cuts or scratches sustained during harvesting can allow fungal and bacterial diseases to enter, unless the kumara are cured by exposing them to high temperatures and high humidity for several days (Helen Leach, *1,000 Years of Gardening in New Zealand,* Auckland, Wellington, 1984, p. 59).
164. Lee and Lam, *Sons of the Soil*, pp. 494–495.
165. Paul Fox, *Clearings: Six Colonial Gardeners*, Carlton, 2004, p. xv.
166. *South Australian Advertiser*, Adelaide, 8 October 1864, p. 2.
167. Alexander Don, Diary, 1 January 1883, p. 126, quoted in Neville Ritchie, 'Archaeology and history of the Chinese in southern New Zealand during the nineteenth century: a study of acculturation, adaptation, and change', PhD Thesis, University of Otago, Dunedin, 1986, p. 640.
168. The Wardian case, a miniature greenhouse similar to a modern terrarium, was invented in England around 1829 by Dr Nathaniel Ward. It revolutionised the transport of plants around the globe (Fox, *Clearings*, p. xv).
169. James Beattie, 'Acclimatisation and the "Europeanisation" of New Zealand, 1830s–1920s', *Environment and Nature in New Zealand*, Vol. 3, No. 1, February 2008, pp. 5–6. See also James Beattie, Jasper M. Heinzen and John Adam, 'Japanese gardens and plants in New Zealand, 1850–1950: transculturalism and transmission', *Studies in the History of Gardens and Designed Landscapes*, Vol. 28, No. 2, 2008, pp. 219–236.
170. Paul Fox, 'Keeping aridity at bay: acclimatisation and settler imagination in nineteenth-century Australia', *Australian Humanities Review*, Issue 36, July 2005, ejournal. See also Dawson, *History of Gardening*, pp. 118–120.
171. Fox, *Clearings*, p. 43; and Suzanne Hunt, 'Vegetable plots and pleasure gardens of the Victorian goldfields', in McCalman, Cook and Reeves (eds.),

Gold: Forgotten Histories and Lost Objects of Australia, Cambridge, 2001, pp. 270–272.

172. Eddie Quong interviewed by Diana Geise, 1993, National Library of Australia, TRC 3005.
173. Atkinson, 'Chinese market gardening', p. 47.
174. Diana Madgin, 'Alice Lee: octogenarian market gardener', *New Zealand Gardener,* Vol. 54, Issue 5, May 1998, pp. 75–76.
175. George Sue interviewed by Joanna Boileau, Auckland, 7 April 2012.
176. Lee and Lam, *Sons of the Soil*, p. 390.
177. Noel Kingsbury, *Hybrid, The History and Science of Plant Breeding*, Chicago, London, 2009, p. 177; and Dawson, *History of Gardening*, pp. 103–104. Yates was established in 1883 in Auckland by Arthur Yates, son of a Manchester seed merchant. In 1887 Yates opened a store in Sydney, leaving his brother Ernest to manage the Auckland branch. The business flourished on both sides of the Tasman (Dawson, *History of Gardening*, p. 227; and Yates Australia, 'History of Yates', http://www.yates.com.au/about/history/, accessed 9 April 2012).
178. The Te Aro seed company continued trading until the late 1950s; Ting's sons George, Arthur and Leslie took over the business after he retired (Lee and Lam, *Sons of the Soil*, pp. 123, 277).
179. Colleen Morris, 'Chinese market gardens in Sydney', *Australian Garden History,* Vol. 12, No. 5, March/April 2001, pp. 5–6.
180. Seed advertisements in *Tung Wah Times*: Anderson & Co.: 27 January 1917, p. 12c, 16 February 1918, p. 14c; George Austen & Co.: 9 February 1924, p. 5 c-d; P.L.C. Shepherd & Son: 6 December 1899, p. 3d; D. J. Reid & Co.: 21 February 1920, p. 22b; and W.A. Smith: 28 January 1922, p. 4c.
181. R. A. Nicol Seeds Merchant, advertisements in *Man Sing Times*: 22 August 1921, p. 26b; 31 August 1921, p. 20d; 12 September 1921, p. 25; and 21 September 1921, p. 18b.
182. Seed advertisements in *New Zealand Chinese Growers' Monthly Journal*: Te Aro Seeds Ltd: 1 September 1949, p. 7, 1 October 1949, p. 2, 1 November 1949, p. 6; Arthur Yates and Co.: 1 September 1949, p. 5, 1 October 1949, p. 4, 1 November 1949, p. 2; and F. M. Winstone Seeds Ltd: 1 March 1951, p. 2, 1 April 1951, p. 4; 1 May 1951, p. 3.
183. Lee and Lam, *Sons of the Soil*, p. 74.
184. Ibid., p. 75.
185. Ibid., p. 415.
186. Robert Wing, Adrienne Wing and Vanessa Tanner, pers. comm.
187. Lee and Lam, *Sons of the Soil*, p. 250.
188. Brian Young interviewed by Lily Lee, 12 October 2006, quoted in Lee and Lam, *Sons of the Soil*, p. 254.

189. *Inglewood Advertiser*, December 1863, quoted in David Horsfall, *March to Big Gold Mountain*, Melbourne, 1985, p. 12.
190. Gifford and Eileen Eardley, 'The Chinese market gardeners of St George district', *St George Historical Society Bulletin*, December 1970, p. 75.
191. Joanne Monk, 'The diggers in the trenches: a history of market gardens in Victoria 1835–1939', *Australian Garden History*, Vol. 4, No. 1, July/August 1992, p. 4.
192. *Auckland Star*, 10 April 1894, p. 8.
193. Lee and Lam, *Sons of the Soil*, pp. 324–325.
194. Ibid., pp. 67–68.
195. James Ng, *Windows on a Chinese Past*, Dunedin North, Vol. 3, 1999, p. 211; and Lee and Lam, *Sons of the Soil*, p. 107.
196. Sandra Pullman, 'Along Melbourne's rivers and creeks', *Australian Garden History*, Vol. 12, No. 5, March/April 2001, p. 9; and Thorpe et al., 'Otaki's market gardens', p. 27.
197. Lee and Lam, *Sons of the Soil*, p. 310.
198. Ibid., pp. 37, 68, 97.
199. Ibid., p. 371.
200. Ibid., pp. 68, 70, 73, 81, 96–97, 129, 159.
201. Ling, 'Chinese market gardening', pp. 63–64.
202. Lee and Lam, *Sons of the Soil*, pp. 341, 358–360.
203. Ibid., p. 403.
204. Peter Lee interviewed by Lily Lee, 11 March 2008, quoted in Lee and Lam, *Sons of the Soil*, p. 81.
205. *Tuapeka Times*, 26 September 1874, p. 2; *Western Australian*, Perth, 30 March 1898, p. 2; *North Melbourne Advertiser*, 4 August 1882, p. 3; *Auckland Star*, 8 March 1889, p. 2; 3 October 1890, p. 3; *Argus*, Melbourne, 15 September 1897, p. 5; *West Australian*, Perth, 28 November 1912, p. 8; 2 November 1928, p. 18.
206. *Mildura Cultivator*, 7 August 1897, p. 2; and *Argus*, Melbourne, 15 September 1897, p. 5.
207. By 1867 Ah Siug Jong was living in the Avoca district, where he was involved in a fracas with other Chinese miners and charged with assault. He was found to be of unsound mind and detained in mental asylums for over 20 years. His reminiscences, both a personal exoneration and a plea for justice, were written from 1872 while he was incarcerated in a lunatic hospital in Yarra Bend. He died in Sunbury Asylum in October 1900 (Ruth Moore and John Tully, 'Introduction', in Ah Siug Jong, *A Difficult Case: An Autobiography of a Chinese Miner on the Central Victorian Goldfields*, Daylesford, 2000, pp. 1–4).

208. The accuracy of this map has been demonstrated. Ruth Moore and John Tully located the site of Jong's dwelling to within three or four metres, and also the remains of the chimney and cook houses as marked on his map (ibid., pp. 136–138).
209. Ibid., pp. 67–69.
210. Daniel Kulp, *Country Life in South China: The Sociology of Familism*, Taipei, 1972, pp. 155–156; and Robert Knapp (ed.), *Chinese Landscapes: The Village as Place*, Honululu, 1992, p. 100.
211. Frost, 'Migrants and technological transfer', p. 127.

CHAPTER 7

The Social Environment

Chinese market gardeners in Australia and New Zealand were part of a transnational Chinese community. In common with diasporic communities around the world, they had to adapt to new social, political and geographical settings and forge new social support groups, and they maintained intricate networks connecting them with their homeland. The changing social environments within which Chinese market gardeners operated in Australia and New Zealand both provided opportunities and imposed constraints upon them. This chapter explores the vital role which social networks played in the development of Chinese market gardening. These included relationships between market gardeners and their compatriots in China, within the overseas Chinese communities in Australia and New Zealand, and between Chinese market gardeners and the predominantly European communities in which they lived. These relationships had both positive and negative impacts upon their lives.

Bonds of Kinship, Language and Native Place

The intricate social networks which market gardeners maintained with their home villages in China and with their compatriots living overseas in Australia and New Zealand helped to sustain bonds of kinship and native place. Through processes of chain migration they ensured the continuity of Chinese culture, language and horticultural skills across the generations. They supported the circulation of knowledge and technology

J. Boileau, *Chinese Market Gardening in Australia and New Zealand*,
Palgrave Studies in the History of Science and Technology,
DOI 10.1007/978-3-319-51871-8_7

across national borders, between market gardeners settled overseas, and between generations. These networks also helped Chinese market gardeners build their businesses and negotiate their relationships with the wider community. At the same time, there were also divisions and tensions within the Chinese community. Like any social group Chinese people had their disputes and there were divisions based on clan and district loyalties.

Chinese Market Gardeners and their Homeland

After the gold mining era, market gardening provided an opportunity for a more settled existence and a base for continued chain migration supplying additional labour for gardens. However, Alexander Don's *Roll of Chinese* demonstrates that market gardeners remained highly mobile, moving around within New Zealand and making return visits to China to visit family and often to marry.[1] In the 1920s and 1930s, young men working overseas were often sent back to China to receive an education in their native country, and also to find wives.[2] These ongoing contacts and experiences were an important avenue for maintaining Chinese traditions and passing on skills from one generation to the next. The overseas Chinese who returned to their home villages conveyed news and information about opportunities overseas, and some who returned to Australia or New Zealand with a Chinese education became invaluable community leaders. One was Andrew Chong, who went on to a university education in New Zealand.[3] He played a key role in establishing the Dominion Federation of New Zealand Chinese Commercial Growers in the 1940s and assisted in negotiations with European growers' organisations.

Overseas Chinese communities contributed to the modernisation of education in China and raised funds to build schools in their home villages. This benefitted both returnees and the local community. Eric Wong Ming, for example, whose father was a gold miner in Australia then became a relatively successful market gardener in New Zealand, attended the new local school in his home village of Gwa Leng between the ages of 10 and 14. The school was built with remittances sent from Gwa Leng expatriates living in Australia and New Zealand.[4]

The maintenance of family ties with China also had its negative effects. Events in China, such as natural disasters and civil war, had a direct impact on market gardeners living overseas as they received appeals for assistance from their home villages, for example to provide funds to build forts to protect communities from bandits. These appeals increased market gardeners' uncertainty about the future of their homeland, their anxieties about their families in China and the pressure on them to support them financially.[5] Such financial demands further limited their capacity to accumulate capital and expand their businesses. As Raymond Chui observes, Chinese in New Zealand were concerned to increase the power and prestige of their family or village in China. Their remittances enabled their families to buy land, open businesses and move up the social ladder.[6] Accordingly, they responded pragmatically to changes in China and New Zealand. They had to weigh up many factors in making decisions, including immigration restrictions, their own financial situation, the need to support newly arrived kin and conditions in their home village.

These anxieties were lessened to some extent by the formation of same place organisations (*tongxianghui*) in Australia and New Zealand. These mutual aid organisations brought market gardeners and their compatriots together, maintaining language and culture and co-ordinating responses to the appeals they received from their home villages.[7] The Chinese community in Sydney established various same place societies, most run by the same merchants who provided day-to-day assistance through their stores.[8] Similar organisations were established in New Zealand between 1900 and 1936.[9] Market gardeners were active members of these organisations.[10]

Market gardeners also joined with their compatriots in political action, signing the petitions which overseas Chinese communities used to voice their concerns before Chinese consular representation was established in Australia or New Zealand. These petitions concerned issues such as the poll tax; the establishment of diplomatic representation; opium smoking and gambling.[11] These petitions are evidence of the growing political activity of Chinese market gardeners and their compatriots which forged links between diasporic communities and China.

While these representations did little to modify immigration restrictions, they resulted in the visits of Chinese commissioners to Australia in 1887 and Australia and New Zealand in 1906–1907; and in the appointment of Chinese consuls in each country in 1909. The first Chinese consuls appointed were Liang Lanxun, based in Melbourne, and Huang Rongliang, based in Wellington.[12] This was an important step. Overseas Chinese communities now had strong advocates for their interests and aspirations, and the consuls promoted the formation of national Chinese organisations in their host countries with varying degrees of success.[13]

Despite efforts to establish national Chinese organisations and break down clan and district divisions within overseas Chinese communities, there is evidence that local loyalties remained strong, particularly during the unsettled period between 1916 and 1928 when the Kuomintang under the leadership of Chiang Kai Shek finally established control over China. Faced with civil war in China and local opposition to their dominance in the market gardening industry, Chinese market gardeners in Australia and New Zealand turned to their traditional sources of support – clan and native place associations. In New Zealand, these networks provided basic support for members and cut across the complex, shifting alliances that emerged in response to the civil war in China.[14] These bonds constituted a fundamental social matrix of protection, status and identity.[15]

During the hard times of the Depression in the early 1930s the Chinese community relied on their local support networks.[16] In New Zealand, Chinese people who were not British subjects were not eligible for social security until 1938, when the Labour government under Michael Joseph Savage passed the Social Security Act.[17] After the Japanese invasion of China in 1935 overseas Chinese communities were united as never before as they worked to raise funds to support the war effort. In New Zealand, a national New Zealand Chinese Association was formed with headquarters in Wellington.[18] However, after the major threat posed by the Japanese had passed, Chinese-New Zealanders were more concerned with getting their families to New Zealand, and local loyalties became paramount once again.[19] Their remittances to China decreased and always upwardly mobile, they used their capital to expand their businesses and improve their standard of living.

Evolving Overseas Chinese Communities

While negotiating their links to their homeland and its changing political and economic climate, Chinese market gardeners in Australia and New Zealand formed part of an evolving local Chinese community. From the early days on the goldfields market gardeners relied on Chinese labour and the maintenance of kin and family ties were central to their financial success. The cooperative work organisation which characterised Chinese market garden enterprises and the social and business ties between market gardeners, fruit and vegetable retailers and general merchants supported the growth of market gardening enterprises. These bonds of kinship helped to shape the opportunities available to Chinese market gardeners within their communities and promoted the exchange of horticultural knowledge and skills.

These relationships entailed reciprocal obligations between partners and the workers they employed on their gardens, who were often related or from the same village or district in China. New arrivals from China received food, shelter and employment, and training in horticultural methods and crops appropriate to Australia and New Zealand as well as managing business relationships with Europeans. The skills they gained equipped them to later establish their own gardens.[20] In return the new arrivals had financial and family obligations to their employers and were expected to contribute to the business through their labour.[21] There were also reciprocal obligations between partners in market garden enterprises who were often responsible for different aspects of the business, such as planning the seasonal range of crops and crop rotations, transport and marketing.

The intricate network of clan and village ties within each market garden enterprise extended to other market gardens within a region, and beyond that to fellow kinsmen and countrymen who owned businesses in other towns and cities. These networks were an efficient, informal support system – a source of labour as well as financial and social capital. Close ties were maintained between relatives and fellow villagers working on gardens in different locations and partners and employees moved between them as the need arose. There was also frequent movement between the main businesses operated by Chinese, particularly fruit and vegetable stores and market gardens.[22]

Market gardening provided for the first time some stability and the possibility of a planned life, with regular savings and flexible business partnerships which allowed visits home. It also entailed a new sense of responsibility since gardens had to be maintained over an extended period of time to ensure a constant return.[23] When market gardeners met at railway stations and central markets while organising the transport and sale of their produce, business opportunities mingled with opportunities to relax and socialise, and share information about daily concerns such as weather, crops and prices. Dow Chung, a market gardener in Otaki in the 1950s, recalls how the women complained that the men were gossiping at the station while their wives worked.[24] These social contacts enabled market gardeners to keep in touch with market trends.

The advent of central produce markets in cities from the late nineteenth century provided gathering places for urban Chinese market gardeners. They were often located close to Chinatowns, so gardeners could visit nearby Chinese restaurants and stores and meet their countrymen after delivering their produce.[25] In New Zealand, Otaki market gardener Kee Yung recalls the noisy chatter as Chinese growers had breakfast at the Great Wall Cafe after unloading their produce at the central markets in Wellington in the 1950s.[26]

Market gardeners maintained regular contacts with Chinese merchants who assisted them in many ways through their business networks.[27] In Sydney, for example, merchants played a key role in housing newly arrived countrymen, providing them with food, lodging and information about job opportunities in Sydney and rural areas. They helped market gardeners negotiate with bureaucracy and acted as banks, providing them with finance for their enterprises, helping them remit money to their families in China and bailing them out if they were arrested for gambling. They also arranged return travel to China and sponsored new migrants, legally and illegally, redressing to some extent the decline in numbers of the Chinese community in Australia and labour shortages on gardens. Market gardens and stores were often linked by common district ties. Kwong War Chong and Co and War Hing and Co in Sydney, for example, were frequented by market gardeners from the home districts of their proprietors.[28]

As in any social group, relations within the overseas Chinese community were complex and not always harmonious. From the late nineteenth century internal political divisions and rivalries between district and dialect groups emerged in response to changing political events in China and also in response to local issues such as gambling.[29] Contemporary newspapers reported disputes over money (for example wages and gambling debts), thefts and assaults.[30]

Nineteenth-century Chinese market gardeners were almost exclusively male, but their social and working environment changed from the early twentieth century as the gender ratio gradually became more balanced. In Australia the females added to the Chinese community largely comprised the offspring of relationships between Chinese men and women of European or Aboriginal descent, rather than new migrants. It was not until 1959 that the wives and children of Chinese-Australian citizens were officially permitted to enter the country permanently.[31] In New Zealand immigration restrictions were relaxed during the Sino-Japanese war of the late 1930s, allowing wives and children to enter the country as refugees.[32]

The arrival of wives and children was a watershed for the New Zealand Chinese community, resulting in a significant shift in social dynamics. The arrival of Chinese growers' wives and children enabled market gardening enterprises to expand and diversify, and was the starting point for many family businesses combining market gardens and fruit and vegetable stores. These businesses were established as family members arrived between the 1940s and 1955, and Chinese people became more settled within local communities. Jean Meng Yee, grand-daughter of Louie Yee Hing, who established a market garden in Gisborne in the late 1890s, describes the changes that occurred by the 1950s:

> The women were the stabilisers. They were the ones that kept the thing going... before the womenfolk came out a lot of the men were drifters, they were just (living) sort of hand to mouth, whereas when the women came it was like building a nest egg, building a home for the children... So when the family was together you had the incentive to work harder.[33]

The major attractions of market gardening were the independence and the opportunity to work flexible hours and bond with family, as gardeners worked with their wives, children and other kin in a shared enterprise. George Sue, who grew up in a market gardening family in Otaki in the 1940s and 1950s and later had his own market garden in Levin, recalls:

> It was a good lifestyle for me, it gave me the opportunity to be with the family, there were times when my kids were in primary school, I could take time off to go to sports days, to be with them, and catch up with work later... you had meals with them, they were able to work alongside you.[34]

The development of a structured Chinese business community, with social and financial networks linking market gardening partnerships, hawkers, fruit and vegetable retailers and general merchants, can be seen as a response to an uncertain political, social and economic environment.[35] The move of Chinese immigrants from gold mining into market gardening and their increasing urbanisation in the late nineteenth century occurred at a time when European society was becoming increasingly antagonistic to their presence.[36] Restrictive immigration policies had a significant social and economic impact on market gardening enterprises from the 1880s until well into the twentieth century. They strengthened existing mutual support networks within the Chinese community and within the European community, while reinforcing the physical and social barriers between them.

Bridging the Cultural Divide

Interactions bridging the divide between Chinese and European communities had positive and negative consequences for Chinese market gardening. The first part of this section discusses relationships between Chinese and Europeans at the broader societal level. Europeans held complex and often contradictory attitudes towards Chinese people. These attitudes could be modified through exchanges related to food and changing dietary patterns, shared desire to develop the land and the contribution which Chinese made to the economy. The second part of this section examines the networks and relationships that market gardeners formed with the wider community at the local level, focusing on the daily interactions which Chinese market gardeners had with the wider community – with neighbours, customers, employees, contractors, produce auctioneers, local government officials and police. These relationships could be positive, fostering understanding and familiarity between cultures, bridging ethnic and social boundaries and promoting business opportunities. They could also be negative, imposing constraints on the successful operation of market gardening enterprises, requiring negotiation and accommodation and at times generating conflict.

Aliens or Law Abiding, Industrious Citizens?

Nineteenth-century European attitudes towards Chinese immigrants in Australia and New Zealand were complex and often ambivalent, including belief in the superiority of the European race and European technology, fear of economic competition and a strong fear of miscegenation.[37] Australia, New Zealand and Canada were part of an international exchange of ideas and people between 'white men's countries' dedicated to maintaining white racial dominance. The increasing ease of communications across the Pacific facilitated the spread of anti-Chinese views. From the 1850s these countries all introduced anti-Chinese immigration legislation, Australia leading the way in 1855. After intercolonial conferences were held on 'the Chinese question' in 1881 and 1888, New Zealand and Canada followed. Their legislation adopted Australia's model, involving payment of a poll tax and restrictions on the number of Chinese immigrants per vessel.[38]These beliefs and attitudes impacted on many aspects of the lives and businesses of Chinese market gardeners, affecting their relations with Europeans.

The comments of the French journalist Edmund Marin La Meslee, who visited Australia in 1883, were typical of the complex racial attitudes that Europeans held in his day. On one hand he wrote of Chinese people: 'They are vegetable purveyors and without them those delicious necessities for European tables would be beyond the reach of most people . . . '. On the other he shared the general anti-Chinese paranoia of the time, commenting that the introduction of cheap Chinese labour threatened the livelihoods of Europeans: 'for sober, patient, industrious Chinese can always undersell the European and so necessarily bring about his downfall'.[39]

The ambivalence of European attitudes towards Chinese people was reflected in their relationships with Chinese market gardeners. They purchased their produce, but were often wary of establishing closer social relationships. Parents would warn their children to keep their distance from Chinese market gardeners and not to enter their houses when they went to buy vegetables. Dorothea Nichols and George Redding, who grew up in Ryde in the 1930s, both recall how they were frightened of the Chinese market gardeners in their neighbourhood even though they gave them presents of sweets and Chinese stamps.[40] At the same time, close relationships between Chinese market gardeners and the local community were often formed.

Chinese people were closely associated with providing food to Europeans, as growers, hawkers and fruit and vegetable retailers. Food was therefore a key bridge between European and Chinese cultures. Chinese market gardeners had a significant influence on the diets of Australians and New Zealanders, stimulating cultural change. In his gastronomic history of Australia, Michael Symons lists the staples of the colonial diet as salt meat, sugar, flour and tea, describing them as the 'rations of an ocean-going empire'.[41] This culinary heritage was remarkably enduring. A feature of the European diet in New Zealand and the Australian colonies throughout the nineteenth century was the high consumption of meat. Mutton and beef was cheap and readily available to all social classes and became the preferred source of protein.[42]

Many pastoral workers and labourers, particularly in the Australian outback, complained of their monotonous diet of meat, tea and damper (unleavened bread cooked on an open fire), and the lack of fresh vegetables. This was exacerbated by the widely held belief that it was not possible to grow vegetables in the outback and that Europeans were unsuited to heavy agricultural work in the tropics. The fact that Chinese market gardeners all over Australia showed that it was possible to grow vegetables in a variety of climates reinforced the demands of pastoral workers for more fresh vegetables in their rations. In Queensland in 1878, a correspondent to the *Brisbane Courier* observed that in the early days of the colony Chinese gardens were unknown and these pastoral workers would have been derided as 'impractical epicures'. However, he said, times had changed and their claims were reasonable:

> There is scarcely any occupied part of the colony to which the industrious Mongolian [gardener] has not penetrated and he is ever ready to follow the bushman to the remotest spot, provided it be provided with permanent water.[43]

By the 1890s dietary attitudes were changing. Commentary in newspapers from the 1870s and 1880s shows there was already public awareness of the health benefits of vegetables and of the role of Chinese gardeners in promoting the health of the community.[44] In 1874, for example, a New Zealand newspaper reported on around 20 Chinese market gardeners in Charleville, Queensland, who also kept boarding houses. Their establishments were well patronised by stockmen coming in from the bush, for they were served ample quantities of fresh vegetables and could 'drink and gamble to their hearts content without the interference of the police'. The editor commented:

> At Jimmy Ah Que's, when a man has spent his last shilling in drink, he is, they say, allowed a week's board free, and the vegetable diet pulls him around in no time.[45]

In 1886 another New Zealand newspaper commented on the laziness of Australians and their neglect of vegetable cultivation, in contrast to the industry of Chinese gardeners:

> Were it not for the indefatigable industry of the Chinese, who have an unsolicited monopoly of market gardening in Victoria and New South Wales, the people of these colonies would have nothing to eat with the meat (of which they eat three times as much each day as they properly

should do) except bread and potatoes and ere long they would be seized with scurvy and die like sheep with the rot.[46]

The growing understanding of nutrition among the medical profession and the efforts of health authorities to raise public awareness of the importance of a balanced diet made inroads into the negative attitudes that shaped social interactions across the cultural divide. In 1893, Doctor Phillip Muskett published *The Art of Living in Australia* in which he deplored the unvarying Australian diet of meat three times a day.[47] He emphasised the health benefits of the many varieties of vegetables which were easily grown but rarely appeared on Australian dining tables, particularly salad greens such as chicory, lettuce, mustard and watercress. Chinese market gardeners were the main suppliers of salad greens and he noted their important role in supplying fresh vegetables to the public. Newspapers in New Zealand also commented on the healthy, economical diet of Chinese people, which included little meat.[48] Thus changes in dietary preferences contributed to the building of relationships across the cultural divide. Chinese market gardeners benefited from increasing demand for their produce and in turn they influenced European diet through making a wide range of fresh vegetables readily available to householders at reasonable prices.

Another potential bridge between cultures which could counter negative attitudes towards Chinese people was a shared commitment, and attachment, to the land. Producing a year-round supply of vegetables for European consumers involved a long-term commitment to the land and maintaining its fertility. Discussing the symbolic attributes of gardens and fields, archaeologist Kathryn Gleason highlights their social and cultural significance – as constructed landscapes and bounded spaces, they shed light on cultural issues such as power relations, colonisation, work and leisure, and subsistence. They are ongoing projects, part of the cycle of daily, seasonal and annual changes, and inspire a host of feelings including delight in nature, fear of crop failure or hope for prosperity and a good harvest.[49] Like all peoples, by engaging with the land European and Chinese settlers legitimised their place in the world, asserted their identity and laid claim to territory.[50]

Relationships with and access to land helped define relationships between Chinese and Europeans. The European view of Chinese people as aliens and sojourners, and in some cases legal restrictions against Chinese owning land, which prevailed well into the twentieth century, restricted their access to land and meant that most Chinese market gardeners were leaseholders. However, Chinese people in Australia are now recognised as pioneers who played a key role in land clearing in North Queensland and western New South Wales.[51] This suggests that Chinese and Europeans shared common attitudes towards the land, in particular their dislike of 'wasteland'. Yuanfang Shen observes that many Chinese immigrants in Australia saw themselves as having a pivotal role in 'building the country'. In her analysis of their narratives she argues that they constructed their identities as pioneers and the equals of Europeans in their endeavours to develop the land. Writing in the 1920s, Taam Tze Pui, who was a pastoral worker in Queensland among other occupations, depicts Australia as an uncultivated realm (in Chinese *huang* (wasteland) or *chunudi* (virgin land)), which was open to both Eastern and Western efforts to develop it.[52] In New Zealand Chinese market gardeners also played a key role in clearing land, such as in the Ohakune district.[53] Meng Foon, a former market gardener and currently Mayor of Gisborne, reflects on the earlier generation of market gardeners:

> The Chinese market gardeners who came to Gisborne were all pioneers, they came with a dream that this was a better place than home. They had literally the shirts on their backs but with hard work, saving all their wages, they slowly started to spread their wings. They were a resilient lot, heads down, arse up, working under the sun, stars and moon, all weathers. They held on to the dream.[54]

Thus both Europeans and Chinese in Australia and New Zealand can be seen as colonists, civilising an uncultivated land.[55] The limited written records that Chinese immigrants left behind in Australia indicate that they saw themselves as contributing to the economic development of the country and at least some wished to settle permanently. In Darwin in 1882, Chinese residents held a banquet for a delegation of South Australian parliamentarians touring the Northern Territory. A speech was read out on their behalf, in which they stated:

> We claim that had it not been for the influx of Chinese to the Territory it would not now present the encouraging position which it maintains... It has frequently been urged that we are a migratory race but no inducement has been held out for us to settle in this country. Had there been, many of us

> before now would have been cultivators of the soil instead of being mere labourers on your goldfields. As we well know that vast portions of this settlement are suitable for agriculture, we would like opportunities of settling and making permanent homes.[56]

Elsewhere in Australia, Chinese community leaders made similar points. In 1879 the Melbourne merchants Lowe Kong Meng and Louis Ah Mouy and the missionary and social reformer Cheong Cheok Hong published a pamphlet arguing against immigration restrictions and pointing out that their countrymen were good colonists. By their use of the term 'colonists' they underlined the commitment of Chinese immigrants to their adopted country, and they specifically referred to the contribution of market gardeners.[57]

Chinese people were willing to share their horticultural skills with Europeans and Europeans were willing to learn from them. In 1899 students and curriculum advisors from a Sydney agricultural college visited a Chinese market garden in Botany where they observed their vegetable growing techniques. The report of their visit in the *Tung Wah Times* noted that Chinese 'do not hold their knowledge exclusively to themselves' and described how impressed the European visitors were: 'they come humbly to learn and respect them (Chinese gardeners) as teachers'.[58]

While Chinese market gardeners in rural areas shared with Europeans a desire to till the soil and make it productive and demonstrated their skills in intensive cultivation, this common link to the land did not operate as strongly in urban areas. However, in cities other social factors operated to create bridges between cultures. The Chinese community congregated in Chinatowns in cities such as Melbourne, Sydney, Auckland and Wellington, and in camps and Chinese quarters on the fringes of country towns, where they found mutual support. The clear social and class divisions which remained within European society well into the twentieth century meant that Chinese people lived side by side with the poor and the working classes in inner city suburbs, providing opportunities for contacts between cultures.

In the latter half of the nineteenth century entrenched inequalities between rich and poor were reinforced by anti-Chinese attitudes, particularly in the growing cities. Melbourne and Dunedin, for example, expanded rapidly after the gold rushes but side by side with grand civic buildings such as banks, hotels and theatres were pockets of poverty. In these inner city suburbs cheap housing, stables, businesses and factories coexisted with rural land uses such as poultry raising. They developed in an era before town planning and the introduction of building and zoning regulations, and suffered from overcrowding, poor drainage and inadequate sewerage systems.[59] Here, Chinese market gardeners who established gardens in outer suburbs regularly visited to socialise, purchase goods from Chinese stores and on market days often stayed overnight in cheap lodgings operated by their compatriots.[60]

Alan Mayne explores the myths surrounding inner city 'slums'. They were characterised as poverty traps and sources of crime, prostitution, gambling and disease. Chinese were stereotyped as disease and vice ridden, and class, race and gender differences were conflated.[61] Smaller Chinatowns in rural centres were also often located at the 'poor' end of town. Janis Wilton comments that while market gardens in regional New South Wales were usually located on the edges of towns near water sources and on suitable alluvial soil, this also placed Chinese people on the periphery of white social and working life.[62]

At the same time Chinese immigrants were often described as sober, hardworking, industrious and law abiding, attributes which promoted their acceptance by the wider community.[63] Prevailing anti-Chinese attitudes were modified by the significant contribution Chinese market gardeners made in providing fresh produce. In New Zealand, an editorial in the *Star* newspaper in 1888 observed that there was another side to the 'Chinese question' in Australia:

> From one end of the great island continent to the other they are the providers of garden produce for a nation, either too careless to attempt the necessary task of providing it for themselves, or incapable of the ceaseless, tireless industry by which alone, in most portions of that arid country, it is possible to accomplish it.[64]

Despite the complex and often ambivalent attitudes that underpinned relationships between Europeans and Chinese, there were a number of factors that operated as bridges across cultures. These included the significant contribution which Chinese market gardeners made in providing fresh produce at reasonable prices and enhancing European diets; a shared commitment to the land and making it productive; and shared spaces in inner city suburbs. The following section turns to the day-to-day relationships between Chinese market gardeners and Europeans at the local level.

Interactions in the Local Community

Many historians comment on the social isolation of Chinese market gardeners, for example in studies of Chinese market gardeners in Perth, Sydney and New Zealand.[65] But Chinese market gardeners depended on European clientele in order to earn a steady income and they interacted with the local community every day – with neighbours, customers, employees, contractors, produce auctioneers, local government officials and police. These relationships could be both negative, imposing constraints on the successful operation of market gardening enterprises, and positive, promoting business opportunities and building bridges across the ethnic divide. For example, in her study of the Chinese community on the west coast of the South Island in New Zealand, Julia Bradshaw documents the range of contacts between the European and Chinese communities, ranging from cross-cultural marriages and business partnerships to participation in agricultural shows. Chinese market gardeners also made donations to local charities, churches and hospitals. She stresses the complexities of these contacts; the norm was generally uneasy co-existence and limited social interaction.[66]

Good Neighbours

The work of Bradshaw in New Zealand, McGowan in New South Wales, Lancaster in Victoria and Cathie May and Henry Reynolds in northern Australia, supports the contention that the European community was generally tolerant of the Chinese people living among them.[67] This accords with the strand in Chinese historiography which shifts the focus away from the sojourner model to the settled Chinese communities which developed in Australia and New Zealand.[68] Even during periods of strong anti-Chinese sentiment, such as the 1880s, the courtesy and politeness of Chinese market gardeners in their dealings with their customers was noted. As a correspondent to the *Otago Witness,* Vincent Pike, wrote:

> For whatever else he may be – gambler, opium smoker, liar, cheat – and he is roundly accused of being each and all – John Chinaman is never boorish. He may not be so ready to 'savey' (understand) as the amenities of intercourse require, but in his commercial transactions, especially with women, his

> civility is boundless and his smiling forbearance is something marvellous to behold.[69]

Thus the daily dealings which market gardeners had with their customers were the foundation of successful Chinese market gardening enterprises. Market gardeners used dictionaries and phrase books to help them communicate with the wider community. Chinese market gardener Georgie Ah Ling's collection of belongings, housed in the Donald Agricultural Museum in Victoria, includes well-used copies of a Chinese dictionary and a Chinese-English pronunciation guide and phrase book published in Hong Kong in 1923. These books reflect processes of cross-cultural exchange and Georgie's social interactions with his neighbours. The language in the phrasebook is practical and functional, and it is divided into sections relating to daily life in Australia, including business and legal interactions, health, food and clothing. It conjures up the world of Chinese-Australians in the early twentieth century and how they negotiated their dealings with Europeans. It also reflects Georgie's personal situation. Despite his poor English, he made long-term friendships with many Donald residents.[70] Another example is an English-Chinese word and phrase book published in Sydney in the 1880s which illustrates the extensive business and personal relationships that Chinese maintained with Europeans.[71] Its contents cover the main areas of Chinese commercial activity in Sydney at the time, notably fruit and vegetable growing and retailing, store keeping, domestic service, and furniture making.

With the advent of central city markets and the auction system from the late nineteenth century market gardeners had regular dealings with produce agents and auctioneers, both European and Chinese. These relationships always had to be negotiated, the grower wanting the best price for his produce by selling at the most advantageous time, the auctioneer requiring a reliable supply of good quality produce. In New Zealand, Chinese growers in Otaki developed relationships with individual auctioneers in the main produce firms operating in Wellington, directing different crops to their preferred firm. Discussing the produce merchants he used, Andrew Yung recalls:

> I used to be a Leary one – . . . lettuce, beans to Leary's . . . because the lettuce auctioneer's always looking after me. When the lettuce price, he reckon it will be cheap, he sold mine first. When he reckoned the price was going up, he sold mine last . . . All my tomatoes – Market Gardeners . . . I gave the

> whole lot to Market Gardeners, not even one case to other markets, and so Alan was always looking after me.[72]

To promote long-term business relationships with their customers, many market gardeners were generous in giving credit. In New Zealand, market gardeners and vegetable hawkers kept track of the credit owed them by marking the sums owed by customers on convenient doorposts or fences.[73] The amount of this credit and the terms of repayment were undoubtedly a matter for negotiation between gardeners and their customers. There are many accounts of Chinese market gardeners giving gifts of sweets, fresh fruit and vegetables to the local children they met on their rounds in cities and in country towns. Europeans shared with Chinese a passion for ginger preserved in syrup; at Christmas or Chinese New Year Chinese gardeners often gave jars of preserved ginger or lychees to their regular customers. In New South Wales, for example, these stories come from the Sydney suburbs of Botany, Ryde and Rockdale, and also from the towns of Dubbo, Murwillumbah, Narranderah and Narrabri.[74] In the Sydney suburb of Willoughby, cicadas formed a bridge between cultures, when Chinese market gardeners collected a birdcage full of cicadas for the local children.[75]

There was also reciprocity in these relationships. Owen Bennett, who grew up in Ryde, recalls that as children they would catch eels in the creeks and exchange them with the local Chinese market gardeners for vegetables.[76] In an earlier, more systematic example of reciprocity, in 1895 the *Northern Territory Times and Gazette* reported on illegal traffic in beef in Borroloola near the Gulf of Carpentaria. Cattle poachers traded the beef with Chinese market gardeners in exchange for fresh vegetables and other food supplies. Three gardeners, Ah Song, Ah Que and Hang Yee, were charged with being receivers and each fined £10.[77]

Adding to the networks created between the Chinese and European communities were the regular dealings between market gardeners and their European landlords and contractors who assisted them with ploughing and repairing machinery. In the Melbourne suburb of Brighton, for example, a stable, relatively harmonious community of European and Chinese market gardeners developed in the 1880s and 1890s. Many Chinese market gardeners leased land from Europeans who were themselves market gardeners, which meant that they had shared interests and concerns.[78]

In New Zealand many Chinese market gardeners leased land from Pākehā and Māori landowners. In the Otaki district, for example, Chinese gardeners leased land from Māori at Te Horo, around Otaki township and Ohau. According to Lily Lee and Ruth Lam and Manying Ip, relations between Māori and Chinese were generally good – they were friends, neighbours and relatives through intermarriage, and shared common cultural traits such as close attachment to the land, respect for ancestors and strong emphasis on kinship and extended family.[79]

Chinese market gardeners participated in the activities of the communities in which they lived, making donations to hospitals, schools, churches and other charities and participating in local agricultural shows and celebrations. For example, in Greymouth on the West Coast of the South Island, Chinese market gardeners donated vegetables to the Grey Valley Hospital.[80] In the 1870s and 1880s they regularly won prizes in the Greymouth Horticultural Society exhibitions. For example, Ah Kong won prizes in 1874 and 1875 and Ah Kum won prizes in 1884, 1885 and 1886.[81]

There are many similar examples of Chinese market gardeners in Australian cities and country towns supporting local charities. In 1909 the *Tung Wah Times* reported that Chinese market gardeners in Bendigo had made donations to the local hospital.[82] In New South Wales, Chinese market gardeners Tip Nooey in Narrandera and Mow Hee in Dubbo were regular donors to their local hospital funds in the late nineteenth and early twentieth centuries.[83] In Sydney in the early 1920s, Chinese market gardeners supplied free vegetables to unemployed workers and the Board of the Unemployed Workers Association published a letter of thanks in the *Tung Wah Times.*[84]

Chinese market gardeners became involved in other community organisations such as sporting groups. This was indicative of a reduced social distance between them and their European neighbours, as overseas Chinese communities became increasingly Westernised from the early twentieth century, influenced by political and economic changes in China.[85] Chinese in Australia and New Zealand adopted European dress and engaged in European sports and pastimes, contributing to the breaking down of social barriers and the building of social networks between the two communities. The Chinese community in New Zealand, for example, participated in sports such as cricket, football, athletics, horse racing, and by the 1920s, car racing.[86]

These exchanges and interactions contributed to better understanding between Chinese market gardeners and the wider community and promoted business relationships. They countered the general hostility towards Chinese at the political level and provide some evidence of cooperative relationships. However, these cross-cultural relationships had to be nurtured and worked at, and were complicated by the multi-racial nature of the wider societies in which Chinese immigrants lived, with their ethnic and class divisions.

Conflict, Cooperation and Circumvention

Cross-cultural relationships were not always harmonious. Chinese gardeners were often made to feel unwelcome in the communities in which they lived, as anti-Chinese sentiments in the wider society fuelled anti-social behaviour. They were attacked by groups of larrikins or neighbourhood children who threw stones, subjected them to racist taunts or stole vegetables. In New Zealand newspapers carried regular reports of thefts of produce, money and goods from Chinese market gardens and huts, and physical attacks on them, particularly in the late nineteenth and early twentieth centuries.[87] The frequency of attacks on Chinese market gardeners provoked comment in both the mainstream media and church publications. In 1889 the *New Zealand Presbyterian* reported: 'A Chinaman living alone cannot leave his house or garden on Sundays (to attend church) for fear of his windows being broken or his garden beds spoiled in his absence'.[88] However, as Ng points out, in New Zealand serious incidents of violence between Europeans and Chinese were widely scattered in time and place and relations between the two races improved from the 1920s with the rising prestige of China.[89]

As Chinese market gardeners established businesses in the growing cities in the later nineteenth century, fences were important in marking the boundaries between them and their European neighbours. Fences signified ownership, marking the boundary between the spaces occupied by Chinese gardeners and the sometimes hostile European community around them. These boundaries were often transgressed. Newspapers in Australia, including Chinese language newspapers such as the *Tung Wah Times,* carried regular reports of larrikins breaking into market gardens, particularly in urban areas. One of the most serious incidents occurred in the Sydney suburb of Waterloo in 1879. Fifty European larrikins pulled down and burnt the fence surrounding Tat Lim's market garden and turned in horses to destroy his crops. They assaulted Lim and his co-workers and broke the windows of their house.[90] These negative interactions and the unpredictable behaviour of Europeans contributed to a lack of trust between them and Chinese market gardeners.

Market gardeners also had regular interactions with the police and the justice system. Although discriminatory legislation was enacted against them in most colonies, at least in Australia Chinese people had in principle equal rights before the law, and could seek redress through the court system. This was facilitated by the early establishment of policing on the goldfields.[91] Although many incidents probably went unreported, contemporary newspapers in both Australia and New Zealand report many cases in which market gardeners who had been subjected to racist attacks or thefts went to the police and the offenders were brought before the courts and fined, for example in Auckland in 1898 and Masterton in 1909.[92] In some areas Chinese market gardeners developed a long-term relationship with local police. In South Dunedin in 1909, Chinese market gardeners presented Sergeant Conn with a pair of vases on his retirement, to show their gratitude for his protection over the years.[93]

Mirroring their relationships with the general European community, the relationships of Chinese market gardeners with European market gardeners varied from conflict to cooperation, as social, economic and political conditions changed.[94] Despite efforts to boycott Chinese-grown vegetables and the establishment of European growers associations in Australia and New Zealand, these movements generally met with public apathy. The ready supply and reasonable prices offered by Chinese gardeners ensured they continued to have a market for their produce. The reality was that few Europeans wanted to take up market gardening.[95] In Australia local associations of European market gardeners were established in various states, for example in the Cooks River district in Sydney in the 1880s, in Melbourne in 1900 and in Adelaide in 1907, but most were short lived and no national organisation was established.[96] In 1891 in Queensland, a deputation from the Labourers Union travelled to Barcaldine to enlist local support for a boycott of Chinese storekeepers, cooks and gardeners. Many residents were willing to consider action against cooks and storekeepers but, as the *Brisbane Courier* reported, they could not do without the vegetables supplied by gardeners.[97]

In New Zealand, the Canterbury European Market Gardeners Association in Christchurch and the Hutt Farmers Producing Association in the Hutt Valley were both established in 1895.[98] Other anti-Chinese commercial growers groups were established in Auckland in 1908. But by the early twentieth century their focus had shifted from a strong anti-Chinese platform towards promoting the interests of growers more generally and they remained regionally focused until well into the twentieth century. The first national organisation of market gardeners, the Dominion Council of Tomato, Stone Fruit and Produce Growers, was not established until 1929.[99]

From the 1920s there is evidence of cooperation between Chinese and European market gardeners, at least in New Zealand. In 1923 Chinese and European market gardeners in the Hutt Valley established a cooperative fruit and vegetable marketing firm in opposition to Wellington retailers who threatened to force them to buy back the containers they used to ship produce. In Auckland, the marketing company Produce Markets, established in 1929, had a large number of Chinese shareholders and in 1930 three Chinese directors were appointed – Thomas Doo Senior, his son Thomas Doo Junior and Fong Foo Foy.[100] During World War II, when Chinese growers played a major role in supplying the armed forces in the Pacific, attitudes towards them improved. Their standing in the community rose and they were complimented for their patriotic effort in Parliament.[101] These cooperative ventures helped market gardeners make their operations more cost effective through improved marketing and reduced costs of inputs such as transport and fertilisers, and promoted their joint interests. The exchange of knowledge and information between Chinese and European market gardeners also facilitated technological change, as Chinese market gardeners adopted the latest developments in irrigation systems, fertilisers, pesticides and agricultural equipment.

During the war legislation was passed governing registration of the market gardening industry and greater efforts at cooperation between the Dominion Council and the newly formed Dominion Federation of New Zealand Chinese Commercial Growers, established in 1943, were made.[102] In Otaki, the Otaki District Commercial Gardeners Society formed a united organisation with Chinese growers in 1942, named the Otaki District Commercial Gardeners Association. The Association played an important role in protecting gardeners' mutual interests and reducing costs by purchasing fertilisers and seeds in bulk and standardising transport rates.[103]

Other significant interactions across the cultural divide which impacted on the daily lives of Chinese market gardeners were their dealings with state and local governments. The employees of these bodies represented a powerful 'outside group' with the backing of government authority. Like their European neighbours, from the late nineteenth century Chinese market gardeners became increasingly reliant upon collectively provided services such as water, power, roads, sewerage and waste disposal and were subject to an increasingly complex set of local regulations, for example regarding sanitation and building standards.[104] As early as 1874 residents in Wellington, New Zealand, complained about 'sanitary reformers' from the city council forcing householders to buy closet boxes. In a letter to the *Evening Post*, one resident argued that household manure was a valuable fertiliser for gardens, citing the example set by Chinese market gardeners.[105] However, in later years local governments became more aware of public health issues and many enacted bylaws prohibiting the use of nightsoil on market gardens, whether European or Chinese. By the late nineteenth century Chinese communities were becoming increasingly urbanised, and as governments became more concerned with solving emerging social and environmental problems in cities, market gardeners required a more sophisticated understanding of the legal environment in which they were operating their businesses, including local bylaws.

Market gardeners were subject to visits by a variety of local government officials including sanitary inspectors, health officers, building inspectors and engineers. Municipal councils in Australian cities such as Brisbane, Sydney, Melbourne, Perth and Bendigo made inspections of Chinese market gardens and conducted campaigns against practices such as washing vegetables in polluted water, diverting water from street drainage to irrigate gardens and carting vegetables in the same vehicles used to transport manure without cleaning them.[106] In Melbourne, for example, after complaints from local residents in Richmond, health authorities instituted proceedings against two Chinese market gardeners for keeping unsanitary premises. They were each fined £5 and costs.[107]

Since Chinese market gardeners were a major presence in Australian cities in the early twentieth century and highly visible because of their race it is probable that they were more likely than European market gardeners to be prosecuted for infringing regulations. Their interactions with local government authorities could also be hampered by language and cultural barriers. As many gardeners did not speak English well they were unaware of many regulations until they were visited by inspectors. Some local governments made efforts to bridge the language barrier. In the Melbourne suburb of Brighton in the early 1900s, the local Board of

Health took the sensible step of translating health regulations into Chinese and circulating them among Chinese market gardeners.[108] Similarly in Brisbane in 1899, the Central Board of Health recommended the distribution of notices in English and Chinese to all market gardeners setting out hygiene regulations.[109]

Chinese language newspapers also played a role in communicating local government regulations to market gardeners. In 1899 the *Tung Wah News* published an article advising Sydney market gardeners about the often bewildering array of regulations of which they should be aware. They were advised to get a certificate for their horse, not to ride an injured horse and to tie up their carts so their horse would not run away, otherwise they might be fined.[110] In the 1920s the *Tung Wah Times* (successor of the *Tung Wah News*) carried regular warnings to Chinese market gardeners to abide by council regulations regarding washing vegetables with clean water and caring for their horses properly.[111] Market gardening operations, both Chinese and European, were certainly circumscribed by such regulations. However, by raising awareness among the Chinese community of local government regulations and bylaws and the penalties for infringing them, the Chinese language media increased their familiarity with the society in which they lived.

In New Zealand, municipal authorities in Auckland, Christchurch, Invercargill and Gisborne received similar complaints about such practices as using nightsoil as fertiliser and washing vegetables in polluted water, and inspected Chinese market gardens. In most cases the offenders were prosecuted and fined.[112] It was acknowledged in some cases that Chinese market gardeners had improved the condition of formerly unusable swampy land and brought it into productive use. Reporting on his inspection of Chinese market gardens near Western Springs (then Auckland's major water supply) in 1887, Mr J. Pond noted:

> I passed through the Chinese garden and found nothing which could in any way injure the purity of the water supply, and I consider the high state of cultivation to which they have brought this ground greatly preferable to the swampy state it was in before.[113]

One response of Chinese market gardeners to the maze of regulations surrounding them was to find creative ways to circumvent them. They actively subverted government regulations, sometimes laying claim to key resources such as land or water that were not legally theirs. In Cooktown in North Queensland, Chinese gardeners sometimes extended the high wattle fences around their garden plots to encroach onto road reserves or to maintain exclusive access to water sources.[114] In another case Ling Fong, owner of a large market garden in the Perth suburb of Maylands in 1926, extended his land into a road reserve and erected a fence to keep stock out.[115] After local councils introduced charges for water usage, some Chinese market gardeners found ways to evade paying them. In 1899, inspectors from the Melbourne Metropolitan Board of Works visited the premises of Ah Lip and Ah Cheong, Chinese gardeners in Coburg, after unusually low levels of water usage had been recorded. After encountering some resistance from the gardeners, the inspectors discovered that they had tampered with the water meter.[116]

Another regulation affecting market gardening operations was the prohibition on working on Sundays, particularly in the period from the 1880s to World War I. The Labour movements in Australia and New Zealand strongly supported reducing the working hours of Chinese in all their occupations, in reaction to what was seen as unfair competition.[117] New Zealand newspapers often reported prosecutions of Chinese market gardeners for infringing such bylaws, for example in Wellington in 1887 and in the Te Matai Valley near Napier in 1909.[118] Restrictions on working on Sundays particularly affected larger enterprises such as Chan Dah Chee's in Auckland, which had contracts to supply set quantities of vegetables to shipping companies, clubs and hotels. In 1907 five of Chee's employees were charged in the Police Court with gardening on Sundays. The defence argued that Chee had about 50 contracts to fill on the Monday morning for the supply of vegetables to steamers and clubs and it was impossible to meet these orders without working on Sunday.[119] Chinese market gardeners in Australia also faced restrictions on working on Sundays, as they were reminded by the *Tung Wah Times*.[120]

Linked to these restrictions were concerns to regulate the activities of Chinese market gardeners during their limited leisure hours, particularly gambling and opium smoking. Police on both sides of the Tasman raided market gardens in search of illegal immigrants and evidence of opium smoking. In Australia in 1931, Ah Wah and Lee Tiy, Chinese gardeners, were fined £5 in Redfern Police Court for smoking opium in Tuck Wah's garden in Botany, following a police raid.[121] In 1919, Yee Kum and Ah Mon, who had been smuggled into New Zealand the previous year, were discovered on a market garden in Auckland.[122] In Sydney in 1927, police raided market gardens in Liverpool, Mascot, Botany and other locations in search of 12 illegal immigrants said to have been clandestinely landed from a steamer. Ten arrests were made.[123]

Certainly immigration rackets and smuggling rings existed, but such stories were readily sensationalised by the popular press. Smuggling people into the country on board ships was the most common method the Chinese community used to defy immigration restrictions and added to the Chinese population of New South Wales more than any other State after Federation.[124] As Jane Lydon and Shirley Fitzgerald observe in their studies of the Chinese community in Sydney, circumventing regulations and bending rules was an elaborate game on both sides. Police and customs officials were often open to bribery, while in Chinese culture these informal exchanges with officialdom were a part of *guanxi*, the custom of giving gifts in exchange for material gain or to strengthen personal relationships.[125]

These raids also caught up Europeans in the authorities' net, indicating that market gardens were important loci of cross-cultural interaction. For example, in New Zealand in 1905, police searching a Chinese market garden in Palmerston North discovered a European smoking opium.[126] In Australia there were also raids targeted at finding 'fallen' women living with Chinese men.[127] These inspections and raids indicate the level of surveillance under which Chinese market gardeners lived, and also the efforts they made to circumvent the legal restrictions they encountered.

Cross-Cultural Relationships on the Margins

Often situated on the margins of European society, Chinese market gardeners formed relationships with other marginalised social groups, including indigenous people. Chinese market gardeners' need for additional labour contributed to the development of relationships across cultures, particularly between Chinese and Māori in New Zealand from the early twentieth century. Many Chinese market gardeners employed Māori workers, and this was a mutually beneficial relationship. For Chinese market gardeners, Māori provided a valuable additional labour force. For Māori, particularly those living in rural areas where employment opportunities were limited, market gardens provided a much needed source of income, food and accommodation. During the 1930s Depression Māori were initially excluded from unemployment relief and, with many of their menfolk unemployed, Māori women found work on market gardens to support their families.[128] They were often assisted by their relatives, as

Chinese market gardeners commonly let a contract to the head of an extended family and provided them all with accommodation.[129]

In market gardening centres such as Otaki, Ohakune and Pukekohe, and in many other areas of the North Island, Māori formed the core of the market gardening workforce.[130] After World War II there was a major movement of the Māori population to the cities in search of work. Urban market gardens, for example in Mangere, Tamaki and Pukekohe in Auckland, provided employment for Māori families. Chinese market gardeners also moved from rural areas to the city, closer to their markets, and some took their Māori workers with them. Ng Bing Wai, a prominent grower in Taranaki in the 1940s and early 1950s, moved to Auckland in 1954 and took Māori workers with him to work on his market garden in Mangere.[131]

These interactions led to personal relationships, ranging from casual liaisons to enduring marriage and business partnerships. Chinese market gardeners leased land from Māori tribes and Māori wives worked side by side with their husbands on the garden as well as caring for children and preparing meals for employees. An example was Chong Lee, known as the 'carrot king', who married a local Māori woman, Eva Te Hiwi. Together they gardened on Eva's ancestral land north of Wellington.[132] Thus market gardens were important loci of interaction between the Chinese and Māori communities. Their lives were entwined, fostering close relationships that bridged the ethnic divide. Where Māori families lived on the gardens, Māori and Chinese children played together and went to school together. They all helped their parents, working on the gardens after school and at the weekends.

The post-war movement of Māori from small rural communities to cities brought opportunities for employment and economic advancement, as well as threats of social dislocation and separation from family and tribal support networks.[133] Concerns about the living and working conditions of Māori on market gardens first arose in the late 1920s when the Māori population was increasing and their aspirations for self-determination and recovering their traditional lands which had been appropriated by Europeans were growing.[134] These concerns resurfaced in the 1950s. The burden of providing accommodation for their workers often fell on Chinese market gardeners and they could not cope with the arrival of large numbers of extended family members.[135] Despite efforts by the Government to address overcrowding and substandard living conditions on market gardens, living conditions remained spartan, often for Chinese gardeners as well as their Māori workers.[136] The facts remained that Māori were the lifeblood of the market gardening industry, particularly during a period when generational changes in the Chinese community meant that sons and daughters of Chinese market gardeners were moving into other careers, and market gardeners provided Māori families with much-needed employment.

Aboriginal people were employed on Chinese market gardens in Australia, but not to the same extent as Māori in New Zealand. With their multiplicity of tribal groups and languages and small, mobile communities scattered across vast distances, Aboriginal people had greater difficulties in organising politically. As the original custodians of the land, their relationships with later settlers, both European and Chinese, were characterised by ignorance, fear and mistrust. Like European settlers, early Chinese immigrants feared 'the wild black people with spears' and saw them as 'primitive' and lacking civilisation.[137] Certainly there were incidents where Chinese immigrants were killed by Aboriginal people, for example on the Palmer River goldfields in North Queensland in the 1870s.[138] However, tensions eased by the late nineteenth century, as restrictive legislation affected the lives of both groups and they were regarded as second-class citizens. Chinese market gardeners developed strong, enduring ties with Aboriginal people, particularly in rural areas, and oral history accounts include many stories of Chinese and Aboriginal people working together, supplying outback towns with vegetables.

Many Chinese market gardeners in northern Australia and outback areas of the southern states employed Aboriginal workers. According to several commentators in Queensland in the 1890s, including the police inspector at Mackay, local police on the Atherton Tableland and the northern Protector of Aborigines, Dr Walter Roth, Aboriginal people were generally treated more fairly by Chinese than by Europeans and they preferred to work for Chinese employers.[139]

In Queensland the lives of Aboriginal people were particularly closely controlled. The 1897 Aboriginal Protection and Restrictions of the Sale of Opium Act was far more restrictive than any similar legislation operating in New South Wales or Victoria. This legislation encapsulated two major preoccupations of colonial authorities: the conflict-ridden relations between white settlers and Aboriginal communities and anxieties about the Chinese presence in Australia, expressed through pejorative views about the 'moral vice' of opium. The Act implemented a system of closed reserves under the direct control of government officials and gave legal sanction to the forced removal and dispersal of Aboriginal people from their traditional lands. As Aboriginal people were a source of cheap labour for white employers, the 1897 Act also empowered the Protectors of Aborigines to regulate the living and employment conditions of individuals who remained in contact with white society. The legislation was justified by the desire to protect indigenous people from exploitation by both white and Chinese employers, but it is debatable to what extent it was motivated by any real concern for their well-being.[140] Thus Chinese market gardeners, like European employers, were constrained in their access to indigenous labour and subject to surveillance by European authorities. In 1927, for example, the Queensland market gardener Sam Wah was fined £10 and costs in the Gordonvale Police Court for employing Aboriginal people contrary to the Aboriginal Protection Act.[141]

In New South Wales the reserve system was less strictly controlled and mission dwellers were allowed to work outside the missions. Like Māori people in New Zealand, indigenous families had the chance to work together outdoors on market gardens and earn much-needed income. For example, Ah Yook, a Chinese market gardener in Wellington, New South Wales employed Aboriginal men, women and children from the nearby mission, collecting them in his truck each day.[142]

The recollections of Norman Mitchell, a Kuku Yalanji man born around 1900 near Mount Carbine on the Cape York Peninsula, give a rare insight into Chinese-Aboriginal relationships from the Aboriginal perspective. Mitchell spent much of his early life in Maytown, which had been the centre of the Palmer River gold rushes between about 1873 and 1885. He vividly recalls the Chinese people who remained after the gold rushes and set up business as market gardeners, storekeepers and carters in the region. For example, he describes helping a Chinese market gardener weed his plots and cooking favourite Aboriginal bush tucker in the Chinese style:

> We used to go up there . . . don't know whether Dad used to take the opium or not . . . but we used to often go up there to Chinaman's garden, you know. Give the Chinaman a hand, shiftin' grass, weed. Had a big garden growin' there and so forth. Dad used to bring 'em great big sand goanna. He used to grab 'em by the tail and slap 'em on the ground...Chinaman would eat that *minya* [meat food] then. They use that fat part, oil, too, for all kinda cookin' and so forth. And they mixed in that *saw-su*, Chinee rum they call it. They used to have it for purpose of medicine too.[143]

One urban area where close relationships developed between Chinese market gardeners and Aboriginal people was La Perouse, in Sydney.[144] In 1883 a camp was established at La Perouse under the Aborigines Protection Board. Many of the inhabitants came from rural areas and over time the settlement was run by a variety of church and welfare groups. During the Depression of the 1930s unemployed people built makeshift houses in two settlements called Happy Valley and Hill 60 north of the Aboriginal mission. There was extensive interaction between all these communities and the Chinese market gardeners in the area, who gave them food.

As in New Zealand, a natural outcome of the daily social interactions between Chinese market gardeners and the local community in Australia was the formation of intimate relationships between Chinese men and women of other ethnic backgrounds, both indigenous and European. It is difficult to accurately determine the prevalence of these relationships. Researching Chinese families in Western Queensland, Sandi Robb has documented Chinese and Aboriginal unions in the market gardening precinct in Cloncurry in the early twentieth century; the majority of wives were Aboriginal.[145] The children of Chinese-Aboriginal relationships had to negotiate their dual cultural heritage. Jennifer Martiniello, the granddaughter of the market gardener Ned Chong in Oodnadatta, who married an Arrente woman, Minnie Bell, in 1913, describes how her grandfather dictated that only English, 'the language of power', be spoken in the household. This is an indication of Chong's desire to succeed in his adopted country, on European terms. Despite this, Minnie passed on Arrente languages and traditional cultural knowledge to all her children.[146]

Jennifer Martiniello inherited her grandfather Ned Chong's fob watch and chain. Well worn, dented in places, the watch is made of Mexican silver; the chain is inscribed with simple linear motifs along the links. As symbols of Western affluence, they give insights into Chong's life and character.[147] The watch and chain represent his business success, his position in the community and his ability to bridge the gap between the dominant European culture and marginalised groups, both Chinese and Aboriginal. They also symbolise Chong's upward mobility; he was a successful entrepreneur whose market garden was a springboard for expansion into other business enterprises. An 1880s photograph of him survives, dressed in Chinese cap, western waistcoat, shirt, tie and jacket, hair tied back in a pigtail, trousers and boots scuffed and worn. Across his waistcoat he wears his fob watch and chain. It is an eloquent portrait of a hard-working Chinese immigrant of the late nineteenth century, living between two worlds.[148]

According to Jennifer, her father and his siblings were able to gain an education generally reserved only for whites because of her grandfather's standing and contribution to the community.[149] Today, the watch and chain has great personal significance for Jennifer and her son, as part of their family heritage. Jennifer has modified the watch chain to wear as a bracelet, adding an eagle's claw, while her son has the watch. Thus the fob watch and chain show how artefacts give material form to patterns of social relationships, and can mediate between different social worlds. As they have passed down the generations, the watch and chain have been transformed and taken on additional layers of meaning and significance.

Recent research in Australia has expanded and complicated our knowledge of Chinese families, particularly concerning the extent to which Chinese men and white women formed relationships. Kate Bagnall highlights the relationships that developed on the thresholds of European homes, between Chinese gardeners and vegetable hawkers and their regular customers.[150] These customers were usually the women of the household; their encounters involved cooperation, negotiation, exchange and friendship. These daily interactions counter more negative pictures of inequality and conflict – they allowed white wives and servants to get to know Chinese people as individuals, rather than part of an anonymous group. Through these encounters European women modified their attitudes towards Chinese people, developing ideas potentially contradictory to those of the men of their household. These were relationships of mutual advantage: housewives had a regular, convenient supply of fresh vegetables at reasonable prices while Chinese gardeners developed a regular clientele for their produce and were able to earn a living in a country where they were excluded from finding work in many other fields.

Conclusion

Chinese market gardeners in Australia and New Zealand not only adapted their technology and work practices to foreign and often harsh physical environments, they also had to adapt to foreign and often hostile social, cultural and political environments. As Warwick Frost observes in Australia, they created not only a Chinese-Australian style of farming, but a Chinese-Australian culture.[151]

This chapter has explored the opportunities and constraints offered by the complex networks of relationships that Chinese market gardeners maintained, with their kin and home villages in China, within their own overseas Chinese communities and with the wider community. These networks extended across national borders, between Chinese market gardeners living overseas and their kin and native places in China, and between the overseas Chinese communities in Australia and New Zealand. These relationships sustained them, maintaining the bonds of family, kin and native place, and ensuring the continuity of Chinese culture, language and horticultural skills across the generations. They therefore played a key role in the exchange of horticultural technology between China and Australia and New Zealand. They also formed a ladder for upward social and economic mobility, helping Chinese market gardeners build their businesses and negotiate their relationships with the wider community.

Importantly, the bonds within Chinese communities were sometimes fractured and were not always sustained by harmonious relations; there was friction and conflict among Chinese people, as could occur in any human group. At the same time the bonds within groups in the wider community could work against the interests of Chinese market gardeners. These included informal gangs of European larrikins banding together to attack gardens as well as more formal anti-Chinese associations organising boycotts against market gardeners or lobbying for restricting Chinese immigration.

The relationships forged across community boundaries broke down barriers of language, culture and ethnicity. Anti-Chinese attitudes at the broader societal level could be moderated through exchanges related to food and changing dietary patterns; shared desire to develop the land and make it productive; and the contribution which Chinese market gardeners made to the economy. At the local level, the diverse relationships that market gardeners formed in their daily interactions with the wider community fostered understanding between cultures, promoted business opportunities and helped to counter broader anti-Chinese attitudes. They ranged from business partnerships to marriage and *de facto* relationships between Chinese men and women of other ethnic backgrounds, and were the foundation for building successful market gardening enterprises.

These relationships could also be negative, imposing constraints on the successful operation of market gardening enterprises, for example the activities of local government officials implementing the maze of local government regulations that circumscribed the operation of market garden enterprises and police raids on market gardens in search of opium and illegal immigrants. They involved conflict, cooperation, negotiation and in some cases circumvention. Chinese market gardeners were active agents in their own history, as they negotiated their positions in relation to the restrictive immigration policies which dominated their lives from the late nineteenth century until well into the twentieth century, as well as changing societal attitudes. At the local level, they mitigated the constraints posed by institutional racism, government regulations and official surveillance through building personal relationships with the people from all walks of life with whom they came in daily contact. They actively manipulated these relationships to their mutual advantage, often circumventing the maze of regulations which surrounded them.

Market gardens were thus important loci of cross-cultural exchange. Often situated on the margins of European society, in inner city industrial areas or on the fringes of country towns, Chinese market gardeners developed close living and working relationships with other marginalised social groups, in particular with Māori in New Zealand and Aboriginal people in Australia. These included intimate sexual relationships, with all their complexities, ranging from enduring marriage and business partnerships to more temporary liaisons. The children of these relationships had to negotiate their mixed cultural heritage and in turn contributed to the building of bridges between cultures.

Notes

1. Brian Moloughney, Tony Ballantyne and David Hood, 'After gold: reconstructing Chinese communities 1896–1913', in Henry Johnson and Brian Moloughney (eds.), *Asia in the Making of New Zealand*, Auckland, 2006, pp. 65–69; and James Ng, *Windows on a Chinese Past*, Vol. 1, Dunedin, 1993, pp. 87–89.
2. This practice continued until the Sino-Japanese war disrupted shipping (Charles Sedgwick, 'The politics of survival: a social history of the Chinese in New Zealand', PhD Thesis, University of Canterbury, Christchurch, 1982, pp. 65, 325).
3. Born in Wanganui in 1907, Chong was educated in China; he returned to New Zealand in 1929 (Nigel Murphy, *Success through Adversity: A History of the Dominion Federation of New Zealand Chinese Commercial Growers*, Pukekohe, 2012, pp. 16, 21).
4. Henry Chan (ed.), *Zengcheng New Zealanders*, Katoomba, 2007, p. 43.
5. Sedgwick, 'Politics of survival', p. 327.

6. Raymond Chui, 'Transnationalism and migration: Chinese migrants in New Zealand', *Global Asia Journal*, Paper 4, http//digitalcommons/pace.edu/Global_Asia_Journal_4, 2008, p. 6, accessed 27 May 2013.
7. Projects funded by overseas Chinese associations in Australia and New Zealand included modern schools and other infrastructure, defensive walls and watch towers to guard against bandits and safe houses in Guangzhou for family members (Chan (ed.), *Zengcheng New Zealanders,* pp. 20, 111; and Michael Williams, 'Brief sojourn in your native land', M. Letters Thesis, University of New England, Armidale, 1998, p. 53).
8. Generally each district had its own society. In 1907 people from the neighbouring Zengcheng and Dongguan counties established a combined organisation, the Loong Yee Tong (Williams, 'Brief sojourn', pp. 19–20).
9. These included: in Auckland, the Kwong Chew Club for people from Siyi county (1920); and in Wellington, the Poon Fa Association for people from Panyu county (1916); the Tung Jung Association for people from Zengcheng (Jung Seng) and Dongguan counties (1924); and the Seyip Association for people from Siyi county (1936) (Sedgwick, 'Politics of survival', p. 588).
10. See, for example, in New Zealand, Sedgwick, 'Politics of survival', pp. 297, 309, 656–658; and in Australia, Shirley Fitzgerald, *Red Tape Gold Scissors: The Story of Sydney's Chinese*, Sydney, 1996, pp. 45, 109–110.
11. Ng, *Windows on a Chinese Past*, Vol. 3, p. 113; Sedgwick, 'Politics of survival', pp. 656–658.
12. Fitzgerald, *Big White Lie*, p. 111; Sedgwick, 'Politics of survival', pp. 172, 264; and William Tai Yuen, *China's Awareness of New Zealand,* Auckland, 2005, pp. 80–81, 85.
13. Two attempts were made to establish national Chinese organisations in New Zealand, in 1909 and 1928, promoted by the Chinese consuls of the time, but both lapsed (Ng, *Windows on a Chinese Past*, Vol. 3, pp. 156–157).
14. Sedgwick, 'Politics of survival', pp. 588–589.
15. Jane Lydon, *Many Inventions: The Chinese in the Rocks 1890–1930*, Clayton, Victoria, 1999, p. 66.
16. Ng, *Windows on a Chinese Past*, Vol. 3, p. 160.
17. Raymond Richards, *Closing the Door to Destitution: The Shaping of the Social Security Acts of New Zealand and the United States*, University Park, Pennsylvania State University, 2010, pp. 32, 71.
18. Ng, *Windows on a Chinese Past*, Vol. 3, pp. 169–170, 177–178. See also Sedgwick, 'Politics of Survival', p. 286.
19. Sedgwick, 'Politics of survival', p. 591.
20. Lily Lee and Ruth Lam, *Sons of the Soil*, Pukekohe, 2012, p. 59.
21. Sedgwick, 'Politics of survival', pp. 285–286, 320.

22. It is easier to trace these networks and movements in New Zealand, where the Chinese population was much smaller than in Australia and distances not as great.
23. Sedgwick, 'Politics of survival', p. 316.
24. Lee and Lam, *Sons of the Soil*, p. 158.
25. For example, in Perth and Sydney. See Anne Atkinson, 'Chinese market gardening in the Perth metropolitan region, 1900–1920', *Western Geographer*, Vol. 8, 1984, p. 48; and Shirley Fitzgerald, *Red Tape Gold Scissors: The Story of Sydney's Chinese*, Sydney, 1996, p. 68.
26. Kee Yung, interviewed by Anne Thorpe, 17 December 2005, Otaki District Commercial Gardeners Society Oral History Project, Alexander Turnbull Library, OHA-5505, Tape 2.
27. Michael Williams, *Chinese Settlement in NSW: A Thematic History*, Sydney, 1999, pp. 7, 16, 23; Lydon, *Many Inventions,* pp. 80–84; and Fitzgerald, *Red Tape Gold Scissors,* pp. 67–68.
28. Kwong War Chong and Co was run by several partners from Zhongshan district in Guangdong and used by market gardeners from that district, while War Hing and Co. was the meeting place of market gardeners from the Yiu Ming Hung Fook Tong Society, which had its roots in the Gao Yao and Gao Ming districts (Williams, *Chinese Settlement in NSW*, p. 21; and Fitzgerald, *Red Tape Gold Scissors*, p. 163).
29. Sedgwick, 'Politics of survival', p. 149.
30. For example in Australia, *Tung Wah Times,* 12 August 1899, p. 2 c-d; in New Zealand, *Southland Times,* 12 September 1904, p. 4.
31. Jupp (ed.), *The Australian People: An Encyclopedia of the Nation, its People and their Origins*, Cambridge, 2001, pp. 201, 863.
32. *Windows on a Chinese Past*, Vol. 3, p. 183. Australia's response to Japanese expansionism was relatively low key, moderated by admiration for Japan's economic achievements and its status as a major trading partner (Fitzgerald, *Red Tape Gold Scissors*, p. 127).
33. Lee and Lam, *Sons of the Soil,* p. 284.
34. George Sue interviewed by Joanna Boileau, Auckland, 7 April 2012, recording held by interviewer.
35. Sedgwick, 'Politics of survival', p. 318.
36. Recent scholarship places the imposition of discriminatory immigration policies in Canada, the United States, Australia and New Zealand in a transnational context, highlighting the political and intellectual exchanges of ideas between white settler nations (Charles Price, *The Great White Walls are Built*, Canberra, 1974, pp. 3–4, 23; Marilyn Lake and Henry Reynolds, *Drawing the Global Colour Line: White Man's Countries and the International Challenge of Racial Equality*, Cambridge, 2008; Fitzgerald, *Big White Lie* p. viii).

37. Henry Reynolds, for example, places Australia's evolving policies towards indigenous people between 1850 and 1950 in the context of international thinking on race and human evolution (Henry Reynolds, *Nowhere People: How International Race Thinking Shaped Australia's Identity*, Camberwell, 2005, pp. 100–101).
38. Lake and Reynolds, *Drawing the Global Colour Line*; Nigel Murphy, *The Poll Tax in New Zealand*, Wellington, 1996, pp. 13–15.
39. Edmond Marin La Meslee, *The New Australia* (1883), London and Melbourne, 1973, pp. 85–86, 204–206.
40. Dorothea Nichols interviewed by Pauline Curby, 29 July 1997, City of Ryde Library Services; and George Redding interviewed by Pauline Curby, 1998, City of Ryde Library Services.
41. Michael Symons, *One Continuous Picnic: A Gastronomic History of Australia*, Melbourne, 2007, p. 14.
42. Ibid., p. 67; Phillip Muskett, *The Art of Living in Australia* (1893), Adelaide, 2005, n.p. See also David Burton, *200 Years of New Zealand Food and Cookery*, Wellington, 1982, p. 25; Tony Simpson, *A Distant Feast*, Auckland, 2008, pp. 135, 149; and David Veart, *First Catch Your Weka: A Story of New Zealand Cooking*, Auckland, 2008, pp. 4, 29.
43. *Brisbane Courier*, 21 October 1878, p. 2.
44. See, for example, *Brisbane Courier*, 7 January 1887, p. 5.
45. *Grey River Argus*, 19 November 1874, p. 3.
46. *Wanganui Herald*, 4 November 1886, p. 2. The Barcoo River in western Queensland gave its name to 'barcoo rot', a form of scurvy characterised by chronic skin sores, caused by lack of fresh vegetables in the diet (Australian National Dictionary Centre, http://andc.anu.edu.au/australian-words/meanings-origins? accessed 14 November 2012).
47. Muskett, *Art of Living in Australia*, n.p.
48. *Southland Times*, 14 May 1899, p. 4.
49. Kathryn Gleason, 'To bound and to cultivate: an introduction to the archaeology of gardens and fields', in Naomi Miller and Kathryn Gleason (eds.), *The Archaeology of Garden and Field*, Philadelphia, 1994, pp. 2, 4.
50. See, for example, Barbara Bender, *Landscape: Politics and Perspectives*, Providence, 1993, pp. 2–3, 9.
51. Cathie May, *Topsawyers: the Chinese in Cairns 1870 to 1920*, Townsville, 1984, p. 48; Barry McGowan, *Dust and Dreams: Mining Communities in South-East New South Wales*, Kensington, 2010, pp. 123–124.
52. Yuan Fang Shen, *Dragon Seed in the Antipodes; Chinese-Australian Autobiographies*, Carlton North, 2001, p. 50.
53. Warwick Roger, 'True builders of the Waimarino', *Weekly News*, 12 July 1971, p. 10.
54. Lee and Lam, *Sons of the Soil*, p. 295.

55. As Kylie Mirmohamadi observes, in the process of harnessing the land for 'productive' use, successive waves of settlers swept away native flora and fauna, dispossessed the indigenous inhabitants of their country and imposed new colonial spaces. Aboriginal people were placed in the realm of nature rather than culture (Kylie Mirmohamadi, 'Wog plants go home: race, ethnicity and horticulture in Australia', *Studies in Australian Garden History*, Vol. 1, 2003, p. 93).
56. William J. Sowden, *The Northern Territory as it Is: A Narrative of the South Australian Parliamentary Party's Trip and Full Description of the Northern Territory, its Settlements and Industries, (1882),* Darwin, 1984, p. 179.
57. Lowe Kong Meng, Cheok Hong Cheong and Louis Ah Mouy (eds.), *The Chinese Question in Australia, 1878–1879*, Melbourne, 1879, pp. 7–8.
58. *Tung Wah News,* 30 August 1899, p. 2 f.
59. J. M. Tuck, 'The Devils Half Acre 1900–1910', BA Honours Thesis, University of Otago, Dunedin, 1983, p. 1.
60. Fitzgerald, *Red Tape Gold Scissors,* pp. 67–68; Atkinson, 'Chinese market gardening', p. 48; and Lee and Lam, *Sons of the Soil*, pp. 315–316.
61. Alan Mayne, *The Imagined Slum: Newspaper Representation in Three Cities 1870–1914, London,* 1993, p. 3. See also Graeme Davison, 'Introduction', in Graeme Davison, David Dunstan and Chris McConville (eds.), *The Outcasts of Melbourne: Essays in Social History,* Sydney, 1985, p. 3; and Lydon, *Many Inventions,* p. 10.
62. Janis Wilton, *Golden Threads: The Chinese in Regional New South Wales 1850–1950*, Sydney, 2004, p. 29.
63. Rod Lancaster, 'European-Chinese interaction: a pre-Federation rural Australian setting', *Rural Society,* Vol. 10, No. 2, 2000, pp. 229–241.
64. *Star,* Canterbury, 26 March 1888, p. 2.
65. Anne Atkinson, 1984, 'The socio-economic aspects of the Chinese community in Perth, 1900–1920', *Journal of the Royal Western Australian Historical Society*, Vol. 9, No. 2, p. 97; Randwick Municipal Council, *Randwick: A Social History, Kensington,* 1985, p. 157; and Mohommod Taher, 'Asians in New Zealand A Geographical Review and Interpretation', PhD Thesis, University of Auckland, Auckland, 1965, p. 156.
66. Julia Bradshaw, *Golden Prospects: Chinese on the West Coast of New Zealand*, Greymouth, 2009, pp. 52, 70–71.
67. Bradshaw, *Golden Prospects*, pp. 58, 71; McGowan, 'Adaptation and organization', p. 236; Lancaster, 'European-Chinese interaction', p. 240; May, *Topsawyers*, pp. 143–168; and Reynolds, *North of Capricorn*, pp. 69–70.
68. Keir Reeves and Benjamin Mountford, 'Sojourning and settling: locating Australian history', *Australian Historical Studies*, Vol. 42, No. 1, 2011, p. 124.
69. *Otago Witness,* 1 March 1884, p. 11.

70. Donald History and Natural History Group, *Georgie Ah Ling, Donald's Friend*, Donald, 2008, p. 25.
71. James Hayes, 'Good morning Mrs. Thompson: a Chinese-English word-book from nineteenth century Sydney', in Paul Macgregor (ed.), *Histories of the Chinese in Australasia and the South Pacific*, Melbourne, 1995, p. 113.
72. Andrew Yew Dick Yung interviewed by Anne Thorpe, 31 October 2004, Otaki District Commercial Gardeners Society Oral History Project, Alexander Turnbull Library, OHInt-0819/09. transcript p. 24.
73. Ng, *Windows on a Chinese Past*, Vol. 1, p. 323.
74. Cynthia Foley, 'Chinese in the Community', *Western Connections*, No. 38, December 1994, p. 5; Colleen Morris, 'Chinese market gardens in Sydney', *Australian Garden History*, Vol. 12, No. 5, March/April 2001, p. 7; Owen Bennett interviewed by Lesley Goldberg, 3 November 1998, City of Ryde Library Services; Dorothy Packer interviewed by Pauline Curby, 12 September 1997, City of Ryde Library Services; Gifford and Eileen Eardley, 'The Chinese market gardeners of St George district', *St George Historical Society Bulletin*, December 1970, pp. 75, 78; Joanna Boileau, *Families of Fortune*, Murwillumbah, 2009, p. 26; Bill Gammage, *Narrandera Shire*, Narrandera, 1986, p. 144.
75. Rannard, *Forgotten Gardens*, p. 43.
76. Owen Bennett interviewed by Lesley Goldberg, 3 November 1998, City of Ryde Library Services.
77. *Northern Territory Times*, 10 May 1895, p. 2.
78. Weston Bate, *A History of Brighton*, Melbourne, 1963, p. 357.
79. Lee and Lam, *Sons of the Soil*, p.152; and Manying Ip, *Being Māori-Chinese: Mixed Identities*, Auckland, 2008, pp. viii, 238.
80. They included the market gardeners Harry in 1894 and Ti Li in 1916 (*Grey River Argus*, 3 January 1894 p. 2; and 22 December 1916, p. 2).
81. Bradshaw, *Golden Prospects*, pp. 70–71.
82. *Tung Wah Times*, 25 December 1909, p. 7b.
83. Gammage, *Narrandera Shire*, p. 144; and *Dubbo Liberal and Macquarie Advocate*, 3 June 1908, p. 4 and 13 January 1909, p. 3.
84. *Tung Wah Times*, 19 March 1921, p. 7b.
85. Alexander Don, *Outlook*, 4 October 1910, p. 13.
86. *Evening Post*, Wellington, 23 February 1925, p. 2.
87. For example, attacks in Auckland were reported in 1881, 1892, 1894, 1898 and 1905 (*Hawke's Bay Herald*, 6 June 1881, p. 2; *Colonist*, 29 March 1892, p. 4; *New Zealand Herald*, '100 years ago today', 10 June 1994, p. 9 and 18 June 1998, p. 14; and *Bay of Plenty Times*, 6 November 1905, p. 2).
88. *New Zealand Presbyterian*, 1 Nov 1889, p. 85. See also *Cromwell Argus*, 4 August 1885, quoted in Ng, *Windows on a Chinese Past*, Vol. 3, p. 26, fn. 17a.

89. Ng, *Windows on a Chinese Past*, Vol. 3, p. 79.
90. *Sydney Morning Herald*, 7 January 1879, p. 5.
91. Barry McGowan, 'Ringbarkers and market gardeners: a comparison of the rural Chinese of New South Wales and California', *Chinese America: History and Perspectives*, Annual, 2006, p. 39.
92. *New Zealand Herald*, '100 years ago today', 18 June 1998, p. 14; *Grey River Argus*, 12 February 1909, p. 3.
93. *Fielding Star*, 14 April 1909, p. 2.
94. As discussed in Chapter 4, fear of economic competition and European hostility towards Chinese people was greatest at times of economic hardship, for example during the depressions of the 1890s and the 1930s.
95. Bee Dawson, *A History of Gardening in New Zealand*, Auckland, 2010, p. 202.
96. *Sydney Morning Herald*, 13 August 1886, p. 4; *Argus*, Melbourne, 5 April 1900 p. 3, and *Border Watch*, Mt Gambier, 3 August 1907, p. 3.
97. *Brisbane Courier*, 6 January 1891, p. 3.
98. Murphy, *Success through Adversity*, pp. 13–14.
99. In 1939 the Council changed its name to the Dominion Council of Commercial Gardeners (ibid., pp. 16–17).
100. In the 1930s and 1940s Produce Markets was the main auction company in Auckland, selling 80 percent of the vegetables produced by Chinese growers (ibid., p. 15).
101. New Zealand Parliamentary Debates, Legislative Council and House of Representatives, Wellington, 1944, Vol. 266, pp. 633, 635.
102. Bert Cooksley, president of the Dominion Council from 1938 to 1952, worked hard to foster harmonious relationships with Chinese growers and bring them under the umbrella of the Council. They were invited to form district associations and attend Council conferences (Murphy, *Success through Adversity*, pp. 27, 29).
103. Anne Thorpe, 'Otaki's market gardens', *Otaki Historical Society Historical Journal*, Vol. 30, 2008, p. 24; and Murphy, *Success through Adversity*, pp. 20–21.
104. Andrea Gaynor, *Harvest of the Suburbs: An Environmental History of Growing Food in Australian Cities*, Perth, 2006, pp. 194–195.
105. *Evening Post*, Wellington, 4 November 1874, p. 2.
106. *Brisbane Courier*, 16 January 1889, p. 3, 30 September 1889, p. 7; *Mercury*, Hobart, 15 September 1910, p. 3; *West Australian*, Perth, 28 February 1906, p. 9; *Star*, Ballarat, 9 July 1891 p. 3; *Argus*, Melbourne, 14 February 1911, p. 9, 8 October 1907, p. 6; and *Bendigo Advertiser*, 14 October 1903, p. 2.
107. *Star*, Ballarat, 9 July 1891, p. 3.
108. Bate, *History of Brighton*, p. 360.

109. The proposed regulations included storing vegetables in properly ventilated stores and not in dwelling houses; washing vegetables only in clean water; keeping carts used for transporting vegetables clean; and not placing manure heaps where they could drain into any well or creek. They were enacted and were still in force in 1904 (*Brisbane Courier,* 30 September 1889, p. 7; and 4 November 1904, p. 4).
110. *Tung Wah News*, 19 July 1899, p. 2.
111. *Tung Wah Times,* 10 September 1921, p. 7 a; 3 December 1921, p. 7 b; 8 October 1921, p. 7 a; 6 May 1922, p. 7 a; 12 June 1926, p. 7 b-d; and 26 June 1926, p. 7 a-d.
112. *Auckland Star*, 25 March 1887, p. 2; *Star*, Canterbury, 18 May 1895, p. 7, 4 October 1878, p. 3; *Southland Times,* Otago, 28 April 1886, p. 2; and *Poverty Bay Herald,* Gisborne, 19 August 1913, p. 4.
113. *Auckland Star*, 25 March 1887, p. 2.
114. Kevin Rains, 'Behind the wattle fence: the Chinese market gardeners of early Cooktown', in Kevin Wong Hoy and Kevin Rains (eds.), *Rediscovered Past: China in Northern Australia*, North Melbourne, 2009, p. 25.
115. *Mirror*, Perth, 14 September 1929, p. 5; and 21 September 1929, p. 5.
116. *Barrier Miner*, Broken Hill, 17 February 1899, p. 2.
117. *Nelson Evening Mail*, 2 April 1904, p. 3; *New Zealand Herald*, 6 September 1906, p. 6, 11 May 1908, p. 6; *Auckland Star*, 14 February 1905, p. 2; and *Evening Post*, Wellington, 1 July 1921, p. 11.
118. *Bruce Herald,* Otago, 28 October 1887, p. 3; *Evening Post,* Wellington, 18 January 1909, p. 3. See also Lee and Lam, *Sons of the Soil*, p. 149, and Ng, *Windows on a Chinese Past*, Vol. 3, p. 84.
119. *Bush Advocate,* Hawke's Bay, 5 November 1907, p. 5.
120. *Tung Wah Times,* 12 October 1907, p. 7 a.
121. *Northern Miner*, Charters Towers, 19 January 1931, p. 2; and *Sydney Morning Herald*, 20 January 1931, p. 5.
122. They were charged with failing to register as aliens and were required to pay the poll tax and an additional penalty of £5, plus 9 shillings costs (*Poverty Bay Herald*, Gisborne, 26 June 1919, p. 5).
123. *West Australian*, 14 October 1927, p. 19; and *Sydney Morning Herald,* 13 October 1927, p. 11.
124. Williams, *Chinese Settlement in NSW*, p. 32.
125. Lydon, *Many Inventions*, pp.151–152; and Fitzgerald, *Red Tape Gold Scissors*, p. 31.
126. *Evening Post,* Wellington, 19 January 1905, p. 6.
127. *Tung Wah Times*, 29 March 1930, p. 8 e.
128. Rosalind Worrall-Smith, 'The employment of Maori women by Asiatic market gardeners: an aspect of New Zealand race relations', *University of Auckland Historical Society Annual*, 1970, p. 9.

129. New Zealand House of Representatives, *Report of the Committee on Employment of Maoris on Market Gardens*, Appendix to the Journals of the House of Representatives of New Zealand, 1929, Vol. II, G 11, p. 2.
130. Māori were employed on gardens in Masterton, Greytown, Levin, Wanganui, Hawke's Bay, Hamilton and Auckland (Lee and Lam, *Sons of the Soil*, pp. 142, 144, 164, 208, 264, 305, 397).
131. Ibid., p. 213.
132. Ibid., p. 153.
133. Michael King, *The Penguin History of New Zealand*, Auckland, 2003, pp. 472–473.
134. Ip, *Being Māori-Chinese*, pp. 3–4; Ng, *Windows on a Chinese Past*, Vol. 3, p. 158; and Worrall-Smith, 'Employment of Maori women', pp. 1, 5.
135. Memo from Department of Labour and Employment to District Superintendent, Auckland, 4 August 1953, Archives New Zealand, ACIE 8798, EAW2619/2, Box 2, 30/1/2, Part 1; and Letter from Minister of Labour to Revd W. Divers, Pukekohe, 23 April 1959, Archives New Zealand, ACIE 8798, EAW2619/2, Box 2, 30/1/2, Part 1.
136. The Departments of Housing and Māori Affairs could not keep pace with the demand for housing at a time when there were limited resources and housing shortages in many other areas (Letter from Secretary, Minister of Labour to Minister of Maori Affairs. 27 July 1959, Archives New Zealand, ACIE 8798, EAW2619/2, Box 2, 30/1/2, Part 1; and *Evening Post*, Wellington, 24 April 1959.)
137. Raymond Evans, Kay Saunders and Kathryn Cronin, *Race Relations in Colonial Queensland: A History of Exclusion, Exploitation and Extermination*, St Lucia, 1993, pp. 258–259; Diana Geise, *Astronauts, Lost Souls and Dragons: Voices of Today's Chinese-Australians in Conversation with Diana Giese*, St Lucia, 1997, p. 38; Shen, *Dragon Seed in the Antipodes*, p. 51, and Yuanfang Shen and Penny Edwards, 'United by the sweep of a tarnished brush', Panorama, *Canberra Times*, 18 November 2000, pp. 4–5.
138. Evans, Saunders and Cronin, *Race Relations in Colonial Queensland*, p. 257; *Grey River Argus*, 25 October 1875, p. 2.
139. Reynolds, *North of Capricorn*, p. 76. See also Li, quoted in Giese, *Astronauts, Lost Souls and Dragons*, p. 39.
140. Andrew Gillett, 'Opium and Race Relations in Queensland', State Library of Queensland, 2010 http://www.slq.qld.gov.au/__data/assets/pdf_file/0020/163028/SLQ_Black_Opium_Andrew_Gillett.pdf, accessed 22 December 2012.
141. *Townsville Daily Bulletin*, 24 August 1927, p. 13, from *Cairns Chronicle*.
142. Ah Yook and Carole Gass interviewed by Joe Eisenberg, August 1999, Golden Threads Papers.

143. Christopher Anderson and Norman Mitchell, 'Kubara: a Kuku-Yalanji view of the Chinese in North Queensland', *Aboriginal History,* Vol. 5, Part I, 1981, p. 29.
144. Migration Heritage Centre, Powerhouse Museum, 'Happy Valley, Chinese market gardens and migrant camps', online exhibition, http://www.migrationheritage.nsw.gov.au/exhibition/atthebeach/happy-valley/, accessed 28 July 2013.
145. These families migrated south from pastoral stations in the northern gulf region between 1910 and 1920. Unlike those on the coast, Chinese communities in western Queensland remained small and scattered. By 1940 few market gardeners remained in major towns (Sandi Robb, 'Beyond the coast: Chinese settler patterns of the western gulf region, north Queensland', in Kevin Wong Hoy and Kevin Rains (eds.), *Rediscovered Past: China in Northern Australia*, North Melbourne, 2009, pp. 16–18).
146. Jennifer Martiniello, 'As strands of plaited music: my Chinese-Aboriginal-Anglo heritage', in Penny Edwards and Yuanfang Shen (eds.), *Lost in the Whitewash: Aboriginal-Asian Encounters in Australia, 1901–2001*, Canberra, 2003, p. 28.
147. Ibid., p. 27.
148. The work of Sophie Couchman has informed my analysis here; she examines photographs of Chinese Australians in multifaceted ways, illuminating the relationships between colonisers and colonised (Sophie Couchman, 'Then in the distance Quong Tart did we see', *Journal of Australian Colonial History*, Vol. 8, 2006, p. 161; and 'Chinese-Australian family photographs: Tock Family and friends revealed through the camera's lens', *Descent*, Vol. 35, No. 2, 2005, pp. 66–68).
149. Martiniello, 'As strands of plaited music', pp. 28, 30.
150. Bagnall, 'Across the threshold', pp. 12, 24–25, 29.
151. Frost, 'Migrants and technological transfer', p. 128.

CHAPTER 8

Conclusion

This book has explored the history of Chinese people in Australia and New Zealand through the lens of market gardening, the main occupation which Chinese immigrants took up after the gold rushes. It has followed the journeys taken by thousands of men from southern China across the Pacific Ocean in search of gold – their struggle for survival in new lands, their travels back to China to maintain and strengthen the bonds of kin and family. Many of the first generation of gold seekers who took up market gardening achieved their goal of building their family's fortunes and returning home. Others remained and formed enduring communities. For later generations, the Japanese invasion of China, World War II and the rise of the Communist party made it difficult to maintain links with their families and home villages in China. If they were fortunate, they were able to bring wives and children to join them. These reunited families worked hard to build their market gardening enterprises and many expanded into related businesses such as fruit shops. The following generation born in Australia and New Zealand in the 1940s and 1950s had more freedom of choice and were not restricted to the limited range of occupations open to their parents and grandparents. They were able to take full advantage of the stable family environment established by their hard-working parents and grandparents, greater educational opportunities and a range of career choices. Many took up other occupations, including professions such as medicine, law, architecture and accountancy. Some,

J. Boileau, *Chinese Market Gardening in Australia and New Zealand*,
Palgrave Studies in the History of Science and Technology,
DOI 10.1007/978-3-319-51871-8_8

particularly in New Zealand, decided to carry on the family business as commercial growers.[1]

A key theme running through the lives of the individual market gardeners whose stories are told in this book is mobility – not only the mobility of people but the mobility of their modes of living and working, ideas, skills, knowledge and technology. Chinese market gardeners in Australia and New Zealand were an integral part of the population movements sparked by the gold rushes of the 1850s and 1860s, the opening up of the land for agriculture and pastoralism and the growth of towns and cities following the gold rushes. Thus this book is about the mobilisation of knowledge across time and space. It enlarges our understanding of processes of technological change, highlighting the mobility of migrants as a key element in the mobility and adaptation of technologies. It also highlights the mobility of the technologies themselves, taking technology in its broadest sense to include business structures, capital sharing and marketing strategies as well as horticultural skills, techniques and equipment.

Another key theme is the agency of Chinese people themselves as they followed their aspirations in their journeys across the globe and sought to achieve their goals through their own labour and the support of kin and compatriots. Taking a transnational perspective, this book highlights the fundamental importance of community and the maintenance of ongoing ties to kin and native place in China as the foundation for successful market gardening enterprises in Australia and New Zealand, and as a source of support for individuals adopting new ideas.

This book has examined the opportunities and constraints offered by the complex networks of relationships that Chinese market gardeners maintained: with their kin and home villages in China, within their own overseas Chinese communities and with the wider community. It illustrates how the ties within and between social groups can bring people together, build bridges between cultures and mobilise solidarity. Chinese market gardeners negotiated their positions in relation to the restrictive immigration policies which dominated their lives until well into the twentieth century. At the local level, they built relationships with people from all walks of life with whom they came in daily contact and manipulated these relationships to their mutual advantage. At the broader societal level they actively lobbied governments against the restrictive legislation imposed upon them, highlighting the contribution which they made to the economies and the national diets of Australia and New Zealand.

In this study I have tried to place the details of the lives of individual market gardeners against the backdrop of longer-term processes of social, economic and environmental change. Market gardeners responded to the often hostile physical, economic, political and social environments they encountered in Australia and New Zealand, by adapting the agricultural skills and practices and social and business customs they brought from their homelands, innovating and experimenting. Thus the history of market gardening in Australia and New Zealand was shaped by political and legal institutions as well as organisational structures.

Viewing technological exchange as a dynamic process involving ongoing adaptation of innovations to suit local conditions reveals that Chinese market gardening, as an agricultural system, was not transferred unchanged to Australia and New Zealand. Ongoing interactions occurred between different knowledge systems and horticultural traditions – indigenous, European and Chinese. Chinese market gardeners adapted their skills and technology to novel environments and novel crops and adopted developments in European technology. They effectively demonstrated the application of intensive cultivation techniques and irrigation in inhospitable environments, particularly in arid regions of Australia and Central Otago in southern New Zealand. Some Europeans readily acknowledged their superior horticultural skills and were willing to learn from them.

Thus market gardens can be seen as places of environmental and social engagement, where Chinese settlers and their plants continued traditions from their homelands and accommodated to new physical and social environments. In this analysis the skills of Chinese immigrants lay not only in the agricultural techniques they brought from China but also in their entrepreneurship and innovation, adapting their techniques to new environments. Chinese market gardeners are revealed as active agents in history and as innovators and risk takers, developing flexible strategies to cope with complex and changing business and social environments and the vagaries of climate and the changing seasons.

Taking a comparative and transnational approach highlights some of the commonalities of the experiences of Chinese market gardeners in Australia and New Zealand. In both countries Chinese immigrants began gardening on the goldfields to supplement their incomes from mining. They faced similar constraints of institutional racism in their adopted countries, which had close social, economic and political ties as British settler societies in the southern hemisphere. Barriers of language and culture, racist attitudes, restrictive immigration policies, discriminatory employment practices and unpredictable business environments limited their occupational and social mobility and encouraged them to enter niche occupations such as market gardening. They also found similar opportunities. The rapid urbanisation of Australians and New Zealanders following the gold rushes and the high standards of living they enjoyed underpinned the growth of market gardening. There was a sustained demand for fresh produce and Chinese gardeners took advantage of the economic opportunities created in the wake of the expanding population. At the peak of market gardening in the 1920s and 1930s there were Chinese market gardens serving country towns and cities across Australia and New Zealand.

This common history gave rise to similar patterns of economic and social adaptation. Market gardeners drew upon a range of economic resources which enabled them to successfully adapt traditional Chinese agricultural practices to new societies. These included flexible business partnerships, the integration of the various occupations involved in food supply, and social and business networks based on common ties of kinship, language and allegiance to village and district in China. In both countries market gardening provided an economic base for the transition from sojourners to settlers, the formation of enduring Chinese communities and a springboard for upward mobility and expansion into other businesses.

This comparative approach also highlights points of difference in the histories of Chinese market gardening in Australia and New Zealand. Firstly, the pre-eminence of Chinese market gardening lasted longer in New Zealand than in Australia, into the 1960s and 1970s, whereas in Australia this leading position was in decline by the 1950s. This was largely due to the post-war influx of migrants from southern Europe into Australia, who supplanted Chinese people in market gardening.

Secondly, there were differences in relations between Chinese gardeners and indigenous peoples. Unlike indigenous people in Australia, Māori in New Zealand had a long tradition of gardening, based on the tropical crops they brought with them in their voyages of migration from the Pacific islands. From the early twentieth century they played an important role in Chinese market gardening, as both landlords and employees of Chinese market gardeners. Their lives were closely entwined, leading to intermarriage and often close business and family ties between the two communities. In comparison there is more limited evidence of Aboriginal people working for Chinese market gardeners in Australia.

Finally, in New Zealand, with its temperate climate and generally reliable rainfall, there was less emphasis on large-scale irrigation schemes. Less attention was paid to the irrigation techniques of Chinese market gardeners and they had less direct influence on the development of agricultural policy than in Australia. In Australia the proponents of large-scale irrigation schemes and advocates of closer settlement held up the successes of Chinese market gardeners in irrigation and intensive horticulture as models for European settlers to aspire to.

The traditional horticultural methods Chinese immigrants brought with them had remarkable continuity in their adopted countries. At the same time Chinese market gardening underwent ongoing technological change and adaptation. Chinese gardeners introduced Chinese crops to Australia and New Zealand, grew the familiar crops favoured by European consumers and experimented with new crop varieties. They readily adopted advances in European technology which were appropriate to their intensive cultivation methods. Technological change was stimulated by developments in technology, industry, transport and communications, which flowed from population growth and urban expansion. These developments were key factors in the successful adaptation of Chinese horticultural systems, along with economic resources such as labour, capital and markets. The social resources provided by the networks of relationships within and without the Chinese community were also key factors in the success of market gardening enterprises.

Chinese market gardeners in Australia and New Zealand played a key role in the supply of fresh fruit and vegetables for at least 50 years. They could be found in the most marginal areas for agriculture, from the rugged ranges of Central Otago to the deserts of Australia. Adapting their traditional skills in water management and intensive cultivation, they successfully turned the challenges of life in such environments to their advantage. They made a significant contribution to the economies of Australia and New Zealand, fuelling rural expansion and the rapid growth of country towns and cities as well as enhancing diets and public health with a wide variety of reasonably priced fruit and vegetables. Unlike Europeans who dreamed of becoming yeoman farmers, they went against the trend of broadacre farming and specialised in labour intensive production on small acreages. Countering the stereotype of Chinese market gardeners as uneducated peasant farmers, conservative and resistant to change, they emerge as hard working and entrepreneurial, adept at identifying niche markets in land rich but labour poor Australia and New Zealand. In

Michael Symons' words: 'the Chinese became the model of the immigrant group who, drawn by the hope of riches, filled a gap in our horticulture and our diet'.[2]

There was a complex process of ongoing exchanges of human and intellectual capital between China, Australia and New Zealand, and continual adaptations to fit into the economies and societies of the countries in which Chinese immigrants settled. These exchanges included technology, knowledge, skills and novel forms of labour organisation. Thus this book sheds new light on the Chinese diaspora, through its focus on the exchange of horticultural skills and technology, and contributes to broader understanding of Chinese migration, settlement and cross-cultural exchange in Australia and New Zealand. I hope it will provide a stimulus for similar studies in other regions such as Papua New Guinea, the islands of the Pacific, Canada and the United States.

There are a number of other questions for future research. More detailed study of Chinese entrepreneurship is needed, in particular the business structure of Chinese market gardening enterprises and their linkages with other Chinese businesses, including fruit and vegetable retailing and general retailing, and including connections across national borders – between New Zealand and Australia and beyond. Further attention could also be given to the role of clan relationships in shaping and sustaining the industry, and developing a methodology for quantifying the contribution of Chinese market gardeners to the economies of Australia and New Zealand. Further research into the environmental impact of Chinese market gardening and its sustainability is also required; for example, the impact of intensive agricultural practices on soil fertility and the extent to which traditional manuring practices countered this effect. This book has also touched on the question of the psychological aspects of Chinese market gardens and the relationship of Chinese immigrants to the land and to the natural world, an area that has been little researched.[3]

In 2001 archaeologist Ian Jack commented that the rich heritage and history deriving from the pre-eminence of Chinese in market gardening and irrigation in Australia was in need of urgent synthesis and fieldwork.[4] This book goes some way towards filling this gap. I hope this broad study will provide an impetus for more detailed local and regional studies and fieldwork. There is still much work to be done to record the remaining physical evidence of Chinese market gardens, particularly in rural areas. As Barry McGowan has shown, many traces of Chinese gardening activity survive in regional New South Wales, including water races, garden layouts, fruit trees and horticultural equipment. Market gardens are endangered landscapes and have significant heritage value. They have high technological significance, revealing the adaptability and inventiveness of Chinese settlers and their skills in horticulture and water management as well as high cultural significance as tangible reminders of primary production in urban areas and the history of settlement and suburban expansion.[5] Unfortunately the majority of market garden sites in urban areas, as well as many regional centres, have been engulfed by industrial and residential development. This makes it all the more important to preserve the few surviving market garden sites in Australia and New Zealand.

Notes

1. Lily Lee and Ruth Lam, *Sons of the Soil*, Pukekohe, 2012, p. 506.
2. Michael Symons, *One Continuous Picnic*, Melbourne, 2007, p. 84.
3. Zvonkica Stanin touches on these issues in her study of Chinese market gardening on the Loddon River in Victoria, interpreting the open layout of gardens as evidence of communal ownership, cooperative labour and care for the land by generations of gardeners (Zvonkica Stanin, 'From Li Chun to Yong Kit: A Market Garden on the Loddon, 1851–1912', *Journal of Australian Colonial History,* Vol. 6, 2004, p. 31).
4. Ian Jack, 'Some less familiar aspects of the Chinese in 19th century Australia', in Henry Chan, Ann Curthoys and Nora Chiang (eds.), *The Overseas Chinese in Australasia: History, Settlement and Interactions,* Canberra, 2001, pp. 44–54.
5. Joanne Monk, 'The diggers in the trenches: a history of market gardens in Victoria 1835–1939', *Australian Garden History*, Vol. 4, No. 1, July/August 1992, p. 5; and Barry McGowan, 'Chinese market gardens in Southern and Western NSW', *Australian Humanities Review*, Issue 36, 2005, n.p.

Bibliography

Primary Sources

Publications

Australian Bureau of Statistics, Commonwealth Censuses, 1911 and 1921, 2112.0. Vol. III Part XII, Occupations (1911) and Vol. II Part XVII, Occupations, (1921), http://www.abs.gov.au/websitedbs/censushome.nsf/home/historicaldata?, accessed 3 November 2013.

———*Queensland Year Book*, Canberra, Government Printer, 1910–1930.

Australian National Dictionary Centre, http://andc.anu.edu.au/australian-words/meanings-origins?, accessed 14 November 2012.

Bloomfield, G. T., *New Zealand: A Handbook of Historical Statistics*, Boston, G. K. Hall, 1984.

Brady, Edwin, *Australia Unlimited*, Melbourne, G. Robertson, 1919.

Brooks, John, 'The last of the Chinese gardeners at Narrabri', unpublished ms., 1998, Golden Threads Papers, Armidale, University of New England.

Clacy, E., *A Lady's Visit to the Gold Diggings of Australia in 1852–53*, London, Hurst and Blackett, 1853.

Clarke, W. H., 'The fruit and vegetable industries', *Agricultural Gazette of NSW*, December 2 1904, pp. 1183–1204.

Don, Alexander, 'Account of travels in Guangzhou: Kwong Chayu to Sanning', *New Zealand Presbyterian*, 1 February 1881, pp. 149–51.

———'Chinese Mission Work in Otago: Annual Upcountry Tour' 1890–91; 1892–93; 1894–95; reprinted from *New Zealand Presbyterian* and *The Outlook*, Dunedin, Otago Daily Times.

J. Boileau, *Chinese Market Gardening in Australia and New Zealand*,
Palgrave Studies in the History of Science and Technology,
DOI 10.1007/978-3-319-51871-8

———*Under Six Flags*, Dunedin, Wilkie, 1898.

———'Chinese Mission Work in Otago: Annual Inland Tour', 1900–01; 1905–06; 1907–08; 1909–11, Dunedin, New Zealand Presbyterian Chinese Mission [reprinted in *New Zealand Presbyterian* and *The Outlook].*

———*Roll of Chinese, 1883–1929,* reproduced as Vol. 4 of James Ng, *Windows on a Chinese Past,* Dunedin, Otago Heritage Books, 1999. A digitised version is available at http://www.otago.ac.nz/historyarthistory/don/index.php.

Gelding, John, 'John Chinaman and his garden', *Horticultural Magazine and Gardener's Calendar of New South Wales*, Vol. IV, No. 44, August 1867, p. 192.

Jong, Ah Siug, *Diary*, MS 12994, State Library of Victoria [original ms. 1866–1872].

———*A Difficult Case: An Autobiography of a Chinese Miner on the Central Victorian Goldfields, translated, annotated and historical introduction by Ruth Moore and John Tully*, Daylesford, Jim Crow Press, 2000 [ms. 1872].

Knibbs, George Handley, *Official Year Book of the Commonwealth of Australia,* No. 1, 1908, No. 8, 1915, Melbourne, Commonwealth Bureau of Census and Statistics, 1908, 1915.

La Meslee, Edmond Marin, *The New Australia*, translated and edited by Russell Ward, London and Melbourne, Heinemann, 1973 [first published 1883].

Latrobe, Charles, Latrobe to Grey, 2 March 1852, 'Further papers relative to the recent discovery of gold in Australia', pp. 169–72, Victorian Parliamentary Papers 1852–3, LXIV, Melbourne, Government Printer, 1853.

Latrobe University, Chinese Heritage of Australian Federation Project, updated September 2004, http://arrow.latrobe.edu.au/store/3/4/5/5/1/public/index.htm, accessed 20 August 2010.

Legislative Council of Victoria, *Report of Select Committee on the Subject of Chinese Immigration*, Votes and Proceedings of Legislative Council of Victoria, 1856–7, Vol. 2, Melbourne, Government Printer, 1857.

Meng, Lowe Kong, Cheok Hong Cheong and Louis Ah Mouy (eds), *The Chinese Question in Australia, 1878–79*, Melbourne, Victorian Government Publisher, 1879.

Muskett, Phillip E., *The Art of Living in Australia*, web edn, eBooks@Adelaide, Adelaide, University of Adelaide Library, 2005 [first published 1893].

National Archives of Australia, 'Commonwealth Government records about the Northern Territory', http://guides.naa.gov.au/records-about-northernterritory/part2/chapter10/10.9.aspx, accessed 29 November 2016.

New South Wales Government Statistician, *New South Wales Statistical Register*, Sydney, Government Printer, *1861, 1871, 1881, 1891, 1901.*

New South Wales Legislative Assembly, *Report of the Royal Commission into Alleged Chinese Gambling and Immorality, Votes and Proceedings of the*

Legislative Assembly of New South Wales, 1891–1892, Vol. 8, pp. 497–997, Sydney, Government Printer, 1892.

———*Royal Commission of Inquiry as to Food Supplies and Prices, 1913, Sectional Report on the Supply of Fruit and Vegetables*, Sydney, Government Printer, 1914.

New South Wales Legislative Council, *Interim Reports from the Select Committee on the Conditions and Prospects of the Agricultural Industry and Methods of Improving the Same Together with the Digest of Evidence, Index and Appendices*, New South Wales Legislative Council, Sydney, Government Printer, 1923.

New Zealand Government, Office of Registrar General/Census and Statistics Office, *New Zealand Statistics: Population Census taken for the night of 29 April, 1906*, Wellington, Government Printer, 1907.

———*Results of a Census of New Zealand*, 1916, Wellington, Government Printer, 1916.

———*Dominion of New Zealand Population Census, 1926,* Vol. IX, Industrial and Occupational Distribution, *Wellington,* Government Printer, 1930.

New Zealand Legislative Council and House of Representatives, *New Zealand Parliamentary Debates*, Wellington, Government Printer, 1944, Vol. 266.

Northern Territory Archives Service, 'Chinese People in the Northern Territory', https://dtc.nt.gov.au/arts-and-museums/northern-territory-archives-service/archives-subject-guides/chinese-people-in-nt, accessed 29 November 2016.

Office of the Government Statist, Victoria, *Statistical Register of the Colony of Victoria*, Melbourne, Government Printer, 1861, 1871, 1881, 1891, 1900.

Patterson, J. A., *The Gold Fields of Victoria in 1862*, Melbourne, Wilson and MacKinnon, 1863.

Sowden, William J., *The Northern Territory As it Is: A Narrative of the South Australian Parliamentary Party's Trip and Full Description of the Northern Territory, its Settlements and Industries*, Darwin, Government Printer of the Northern Territory, 1994 [Adelaide, Thomas and Co, 1882].

Wilson, Roland, Commonwealth Statistician, '1947 Census Bulletin No 1: Population of States and Territories, Urban and Rural Population', Canberra, Commonwealth Government Printer, 1947.

Young, W., 'Report on the Condition of the Chinese Population of Victoria', Victorian Parliamentary Papers, Vol. 3, Votes and Proceedings, Melbourne, Government Printer, 1869.

Oral Histories

Bennett, Owen, interviewed by Lesley Goldberg, 3 November 1998, City of Ryde Library Services, Ryde, http://www.ryde.nsw.gov.au/_Documents/Oral+Histories/Owen+ Bennett.pdf, accessed 20 September 2011.

Boon, Norman, interviewed by Pauline Curby, 1998, City of Ryde Library Services, Ryde.

Chew, Wayn, interviewed by Joanna Boileau, Sydney, 3 August 2011, recording held by interviewer.

Fong, Allan, interviewed by Ruth Lam, Pukekohe, 27 September 2007, recording held by interviewer.

Fong, Nellie, interviewed by Diana Giese, 28 December 1996, Post-War Chinese Australians Oral History Project, National Library of Australia, Canberra, ORAL TRC 3543.

Ha, Gordon, interviewed by Joanna Boileau, Botany, 27 July 2011, recording held by interviewer.

Kim, George Lee, interviewed by Paul McGregor, 17 May 1994, Australia-China Oral History Project, National Library of Australia, Canberra, ORAL TRC 3522/4/4.

Lee, Peter, interviewed by Lily Lee, 12 March 2008, Oamaru, recording held by interviewer.

Ng, James, interviewed by Lily Lee, 13 March 2008, Dunedin, recording held by interviewer.

Nichols, Dorothea, interviewed by Pauline Curby, 29 July 1997, City of Ryde Library Services, Ryde, http://library.ryde.nsw.gov.au/docs/oralhistories/nichols1/nichols1.htm, accessed 20 September 2011.

Nomchong, Lionel, interviewed by Matthew Higgins, 3 March 1995, *Post-War Chinese Australians Oral History Project*, National Library of Australia, Canberra, ORAL TRC 3196.

Packer, Dorothy, interviewed by Pauline Curby, 12 September 1997, City of Ryde Library Services, Ryde, http://www.ryde.nsw.gov.au/_Documents/Oral+Histories/Dorothy+Packer.pdf, accessed 20 September 2011.

Quong, Eddie, interviewed by Diana Geise, 7 December 1993, *Post-War Chinese Australians Oral History Project*, National Library of Australia, Canberra, ORAL TRC 3005.

Redding, George, interviewed by Pauline Curby, 1998, City of Ryde Library Services, Ryde.

Ross, Enid, interviewed by Warwick Eather, 18 and 22 February 1987, *New South Wales Bicentennial Oral History Collection*, Mitchell Library, Sydney, MLMSS 5163/Box02.

Smith, Esma, interviewed by Rob Willis, 4 September 2009, *Voices of the Bush Oral History Project*, National Library of Australia, Canberra, ORAL TRC 6125/11.

Sue, George, interviewed by Joanna Boileau, 7 April 2012, Auckland, recording held by interviewer.

Sue, George, interviewed by Lily Lee, 12 November 2006, Levin, recording held by interviewer.

Turner, Jeffrey, interviewed by Ruth Lam, Auckland, 15 May 2008, recording held by interviewer.
Wing, Alan, interviewed by Lily Lee, 22 February 2007, Hastings, recording held by interviewer.
Wong, Bing, interviewed by Lorna Wong, 30 July 2007, Wellington, 21 Voices Project, New Zealand Chinese Oral History Foundation, Auckland City Library, 11-OH-15.
Young, Brian, interviewed by Lily Lee, 12 October 2006, Napier, recording held by interviewer.
Young, William, interviewed by Lily Lee, 16 November 2006, Palmerston North, recording held by interviewer.
Yung, Andrew Yew Dick, interviewed by Anne Thorpe, 31 October 2004, Otaki District Commercial Gardeners Society Oral History Project, Oral History Centre, Alexander Turnbull Library, Wellington, OHInt-0819/09.
Yung, Kee, interviewed by Anne Thorpe, 10, 17 December 2005, Otaki District Commercial Gardeners Society Oral History Project, Oral History Centre, Alexander Turnbull Library, Wellington, OHA-5505.
Yook, Ah and Carole Gass, interviewed by Joe Eisenberg, Golden Threads Project, August 1999, Armidale, University of New England.

Newspapers

Australia

Advocate and the Emu Bay Times, Tasmania
Alexandra and Yea Standard, Gobur Thornton and Acheron Express, Victoria
Argus, Melbourne, Victoria
Bacchus Marsh Express, Victoria
Barrier Miner, Broken Hill, NSW
Bendigo Advertiser, Victoria
Border Watch, Mt Gambier, SA
Brisbane Courier, QLD
Bulletin, Sydney, NSW
Cairns Chronicle, QLD
Canberra Times
Clarence and Richmond Examiner, Grafton, NSW
Daylesford Mercury, Victoria
Dubbo Liberal and Macquarie Advocate, NSW
Inglewood Advertiser, Victoria
Melbourne Age, Victoria
Mercury, Hobart, Tasmania
Mildura Cultivator, Victoria

Mirror, Perth
Northern Advocate, Charters Towers, QLD
Northern Miner, Charters Towers, QLD
Northern Territory Times and Gazette, Darwin, NT
North Melbourne Advertiser, Victoria
Ovens and Murray Advertiser, Beechworth, Victoria
Queenslander, Brisbane, QLD
Register, Adelaide, SA
Riverine Grazier, Hay, NSW
South Australian Advertiser, Adelaide, SA
South Australian Register, Adelaide, SA
Southern Cross, Junee, NSW
Star, Ballarat, Victoria
Sturt Recorder, Milparinka, NSW
Sydney Morning Herald, NSW
Townsville Daily Bulletin, QLD
Traralgon Record, Victoria
Tung Wah News, Sydney, NSW
Tung Wah Times, Sydney, NSW
West Australian, Perth, WA

New Zealand

Ashburton Guardian, Canterbury
Auckland Star, Auckland
Bay of Plenty Times, Bay of Plenty
Bruce Herald, Otago
Bush Advocate, Hawke's Bay
Christian Outlook, Otago
Colonist, Nelson
Cromwell Argus, Otago
Daily Telegraph, Napier
Dunstan Times, Otago
Evening Post, Wellington
Fielding Star, Manawatu - Wanganui
Grey River Argus, West Coast
Hawera Normanby Star, Taranaki
Hawke's Bay Herald, Hawke's Bay
Man Sing Times, Wellington
Nelson Evening Mail, Nelson
New Zealand Chinese Growers' Monthly Journal, Wellington
New Zealand Herald, Auckland

New Zealand Presbyterian, Otago
Otago Witness, Otago
Outlook, Otago
Observer, Auckland
Poverty Bay Herald, Gisborne
Southland Times, Southland
Star, Canterbury
Taranaki Herald, Taranaki
Tuapeka Times, Otago
Wanganui Herald, Manawatu - Wanganui
Weekly News, Auckland
West Coast Times, West Coast

Secondary Sources

Allen, Robert, 'Agriculture during the industrial revolution', in Roderick Floud and Donald McCloskey (eds), *The Economic History of Britain Since 1700*, Vol. 1, 1700–1860, Cambridge, New York, Cambridge University Press, 1994, pp. 96–122.

Anderson, Christopher and Norman Mitchell, 'Kubara: A Kuku-Yalanji view of the Chinese in North Queensland', *Aboriginal History*, Vol. 5, Part I, 1981, pp. 21–37.

Anderson, E. N., *The Food of China*, New Haven, London, Yale University Press, 1988.

Ao, Benjamin, 'History and Prospect of Chinese Romanization', *Chinese Librarianship: An International Electronic Journal*, December 1997, http://www.white-clouds.com/iclc/cliej/cl4ao.htm, accessed 20 September 2011.

Ashkanasy, Neal, Edwin Trevor Roberts and Louise Earnshaw, 'The Anglo cluster: Legacy of the British Empire', *Journal of World Business*, Vol. 37, No. 1, Spring 2002, pp. 28–39.

Atkinson, Anne,'Chinese market gardening in the Perth metropolitan region, 1900–1920', *Western Geographer*, Vol. 8, 1984, pp. 38–52.

———'The socio-economic aspects of the Chinese community in Perth 1900–1920', *Journal of the Royal Western Australian Historical Society*, Vol. 9, No. 2, 1984, pp. 96–104.

Atkinson, Neill, *Trainland: How Railways Made New Zealand*, Auckland, Random House, 2007.

Bader, Hans Dieter and Janice Adamson, 'Kong Foong Yuen, The Garden of Prosperity: Final Report on the Archaeological Excavations at Carlaw Park, Auckland', Prepared by Archaeology Solutions for Haydn & Rollett Construction, Auckland, March 2010.

Bagnall, Kate, 'Across the threshold: White women and Chinese hawkers in the white colonial imaginary', *Hecate*, Vol. 28, No. 2, 2002, pp. 9–32.

Bate, Weston, *A History of Brighton*, 2nd edn, Melbourne, Melbourne University Press, 1983.

Beattie, James, 'Hungry Dragons: Expanding the horizons of Chinese environmental history – Cantonese gold-miners in colonial New Zealand 1860s–1920s', *International Review of Environmental History*, Vol. 1, 2015, pp. 103–145.

———'Acclimatisation and the "Europeanisation" of New Zealand, 1830s–1920s', *Environment and Nature in New Zealand*, Vol. 3, No. 1, February 2008, pp. 1–25.

Beattie, James; Jasper M. Heinzen and John Adam, 'Japanese gardens and plants in New Zealand, 1850–1950: Transculturalism and transmission', *Studies in the History of Gardens and Designed Landscapes*, Vol. 28, No. 2, 2008, pp. 219–236.

Beatty, Don, 'Clutterbuck Oil Engine', *Gawler Machinery Restorers Club Inc. Newsletter*, 18, 1993, pp. 9–11.

Beatty, Edward, 'Approaches to technology transfer in history and the case of nineteenth century Mexico', *Comparative Technology Transfer and Society*, Vol. 1, No. 2, August 2003, pp. 167–197.

Belich, James, *Making Peoples: A History of the New Zealanders from Polynesian Settlement to the end of the Nineteenth Century*, Auckland, Allen Lane/Penguin, 1996.

Bender, Barbara (ed.), *Landscape: Politics and Perspectives*, Providence, Berg, 1993.

Bennett, Alex, *A Living From an Acre: How to Defeat the Depression*, Sydney, Finn Bros, 1932.

Binford, Lewis, 'Archaeology as anthropology', *American Antiquity*, Vol. 28, No. 2, 1962, pp. 217–225.

Blainey, Geoffrey, *The Tyranny of Distance: How Distance Shaped Australia's History*, revised edn, Sydney, McMillan, 2001.

Boileau, Joanna, *Families of Fortune: The Chinese in the Tweed Valley*, Murwillumbah, Tweed Shire Council, 2009.

Bolton, Geoffrey, 'Money: Trade, investment and economic nationalism', in Deryk Schreuder and Stuart Ward (eds.), *Australia's Empire*, Oxford, Oxford University Press, 2008, pp. 211–231.

———*A Thousand Miles Away: A History of North Queensland to 1920*, Brisbane, Jacaranda Press in association with the Australian National University, 1963.

Bonyhady, Tim, *The Colonial Earth*, Melbourne, Miegunyah Press, 2000.

Bradshaw, Julia, *Golden Prospects: Chinese on the West Coast of New Zealand*, Greymouth, Shantytown (West Coast Mechanical and Historical Society), 2009.

Bray, Francesca and Joseph Needham, *Science and Civilisation in China, Vol. 6: Biology and Biological Technology, Part II: Agriculture* by Francesca Bray, Cambridge, Cambridge University Press, 1984.

Bromby, Robin, *Unlocking the Land: The Saga of Farming in Australia*, Port Melbourne, Lothian, 1989.

Brooking, Tom, and Eric Pawson (eds), *Seeds of Empire: The Environmental Transformation of New Zealand*, London, I. B. Tauris, 2011.

Broome, Richard, *Coburg: Between Two Creeks*, 2nd edn, Coburg, Coburg Historical Society, 2001.

Buckley, Ken and Ted Wheelwright, *No Paradise for Workers: Capitalism and the Common People in Australia 1788–1914*, Melbourne, Melbourne University Press, 1988.

Bunch, Roland, 'Encouraging farmers experiments', in Robert Chambers, Arnold Pacey and Lori Ann Thrupp (eds.), *Farmer First: Farmer Innovation and Agricultural Research*, London, Intermediate Technology Publications, 1989, pp. 55–60.

Bureau of Meteorology, 'Australia: Climate of our continent', http://www.bom.gov.au/lam/climate/levelthree/ausclim/zones.htm, accessed 10 May 2011.

Burton, David, *200 Years of New Zealand Food and Cookery*, Wellington, Reed, 1982.

Campbell, Susan, *A History of Kitchen Gardening*, London, Frances Lincoln, 2005.

Carney, Judith, 'Landscapes of technology transfer: Rice cultivation and African continuities', *Technology and Culture*, Vol. 37, No. 1, January 1996, pp. 5–35.

Cassidy, Jill, 'Chung Gon, James (1854–1952)', *Australian Dictionary of Biography*, National Centre of Biography, Australian National University, http://adb.anu.edu.au/biography/chung-gon-james-9745/text17213, accessed 3 November 2013.

Cathcart, Michael, *The Water Dreamers: The Remarkable History of Our Dry Continent*, Melbourne, Text Publishing, 2010.

Chambers, Robert, Arnold Pacey and Lori Ann Thrupp (eds.), *Farmer First: Farmer Innovation and Agricultural Research*, London, Intermediate Technology Publications, 1989.

Chan, Henry (ed.), *Zencheng New Zealanders: A History for the 80th Anniversary of the Tung Jung Association of NZ Inc*, Katoomba, Echo Point Press, 2009.

Chan, Sucheng, 'Chinese livelihood in rural California: The impact of economic change 1860–1880', *Pacific Historical Review*, Vol. 53, No. 3, August 1984, pp. 273–307.

———*This Bittersweet Soil: The Chinese in California Agriculture 1860–1910*, Berkeley, University of California Press, 1986.

Chazan, Michael, *World Prehistory and Archaeology: Pathways Through Time*, 2nd edn, Boston, Prentice Hall, 2011.

Chi Wang, Ling, 'The structure of dual domination: Toward a paradigm for the study of the Chinese diaspora in the United States', in Hong Liu (ed.), *The Chinese Overseas*, Vol. 1, Abingdon, London, New York, Routledge, 2006, pp. 279–296.

Choi, C. Y., *Chinese Migration and Settlement in* Australia, Sydney, Sydney University Press, 1975.

Chui, Raymond, 'Transnationalism and migration: Chinese migrants in New Zealand', *Global Asia Journal*, Paper 4, 2008, http://digitalcommons/pace.edu/Global_Asia_Journal_4, accessed 27 May 2013.

Chung Gon, Ted, 'James Chung Gon', Chinese Heritage of Australian Federation, www.chaf.lib.latrobe.edu.au, accessed 20 May 2012. archived, La Trobe University Research Online.

City of Sydney Council, 'History of Sydney City Council', http://www.cityofsydney.nsw.gov.au/__data/assets/pdf_file/0003/65550/hs_chos_history_of_council_1001.pd, accessed 12 September 2011.

———'Sydney Stories: Sydney Horses', http://www.cityofsydney.nsw.gov.au/AboutSydney/HistoryAndArchives/OralHistories/HorsesInnerCity.asp, accessed 12 September 2011.

Clark, Gordon, *Innovation Diffusion: Contemporary Geographical Approaches*, Norwich, Concepts and Techniques in Modern Geography No. 40, 1984, p. 14.

Clark, Manning, *A Short History of Australia*, Camberwell, Vic, Penguin Group, 2006 [first published 1963].

Clay, Karen and Randall Jones, 'Migrating to riches? The evidence of the California gold rush', *Journal of Economic History*, Vol. 68, No. 4, 2008, pp. 997–1027.

Clunas, Craig, *Fruitful Sites: Garden Culture in Ming Dynasty China*, London, Reaktion Books, 1996.

Collins, Jock, 'Chinese entrepreneurs: The Chinese diaspora in Australia', *International Journal of Entrepreneurial Behaviour and Research*, Vol. 8, No. 1/2, 2002, pp. 113–133.

Collins, Jock, Katherine Gibson, Caroline Alcorso, Stephen Castles and David Tait, *A Shop Full of Dreams: Ethnic Small Business in Australia*, Leichardt NSW, Pluto Press Australia, 1995.

Conan, Michel (ed.), *Perspectives on Garden Histories: Dumbarton Oaks Colloquium on the History of Landscape Architecture, 1989*, Washington D.C., Dumbarton Oaks Research Library and Collection, 1999.

———*Gardens and Imagination: Cultural History and Agency*, Washington D.C., Dumbarton Oaks Research Library and Collection, 2008.

Conan, Michel, and W. John Kress (eds.), *Botanical Progress, Horticultural Innovations and Cultural Changes, Dumbarton Oaks Colloquium on the History of Landscape Architecture, 2004*, Washington D.C., Dumbarton Oaks

Research Library and Collection, distributed by Harvard University Press, 2007.

Conan, Michel, and Jeffrey Quilter (eds.), *Gardens and Cultural Change: A Pan-American Perspective: Dumbarton Oaks Colloquium on the History of Landscape Architecture, 2003*, Washington D.C., Dumbarton Oaks Research Library and Collection, distributed by Harvard University Press, 2007.

Connell, John and Angela Ip, 'The Chinese in Sydney: From Chinatown to suburbia', *Asian Profile*, Vol. 9, No. 4, August 1981, pp. 291–308.

Conway, Gordon, *One Billion Hungry: Can We Feed the World?*, Ithaca, London, Cornstock Publishing, Cornell University Press, 2012.

Couchman, Sophie, 'Then in the distance Quong Tart did we see: Quong Tart, celebrity and photography', *Journal of Australian Colonial History*, Vol. 8, 2006, pp. 159–182.

———'Chinese-Australian family photographs: Tock family and friends revealed through the camera's lens', *Descent*, Vol. 35, No. 2, 2005, pp. 66–68.

Crafts, Mark, 'The industrial revolution', in Roderick Floud and Donald McCloskey (eds.), *The Economic History of Britain Since 1700*, Vol. 1, 1700–1860, Cambridge, New York, Cambridge University Press, 1994, pp. 44–59.

Cronin, Kathryn, *Colonial Casualties: Chinese in Early Victoria*, Melbourne, Melbourne University Press, 1982.

Crosby, Alfred, *The Columbian Exchange: Biological and Cultural Consequences of 1492*, Westport, Connecticut, Greenwood Press, 1972.

Curthoys, Ann, '"Men of all nations, except Chinamen": Europeans and Chinese on the gold-fields of New South Wales', in I. McCalman, A. Cook and A. Reeves (eds.), *Gold: Forgotten Histories and Objects in Australia*, Cambridge, Cambridge University Press, 2001.

Darnell, Maxine, 'The Chinese labour trade to New South Wales 1783–1853', PhD Thesis, University of New England, Armidale, 1997.

Davison, Graeme, 'Introduction', in Graeme Davison, David Dunstan and Chris McConville (eds.), *The Outcasts of Melbourne: Essays in Social History*, Sydney, George Allen & Unwin, 1985, pp. 1–28.

Dawson, Bee, *A History of Gardening in New Zealand*, Auckland, Godwit, 2010.

Dawson, Graeme, J.W. McCarty and Ailsa McCleary (eds.), *Australians, A Historical Library: Australians 1888*, Broadway NSW, Fairfax Syme and Weldon, 1987.

Denham, Tim, Mark Donohue and Sara Booth, 'Horticultural experimentation in northern Australia reconsidered', *Antiquity*, Vol. 83, 2009, pp. 634–648.

Department of Conservation, New Zealand, 'Historic Otuataua Stonefields', http://www.doc.govt.nz/conservation/historic/by-region/auckland/central-and-south-auckland/otuataua-stonefields/, accessed 11 February 2013.

Donald History and Natural History Group, *Georgie Ah Ling: Donald's Friend*, Donald, Donald History and Natural History Group, 2008.

Dormer, Marion, *Dubbo, City on the Plains, 1901–1988*, Dubbo, Macquarie Publications, 1988.

Dormer, Marion and Joan Starr, *Settlers on the Marthaguy*, Dubbo, Macquarie Publications, 1979.

Drevins, Desley M. 'Chinese market gardens and market gardeners on Enoggera Creek and Ithaca Creek', presentation given to Ashgrove Historical Society, 1 September 2007, Brisbane, Ashgrove Historical Society.

Eardley, Gifford and Eileen Gifford, 'The Chinese market gardeners of St George district', *St George Historical Society Bulletin*, December 1970, pp. 75–81.

Edmeades, D. C., 'The agronomic effectiveness of lime reverted and dicalcic superphosphates: A review', *New Zealand Journal of Agricultural Research*, Vol. 43, 2000, pp. 1–6.

Edwards, Penny and Shen Yuanfang (eds.), *Lost in the Whitewash: Aboriginal-Asian Encounters in Australia, 1901–2001*, Canberra, Humanities Research Centre, Australian National University, 2003.

Evans, Benjamin, *A History of Agriculture, Production and Marketing in New Zealand*, Palmerston North, Keeling and Mundy, 1969.

Evans, Raymond, Kay Saunders and Kathryn Cronin, *Race Relations in Colonial Queensland: A History of Exclusion, Exploitation and Extermination*, 3rd edn, St Lucia, University of Queensland Press, 1993.

Fahey, Charles, 'Gold and land', in Deborah Gare and David Ritter (eds.), *Making Australian History: Perspectives on the Past*, South Melbourne, Thomson Learning Australia, 2008, pp. 206–212.

Fee, Jeffrey, 'Idaho's Chinese mountain gardens', in Priscilla Wegars (ed.), *Hidden Heritage: Historical Archaeology of the Overseas Chinese*, Amityville, Baywood Publishing, 1993, pp. 65–96.

Finn, A. J., 'Suitable sites for market gardening', *Agricultural Gazette of NSW*, Vol. XXVII, Part 9, 2 September 1916, pp. 633–639.

Fitzgerald, John, *Big White Lie: Chinese Australians in White Australia*, Sydney, University of New South Wales Press, 2007.

Fitzgerald, Shirley, *Red Tape Gold Scissors: The Story of Sydney's Chinese*, Sydney, State Library of New South Wales Press, 1996.

———'Desktop review of heritage significance of market gardens', June 2012, http://laperousemarketgardens.wordpress.com/history/, accessed 4 February 2013.

Flannery, Tim, *The Future Eaters: An Ecological History of Australasian Lands and People*, Sydney, Reed, 1994.

Flood, Josephine, *Archaeology of the Dreamtime: The Story of Prehistoric Australia and its People*, Pymble, NSW, Angus and Robertson, 1995.

Florey, Cecil, *Peninsula City: A Social History of the City of South Perth*, Perth, City of South Perth, 1995.

Foley, Cynthia, 'Chinese in the community', *Western Connections*, No. 38, December 1994, Dubbo and District Family History Society, pp. 4–7.

Fong, Ng Bickleen, *The Chinese in New Zealand*, Hong Kong, Hong Kong University Press, 1959.

Forster, Colin, 'Unemployment and minimum wages in Australia, 1900–1930', *Journal of Economic History*, Vol. 45, No. 2, 1985, pp. 383–388.

Fox, Paul, *Clearings: Six Colonial Gardeners*, Carlton, Victoria, Miegunyah Press, 2004.

Frawley, Kevin, 'Evolving visions: environmental management and nature conservation in Australia', in Stephen Dovers (ed.), *Australian Environmental History: Essays and Cases*, Melbourne, Oxford University Press, 1994, pp. 55–78.

Freedman, Maurice, 'Immigrants and associations: Chinese in nineteenth century Singapore', in Hong Liu (ed.), *The Chinese Overseas*, Vol. 2, London, New York, Routledge, 2006, pp. 104–127.

Frost, Lionel and Tony Dingle, 'Sustaining suburbia: a historical perspective on Australia's urban growth', *Economics and Commerce Discussion Papers*, Discussion Paper No. 23.93, November 1993.

Frost, Warwick, 'Migrants and technological transfer: Chinese farming in Australia, 1850–1920', *Australian Economic History Review*, Vol. 42, No. 2, 2002, pp. 113–31.

———'Australia unlimited?: Environmental debate in the age of catastrophe, 1910–1939', *Environment and History*, Vol. 10, No. 3, August 2004, pp. 285–303.

Furey, Louise, *Maori Gardening: An Archaeological Perspective*, Wellington, Department of Conservation, 2006.

Gammage, Bill, *Narrandera Shire*, Narrandera, Narrandera Shire Council, 1986.

Gaynor, Andrea, *Harvest of the Suburbs: An Environmental History of Growing Food in Australian Cities*, Perth, University of Western Australia Press, 2006.

George, Merrilyn, *Ohakune: Opening to a New World*, Ohakune, Kaipai Enterprises, 1990.

Gerber, James and Lei Guang (eds.), *Agriculture and Rural Connections in the Pacific*, Aldershot, Ashgate Variorum, 2006, pp. 195–234.

Giese, Diana, *Astronauts, Lost Souls and Dragons: Voices of Today's Chinese-Australians in Conversation with Diana Giese*, St Lucia, University of Queensland Press, 1997.

Gifford, Eardley and Eileen, 'The Chinese market gardeners of St George district', *St George Historical Society Bulletin*, December 1970, pp. 75–81.

Gillett, Andrew, 'Opium and race relations in Queensland', State Library of Queensland, 2010, http://www.slq.qld.gov.au/__data/assets/pdf_file/0020/163028/SLQ_Black_Opium_Andrew_Gillett.pdf, accessed 22 December 2012.

Gleason, Kathryn, 'To bound and to cultivate: an introduction to the archaeology of gardens and fields', in Naomi Miller and Kathryn Gleason (eds.), *The Archaeology of Garden and Field*, Philadelphia, University of Pennsylvania Press, 1994, pp. 1–25.

Global Farm Machinery Network, 'Irrigation', http://agmachine.ning.com/group/irrigation, accessed 10 May 2012.

———'Moulding machine', http://agmachine.ning.com/photo/mouldingmachine, accessed 10 May 2012.

Golden threads: 'The Chinese in regional NSW 1950–1950', online database, http://hosting.collectionsaustralia.net/goldenthreads/, accessed 20 January 2011.

Golledge, R. G., 'Sydney's metropolitan fringe: a study in urban-rural relations', *The Australian Geographer*, Vol. 7, 1960, pp. 243–255.

Golson, Jack, 'Aboriginal food plants: some ecological and cultural considerations', in D. J. Mulvaney and J. Golson (eds.), *Aboriginal Man and Environment in Australia*, Canberra, Australian National University Press, 1971, pp. 196–235.

Grace's Guide, 'British industrial history', http://www.gracesguide.co.uk/Richard_Tangye, accessed 11 May 2012.

Grassby, Richard, 'Material culture and cultural history', *Journal of Interdiscplinary History*, Vol. 35, No. 4, Spring 2005, pp. 591–603.

Greasley, David and Les Oxley, 'Globalization and real wages in New Zealand 1873–1913', *Explorations in Economic History*, Vol. 41, No. 1, January 2004, pp. 26–47.

———'Refrigeration and distribution: New Zealand prices and real wages 1873–1939', *Australian Economic History Review*, Vol. 45, No. 1, March 2005, pp. 23–44.

Greasley, David, Jakob Madden and Les Oxley, 'Real wages in Australia and Canada, 1870–1913: globalisation and productivity', *Australian Economic History Review*, Vol. 40, No. 2, July 2000, pp. 178–198.

Griffin, Keith, *The Political Economy of Agrarian Change: An Essay on the Green Revolution*, London, MacMillan Press, 1979.

Grotewold, Andreas, 'Von Thunen in retrospect', *Economic Geographer*, Vol. 35, 1959, pp. 346–355.

Gungwu, Wang, 'The study of Chinese identities in Southeast Asia', in Hong Liu (ed.), *The Chinese Overseas*, Vol. 1, London, New York, Routledge, 2006, pp. 258–278.

Guohong, Zhu, 'A historical demography of Chinese migration', in Hong Liu (ed.), *The Chinese Overseas*, Vol. 1, London, New York, Routledge, 2006, pp. 139–167.

Gwa Leng New Zealand Family History Group, *Gwa Leng Wongs in New Zealand*, Palmerston North, Edmon Wong, 2010.

Haig-Muir, Marnie and Roy Hay, 'The economy at war', in Joan Beaumont (ed.), *Australia's War 1939–1945*, St Leonards NSW, Allen & Unwin, 1996, pp. 107–135.

Harvey, Karen, *History and Material Culture*, London, Routledge, 2009.

Hawke, Gary and Ralph Lattimore, *Visionaries, Farmers and Markets: An Economic History of New Zealand Agriculture*, Wellington, Report to the Foundation for Research Science and Technology, January 1999.

Hayes, James, 'Good morning Mrs Thompson: A Chinese-english word-book from nineteenth century Sydney', in Paul Macgregor (ed.), *Histories of the Chinese in Australasia and the South Pacific*, Proceedings of an International Conference held at the Museum of Chinese Australian History, Melbourne, 8–10 October 1993, Melbourne, Museum of Chinese Australian History, 1995, pp. 113–126.

Head, Lesley, Pat Muir and Eva Hampel, 'Australian backyard gardens and the journey of migration', *Geographical Review*, Vol. 94, No. 3, July 2004, pp. 326–347.

Heinz, William, *Bright Fine Gold: Stories of the New Zealand Goldfields*, Wellington, Reed, 1974.

Helphand, Kenneth (ed.), *Technology and the Garden*, Dumbarton Oaks Colloquium on the History of Landscape Architecture, 2011, Washington, D.C., Dumbarton Oaks Research Library and Collection, 2014.

Hilaire-Perez, Liliane and Catherine Verna, 'Dissemination of technological knowledge in the Middle Ages and early modern era: new approaches and methodological issues', *Technology and Culture*, Vol. 47, No. 3, July 2006, pp. 536–565.

Hiyama, Kaaren, *High Hopes in Hard Times: A History of Grey Lynn and Westmere*, Grey Lynn, Media Studies Trust, 1991.

Hodder, Ian, *Symbols in Action: Ethnoarchaeological Studies of Material Culture*, Cambridge, Cambridge University Press, 1982.

Hoggart, Keith, *Agricultural Change, Environment and Economy: Essays in Honour of W. B. Morgan*, London, Mansell Publishing, 1992.

Horsfall, David, *March to Big Gold Mountain*, Melbourne, Red Rooster Press, 1985.

Hsu, Madeleine Yuan-Yin, *Dreaming of Gold, Dreaming of Home: Transnationalism and Migration between the United States and South China, 1882–1943*, Stanford, Stanford University Press, 2000.

Humphries, Andrea, Michael Wright, Adam Fowler, Paul Tinslay, Denis Gojak and Mark Tooker, 'Rockdale Market Gardens Conservation Management Plan', Sydney, NSW Department of Urban Affairs and Planning, 2001.

Hunt, Donald T., 'Market Gardening in Metropolitan Auckland', MA Thesis, University of New Zealand, Auckland, 1956.

Hunt, John Dixon (ed.), *Garden History: Issues, Approaches, Methods*, Dumbarton Oaks Colloquium on the History of Landscape Architecture, 1989, Washington D.C., Dumbarton Oaks Research Library and Collection, 1992.

Hunt, Suzanne, 'Vegetable plots and pleasure gardens of the Victorian goldfields', in Ian McCalman, Alexander Cook and Andrew Reeves (eds.), *Gold: Forgotten Histories and Lost Objects of Australia*, Cambridge, Cambridge University Press, 2001, pp. 267–284.

Inglis, Christine, 'Chinese in Australia', *International Migration Review*, Vol. 6, No. 3, Autumn 1972, pp. 266–281.

Ip, Manying, *Being Māori Chinese: Mixed Identities*, Auckland, Auckland University Press, 2008.

Jack, Ian, 'Some less familiar aspects of the Chinese in 19 century Australia', in Henry Chan, Ann Curthoys and Nora Chiang (eds.), *The Overseas Chinese in Australasia: History, Settlement and Interactions*, Canberra, National Taiwan University and Australian National University, 2001, pp. 44–54.

Jack, Ian, and Holmes, Kate, 'Ah Toy's Garden: A Chinese market garden on the Palmer River Goldfield, North Queensland', *Australian Journal of Historical Archaeology*, Vol. 2, 1984, pp. 51–58.

Jervis, James (Vince Kelly (ed.)), *The History of Woollahra: A Record of Events from 1788 to 1960 and a Centenary of Local Government*, Sydney, Woollahra Municipal Council, 1965.

Jiao, Tianlong, *The Neolithic of South East China: Cultural Transformation and Regional Interaction on the Coast*, Youngstown, New York, Cambria Press, 2007.

Jones, Paul, *Chinese-Australian Journeys: Records on Travel, Migration and Settlement, 1860–1975*, Canberra, National Archives of Australia, 2005.

Jupp, James (ed.), *The Australian People: An Encyclopedia of the Nation, its People and their Origins*, Cambridge, Cambridge University Press, 2001 [first published 1988].

Karskens, Grace, *The Cumberland Street/Gloucester Street Site, the Rocks: An Historical Discourse*, Sydney, Sydney Cove Authority, 1994.

———'Divergent visions: The archaeology of working people's culture' in *The Cumberland/Gloucester Street Site, The Rocks: Archaeological Investigation Report*, Redfern, NSW, Godden Mackay, 1999, pp. 159–184.

———'Engaging artefacts: urban archaeology, museums and the origins of Sydney', *Humanities Research*, Vol. 9, No. 1, 2002, pp. 36–56.

Karskens, Grace, *The Colony: A History of Early Sydney*, Crows Nest, NSW, Allen & Unwin, 2009.

Kass, Terry, *A Thematic History of the Central West*, Sydney, NSW Heritage Office, June 2003.

King, Franklin Hiram, *Farmers of Forty Centuries*, London, Butler and Tanner, 1949 [first published 1927].

King, Michael, *The Penguin History of New Zealand*, Auckland, Penguin Books, 2003.

Kingsbury, Noel, *Hybrid: The History and Science of Plant Breeding*, Chicago, London, University of Chicago Press, 2009.

Knapp, Robert (ed.), *Chinese Landscapes: The Village as Place*, Honululu, University of Hawaii Press, 1992.

Kulp, Daniel, *Country Life in South China: The Sociology of Familism*, Taipei, Ch'eng Wen Publishing, 1972.

Laird, Mark, '"Much better contrived and built than any other in England": stoves and other structures for the cultivation of exotic plants at Hampton Court Palace, 1689–1702', in Michael Lee and Kenneth Helpland (eds.), *Technology and the Garden*, Washington, D.C., Dumbarton Oaks Colloquium on the History of Landscape Architecture, 2014, pp. 79–107.

Lake, Marilyn and Henry Reynolds, *Drawing the Global Colour Line: White Man's Countries and the International Challenge of Racial Equality*, Cambridge, Cambridge University Press, 2008.

Lancaster, Rod, 'European-Chinese interaction: a pre-federation rural Australian setting', *Rural Society*, Vol. 10, No. 2, 2000, pp. 229–241.

Leach, Helen, *1,000 Years of Gardening in New Zealand*, Auckland, Wellington, A.H. & A.W. Reed, 1984.

———'The terminology of agricultural origins and food production systems – a horticultural perspective', *Antiquity*, Vol. 71, No. 271, March 1997, pp. 135–148.

Lee, Lily and Ruth Lam, *Sons of the Soil: Chinese Market Gardeners in New Zealand*, Pukekohe, Dominion Federation of New Zealand Chinese Commercial Growers, 2012.

Lee, Thong Ling, 'Chinese market gardening in the Auckland Region', MA Thesis, University of Auckland, Auckland, 1974.

Leigh, G. J., *The World's Greatest Fix: A History of Nitrogen and Agriculture*, Oxford, Oxford University Press, 2004.

Liu, Hong (ed.), *The Chinese Overseas*, Vols. 1–4, London, New York, Routledge, 2006.

———'Towards a multidimensional exploration of the Chinese overseas', in Hong Liu (ed.), *The Chinese Overseas*, Vol. 1, London, New York, Routledge, 2006, pp. 1–30.

Long Dodd, Marjorie, *It's a Long Story*, Sydney, self-published, 1997.

Lydon, Jane, *Many Inventions: The Chinese in the Rocks 1890–1930*, Clayton, Victoria, Monash Publications in History, Monash University, 1999.

Macintyre, Stuart, *A Concise History of Australia*, 3rd edn, Port Melbourne, Cambridge University Press, 2009.

Madgin, Diana, 'Alice Lee, octogenarian market gardener', *New Zealand Gardener*, Vol. 54, No. 5, May 1998, pp. 75–76.

Mahaffey, Kath, *Pioneers of the North West Plains*, Vol. II, Moree, Moree and District Historical Society, 1982.

Mangere Historical Society, *Mangere Chronicles*, Mangere, Mangere Historical Society, 1990.

Markus, Andrew, *Fear and Hatred: Purifying Australia and California 1850–1901*, Sydney, Hale and Iremonger, 1979.

Martiniello, Jennifer, 'As strands of plaited music: my Chinese-Aboriginal-Anglo heritage', in Penny Edwards and Shen Yuanfang (eds.), *Lost in the Whitewash: Aboriginal-Asian Encounters in Australia, 1901–2001*, Canberra, Australian National University 2003, pp. 23–36.

Maurya, D. M., 'The innovative approach of Indian farmers', in Robert Chambers, Arnold Pacey and Lori Ann Thrupp (eds.), *Farmer First: Farmer Innovation and Agricultural Research*, London, Intermediate Technology Publications, 1989, pp. 9–14.

May, Cathie, *Topsawyers: The Chinese in Cairns 1870 to 1920*, Townsville, James Cook University, 1984.

Mayne, Alan, *The Imagined Slum: Newspaper Representation in Three Cities 1870–1914*, London, Leicester University Press, 1993.

McCalla, Thomas and Donald Plucknett, 'Collecting, transporting and processing organic fertilisers' in Donald Plucknett and Halsey Beemer (eds.), *Vegetable Farming Systems in China*, Boulder Colorado, Westview Press, 1981, pp. 19–38.

McCarol, Norman, *British History 1815–1906*, Oxford, Oxford University Press, 1991.

McCarty, J. W.,'Australian capital cities in the twentieth century', *Australian Economic History Review*, Vol. X, No. 1, March 1970, pp. 107–137.

———'Australia as a region of settlement in the nineteenth century', *Australian Economic History Review*, Vol. XIII, 1973, pp. 148—167.

McCarty, J. W. and C. B. Schedvin (eds.), *Australian Capital Cities: Historical Essays*, edited by J.W. McCarty and C.B. Schedvin, Sydney, Sydney University Press, 1978.

McCraw, John, *Mountain Water and River Gold: Stories of Gold Mining in the Alexandra District*, Dunedin, Square One Press, 2000.

McGowan, Barry, 'The economics and organisation of Chinese mining in colonial Australia', *Australian Economic History Review*, Vol. 45, No. 2, July 2005, pp. 119–138.

———'Chinese market gardens in southern and western NSW', *Australian Humanities Review*, No. 36, 2005, n. p.

———'Ringbarkers and market gardeners: a comparison of the rural Chinese of New South Wales and California', *Chinese America: History and Perspectives*, Annual, San Francisco, Chinese Historical Association of America, 2006, pp. 31–46.

———'Adaptation and organization: the history and heritage of the Chinese in the Riverina and Western New South Wales, Australia, *Chinese America: History and Perspectives*, Annual, San Francisco, Chinese Historical Association of America, 2007, pp. 233–240.

———*Dust and Dreams: Mining Communities in South-East New South Wales*, Kensington, University of New South Wales Press, 2010.

———*Tracking the Dragon: A History of the Chinese in the Riverina*, Wagga Wagga, Museum of the Riverina, 2010.

McGuire, R. H., 'The study of ethnicity in historical archaeology', *Journal of Anthropological Archaeology*, Vol. 1, No. 2, 1982, pp. 159–178.

McLellan, Alex, 'Market gardening in Otaki', *Otaki Historical Society Historical Journal*, No. 5, 1982, pp. 60–69.

McLure, Margaret, 'Auckland region', *Te Ara–the Encyclopedia of New Zealand*, updated November 2012, http://www.teara.govt.nz/en/auckland-region/, accessed 20 June 2013.

Mei, June, 'Socioeconomic origins of emigration: Guangdong to California, 1850–1882', in James Gerber and Lei Guang (eds.), *Agriculture and Rural Connections in the Pacific*, Aldershot, Ashgate Variorum, 2006, pp. 195–234.

Meisner Rosen, Christine, 'The business–environment connection', *Environmental History*, Vol. 10, No. 1, 2005, pp. 77–79.

Melbourne Markets, 'Melbourne wholesale fruit, vegetable & flower market "History"', http://www.melbournemarkets.com.au/about_us/history, accessed 12 September 2011.

Metherill, Terry, 'Faster: Manly in the 1920s', unpublished manuscript, Ryde, City of Ryde Library Service, 2006.

Miller, Sally, 'European immigrant communities and New Zealand: comparative reflections on a little known history', in Olavi Koivukangas and Charles Westin (eds.), *Scandinavian and European Migration to Australia and New Zealand*, Proceedings of the conference held in Turku Finland, June 1998, Turku, Institute of Migration, Stockholm, Centre for Research in International Relations and Ethnic Relations, 1998.

Migration Heritage Centre, 'Happy Valley, Chinese market gardens and migrant camps', online exhibition, Powerhouse Museum, Sydney, http://www.migrationheritage.nsw.gov.au/exhibition/atthebeach/happy-valley/, accessed 28 July 2013.

Mindfood, 'Traditional Maori recipes', 27 April 2012, http://www.mindfood.com/at-traditional-maori-recipes-global-tastes.seo, accessed 22 February 2013.

Mirmohamadi, Kylie, 'Wog plants go home: race, ethnicity and horticulture in Australia', *Studies in Australian Garden History*, Vol. 1, 2003, pp. 91–108.

Missingham, Bruce, Jacqui Dibden and Chris Cocklin, 'A multicultural countryside? Ethnic communities in rural Australia', *Rural Society*, Vol. 16, No. 2, 2006, pp. 131–150.

Mokyr, Joel, 'Technological change, 1700–1830', in Roderick Floud and Donald McCloskey (eds.), *The Economic History of Britain Since 1700*, Vol. 1, 1700–1860, Cambridge, New York, Cambridge University Press, 1994, pp. 12–43.

Moloughney, Brian, Tony Ballantyne and David Hood, 'After gold: Reconstructing Chinese communities 1896–1913', in Henry Johnson and Brian Moloughney (eds.), *Asia in the Making of New Zealand*, Auckland, Auckland University Press, 2006, pp. 58–75.

Monk, Joanne, 'The diggers in the trenches: a history of market gardens in Victoria 1835–1939', *Australian Garden History*, Vol. 4, No. 1, July/August 1992, pp. 3–5.

Morris, Colleen, 'Chinese market gardens in Sydney', *Australian Garden History*, Vol. 12, No. 5, March/April 2001, pp. 4–7.

Mosely, Henry, *The Mechanical Principles of Engineering and Architecture*, New York, Wiley and Sons, 1869.

Moyal, Ann, *A Bright and Savage Land: Scientists in Colonial Australia*, Sydney, Collins, 1986.

Mullan, Brett, Andrew Tait and Craig Thompson, 'Climate', *Te Ara–the Encyclopaedia of New Zealand*, updated November 2012, http://www.teara.govt.nz/en/climate, accessed 4 May 2011.

Mulvaney, John and Johan Kamminga, *Prehistory of Australia*, Washington, London, Smithsonian Institution Press, 1999.

Murphy, E. and M. Aucott, 'An assessment of the amounts of arsenical pesticides used historically in a geographical area', *Science of the Total Environment*, Vol. 218, No. 2–3, July 1998, pp. 89–101.

Murphy, Nigel, *The Poll-Tax in New Zealand*, Research paper commissioned by the New Zealand Chinese Association, 2nd edn, Wellington, Office of Ethnic Affairs, Department of Internal Affairs, 2002.

———*A Chronology of Events Relating to the History of the Chinese in New Zealand*, Wellington, Alexander Turnbull Library, 2002.

———'A history of the horticultural industry in New Zealand', unpublished manuscript, 2010.

———*Success Through Adversity: A History of the Dominion Federation of New Zealand Chinese Commercial Growers*, Pukekohe, Dominion Federation of New Zealand Chinese Commercial Growers, 2012.

Muscolino, Micah, 'Global dimensions of modern China's environmental history', *World History Connected* 6 (2009): 31 pars http://www.historycooperative.org/journals/whc/6.1/muscolino.htm, accessed January 19, 2012.

Needham, Joseph, (in collaboration with Wang Ling and Lu Gwei Djen) *Science and Civilisation in China, Vol. 4, Part III, Civil Engineering and Nautics*, Cambridge, Cambridge University Press, 1971.

———(in collaboration with Lu Gwei Djen and Huang Tsing-Tsung) *Science and Civilisation in China, Vol. 6, Biology and Biological Technology, Part I, Botany*, Cambridge, Cambridge University Press, 1986.

Needham, Joseph and Bray, Francesca, *Science and Civilisation in China, Vol. 6, Biology and Biological Technology, Part I, Agriculture*, Cambridge, Cambridge University Press, 1984.

Newman, Jacqueline, *Food Culture in China*, Westport, London, Greenwood Press, 2004.

New Zealand Historic Places Trust, 'Lye Bows Historic Area', http://www.historic.org.nz/theregister/RegisterSearch/RegisterResults.aspx?RID=7547, accessed 14 June 2011.

Ng, James, *Windows on a Chinese Past*, 4 Vols, Dunedin, Otago Heritage Books, 1993–1999.

———'The sojourner experience: The Cantonese gold seekers in New Zealand', in Ip (ed.), *Unfolding History, Evolving Identity: The Chinese in New Zealand*, Auckland, Auckland University Press, 2003, pp. 5–30.

Nixon, R. E., 'The Chinese community of Camden', unpublished manuscript, Camden Historical Society, 1976.

O'Brien, Patrick, 'Modern conceptions of the industrial revolution,' in Patrick O'Brien and Roland Quinault (eds.), *The Industrial Revolution and British Society*, Cambridge, Cambridge University Press, 1993, pp. 1–30.

Oxford Dictionaries, www.oxforddictionaries.com, accessed 10 May 2012.

Pacey, Arnold, *Technology in World Civilisation*, Cambridge, Massachusetts, MIT Press, 1993.

Park, Geoff, 'Swamps which might doubtless be easily drained: swamp drainage and its impact on the indigenous', in Eric Pawson and Tom Brooking (eds.), *Environmental Histories of New Zealand*, Auckland, Oxford University Press, 2002, pp. 151–168.

———*Nga Ururora: Ecology and History in a New Zealand Landscape*, Wellington, Victoria University Press, 1995.

Parker, Frances, 'The safe use of farm chemicals by market gardeners of non-English speaking background: developing an effective extension strategy for the Sydney basin', Sydney, Rural Industries Research & Development Corporation, 2000, http://www.rirdc.gov.au, accessed 12 May 2012.

———'Making periurban farmers on the fringe matter', Proceedings of the State of Australian Cities Conference, Sydney, University of New South Wales, 2007, pp. 259–269.

Parr, Tom E., *Reminiscences of a NSW South West Settler*, New York, Carlton Press, 1977.

Pawakapan, Niti, 'The Chinese in Dunedin between the 1920s and the 1930s', MA Thesis, University of Otago, Dunedin, 1987.

Payne, Val, *Celebrating Mangere Bridge*, Mangere, Mangere Historical Society, 2005.

Pearson, Warwick, 'Water power in a dry continent: the transfer of watermill technology from Britain to Australia in the 19 century', PhD Thesis, University of New England, Armidale, 1995.

Pechenkina, Ekaterina, Robert Benfer and Xiaolin Ma, 'Diet and health in the Neolithic of the Wei and middle Yellow River basins, Northern China', in Mark Nathan Cohen and Gillian M. Crane-Kramer (eds.), *Ancient Health: Skeletal Indicators of Agricultural and Economic Intensification*, Gainsville, University Press of Florida, 2007, pp. 255–272.

Peryea, Francis, 'Historical use of lead arsenate insecticides, resulting soil contamination and implications for soil remediation', Proceedings, 16th World Congress of Soil Science, Montpellier, France, 20–26 August 1998, http://soils.tfrec.wsu.edu/leadhistory.htm, accessed 1 March 2012.

Pesticide Action Network, 'Fact sheet: rotenone', http://www.pan-uk.org/pest news/Actives/rotenone.htm, accessed 5 March 2012.

Plucknett, Donald and Halsey Beemer (eds.), *Vegetable Farming Systems in China*, Special Studies in Agricultural Science Policy, Boulder, Colorado, Westview Press, 1981.

Powell, Joe, *Mirrors of the New World: Images and Image-makers in the Settlement Process*, Folkestone/Hamden Connecticut, Dawson Archon Books, 1977.

Powell, J. M., *Plains of Promise, Rivers of Destiny: Water Management and the Development of Queensland 1824–1990*, Bowen Hills, Boolarong Publications, 1991.

———'Thomas Griffith-Taylor', in *Australian Dictionary of Biography*, 1990, http://adb.anu.edu.au/biography/taylor-thomas-griffith-8765, accessed 30 April 2012.

———*Environmental Management in Australia* 1788–1914: *Guardians, Improvers and Profit*, Melbourne, Oxford University Press, 1993.

Price, Charles, *The Great White Walls are Built: Restrictive Immigration to North America and Australia, 1836–1888*, Canberra, Australian Institute of International Affairs in association with Australian National University Press, 1974.

Pullman, Sandra, 'Along Melbourne's rivers and creeks', *Australian Garden History*, Vol. 12, No. 5, March/April 2001, pp. 9–10.

Queensland Department of Agriculture, Fisheries and Forestry 'About taro', http://www.daff.qld.gov.au, accessed 27 March 2013.

Rains, Kevin, 'Beyond the wattle fence: the Chinese market gardeners of early Cooktown', in Kevin Wong Hoy and Kevin Rains (eds.), *Rediscovered Past: China in Northern Australia*, North Melbourne, Chinese Heritage In Northern Australia (CHINA), 2009, pp. 21–30.

Randwick Municipal Council, *Randwick: A Social History*, Kensington, University of New South Wales Press, 1985.

Rannard, Ian, *The Forgotten Gardens: The Story of the Last Market Gardens in Willoughby and Northbridge, NSW*, Douglas Park, Ian Rannard, 2005.

Rawski, Evelyn Sakakida, *Agricultural Change and the Peasant Economy of South China*, Cambridge Massachusetts, Harvard University Press, 1972.

Reeves, Keir, 'Historical neglect of an enduring Chinese community', *Traffic*, No. 3, July 2003, pp. 53—77.

———'Tracking the dragon down under: Chinese cultural connections in gold rush Australia and Aotearoa New Zealand', *Graduate Journal of Asia-Pacific Studies*, Vol. 3, No 1, 2005, pp. 49—66.

Reeves, Keir, and Benjamin Mountford, 'Sojourning and settling: locating Australian history', *Australian Historical Studies*, Vol. 42, No. 1, 2011, pp. 111–125.

Reidy, Jade, *Not Just Passing Through: The Making of Mount Roskill*, Auckland, Auckland City Council, 2007.

Resources For History, R4H, 'Farming in Celtic Britain', http://resourcesforhistory.com/Celtic_Farming_in_Britain.htm, accessed 25 March 2011.

Reynolds, Henry, *North of Capricorn: The Untold Story of Australia's North*, Crows Nest, NSW, Allen and Unwin, 2003.

———*Nowhere People: How International Race Thinking Shaped Australia's Identity*, Camberwell, Vic., Penguin, 2005.

Rhoades, Robert, 'The role of farmers in the creation of agricultural technology', in Robert Chambers, Arnold Pacey and Lori Ann Thrupp (eds.), *Farmer First: Farmer Innovation and Agricultural Research*, London, Intermediate Technology Publications, 1989, pp. 3–8.

Richards, Oline, 'Chinese market gardening: a Western Australian perspective', *Australian Garden History*, Vol. 13, No. 1, July/August 2001, pp. 19–21.

Richards, Paul, *Indigenous Agricultural Revolution: Ecology and Food Production in West Africa*, London, Melbourne, Sydney, Auckland, Hutchinson, 1985.

Ritchie, Neville, 'Archaeology and history of the Chinese in southern New Zealand during the nineteenth century: a study of acculturation, adaptation, and change', PhD Thesis, University of Otago, Dunedin, 1986.

———'Traces of the past: archaeological insights into the New Zealand Chinese experience in southern New Zealand', in Manying Ip (ed.), *Unfolding History, Evolving Identity: The Chinese in New Zealand*, Auckland, Auckland University Press, 2003, pp. 31–47.

Robb, Sandi, 'Beyond the coast: Chinese settler patterns of the western gulf region, north Queensland', in Kevin Wong Hoy and Kevin Rains (eds.), *Rediscovered Past: China in Northern Australia*, North Melbourne, Chinese Heritage In Northern Australia (CHINA), 2009, pp. 24–35.

———'Mar & Mar: a tale of two Chinese gardeners at Winton', in Kevin Wong Hoy and Kevin Rains (eds.), *Rediscovered Past: Valuing Chinese Across the North*, North Melbourne, Chinese Heritage In Northern Australia (CHINA), 2012, pp. 36–44.

Robin, Libby and Tom Griffiths, 'Environmental history in Australasia', http://ceh.environmentalhistory-anz.org/wpcontent/uploads/Environmental_History_in_Australasia2004.pdf, accessed 8 February 2013.

Roche, Michael, 'The state as conservationist, 1920–60: "wise use" of forests, lands and water', in Eric Pawson and Tom Brooking (eds.), *Environmental Histories of New Zealand*, Auckland, Oxford University Press, 2002, pp. 183–199.

Roesch, Frank, 'Irrigating old style: the Skinner system', *Farm Collector*, July 2009, http://www.farmcollector.com/equipment/Irrigating-Old-Style-The-Skinner-System.aspx, accessed 29 January 2012.

Roger, Warwick, 'The true builders of the Waimarino', *Weekly News*, 12 July 1971, pp. 10–13.

Rogers, Everett, *Diffusion of Innovations*, 5th edn, New York, Free Press, 2003.

Rosen, Christine Meisner, 'The business–environment connection', *Environmental History*, Vol. 10, No. 1, 2005, pp. 77–79.

Rosenberg, Nathan, *Inside the Black Box: Technology and Economics*, Cambridge, Cambridge University Press, 1982.

———*Exploring the Black Box: Technology, Economics and History*, Cambridge, Cambridge University Press, 1994.

Ruttan, V. W. and Yujiro Hayami, 'Technology transfer and agricultural development', *Technology and Culture*, Vol. 14, No. 2, Part 1, April 1973, pp. 119–151.

Ryan, Jan, *Ancestors: Chinese in Colonial Australia*, South Fremantle, Fremantle Arts Centre Press, 1995.

Ryder, Leonie, 'Incorrigible colonist: ginger in Australia 1788–1950', PhD Thesis, University of Adelaide, Adelaide, 2010.

Sakiyama, R., 'What is Horticulture?', *Chronica Horticulturae*, International Library for Horticultural Science, Vol. 31, No. 2, 1991, p. 17.

Schedvin, C. B., 'Staples and regions of Pax Britannica', *Economic History Review*, Vol. 43, No. 4, November 1990, pp. 533–539.

Scoones, Ian and John Thompson (eds.), *Beyond Farmer First: Rural People's Knowledge, Agricultural Research and Extension Practices*, London, International Technology Publications, 1994.

Sedgwick, Charles, 'The politics of survival: a social history of the Chinese in New Zealand', PhD Thesis, University of Canterbury, Christchurch, 1982.

Shanahan, Martin and John Wilson, 'Measuring inequality trends in colonial Australia using factor-price ratios: the importance of boundaries', *Australian Economic History Review*, Vol. 47, No. 1, March 2007, pp. 6–21.

Shen, Yuanfang, *Dragon Seed in the Antipodes: Chinese-Australian Autobiographies*, Carlton North, Melbourne University Press, 2001.

Shen, Yuanfang and Penny Edwards, 'United by the sweep of a tarnished brush', *Canberra Times*, 18 November 2000, pp. 4–5.

Sheng, Fei, 'Environmental experiences of Chinese people in the mid-nineteenth century Australian gold rush', *Global Environment, A Journal of History and Natural and Social Sciences*, 7–8, 2012, pp. 99–127.

Simpson, Tony, *A Distant Feast: The Origins of New Zealand Cuisine*, Auckland, Godwit, 2008.

Singhal, Arvind and James Deering (eds.), *Communication of Innovations: A Journey with Ev Rogers*, Sage Publications, New Delhi, 2006.

Smith, Derek, 'Market gardening at Adelaide's urban fringe', *Economic Geography*, Vol. 42, No. 1, January 1966, pp. 19–36.

Stanin, Zvonkica, From Li Chun to Yong Kit: a market garden on the Loddon, 1851–1912', *Journal of Australian Colonial History*, Vol. 6, 2004, pp. 15–34.

Stapleton, Darwin H., *The Transfer of Early Industrial Technologies to America*, Kansas, Allen Press, 1987.

Star, Paul and Lynne Lochhead, 'Children of the burnt bush: New Zealanders and the indigenous remnant, 1880–1930', in Eric Pawson and Tom Brooking (eds.), *Environmental Histories of New Zealand*, Auckland, Oxford University Press, 2002, pp. 183–199.

Stedman, Geoffrey, 'The South Dunedin Flat: a study in urbanisation 1849–1965', MA Thesis, Department of Geography, University of Otago, Dunedin, 1966.

Sturman, Andrew and Nigel Tapper, *The Weather and Climate of Australia and New Zealand*, 2nd edn, South Melbourne, Oxford University Press, 2006.

Svenson, Geoffrey, 'Marginal people: the archaeology and the history of the Chinese at Milparinka', MA Thesis, University of Sydney, Sydney, 1994.

Symons, Michael, *One Continuous Picnic: A Gastronomic History of Australia*, Melbourne, Melbourne University Press, 2007.

Taher, Mohommod, 'Asians in New Zealand: a geographical review and interpretation', PhD Thesis, University of Auckland, 1965.

Tai Yuen, William, *The Origins of China's Awareness of New Zealand, 1674–1911*, Auckland, New Zealand Asia Institute, 2005.

Thick, Malcolm, *The Neat House Gardens: Early Market Gardening Around London*, Totnes, Prospect Books, 1998.

Thirsk, Joan, *Alternative Agriculture: A History*, Oxford, Oxford University Press, 1997.

Thomas, Mark, '"A substantial Australian superiority"? Anglo-Australian comparisons of consumption and income in the late nineteenth century', *Australian Economic History Review*, Vol. 35, No. 2, 1995, pp. 10–38.

Thorpe, Ann, 'Otaki's market gardens', *Otaki Historical Society Historical Journal*, Vol. 30, 2008, pp. 13–35.

Todd, Jan, *Colonial Technology: Science and the Transfer of Innovation to Australia*, Cambridge, Cambridge University Press, 1995.

Tuck, J. M., 'The Devil's Half Acre, 1900–1910', BA Honours Thesis, University of Otago, Dunedin, 1983.

Tyrell, Ian, *True Gardens of the Gods: California–Australia Environmental Reform 1860–1930*, Berkeley, University of California Press, 1999.

Veart, David, *First Catch Your Weka*, Auckland, Auckland University Press, 2008.

Vinning, Grant, *Select Markets for Sweet Potato, Taro and Yam*, Report for Rural Industries Research and Development Corporation, Kingston, ACT, Rural Industries Research and Development Corporation, May 2003.

Von Tunzelman, G.N., 'Technological and organizational change in industry in the early Industrial revolution', in Patrick K. O'Brien and Roland Quinault (eds.), *The Industrial Revolution and British Society*, Cambridge, Cambridge University Press, 1993, pp. 254–282.

———'Technology in the early nineteenth century', in Roderick Floud and Donald McCloskey (eds.), *The Economic History of Britain Since 1700*, Vol. 1, 1700–1860, Cambridge, New York, Cambridge University Press, 1994, pp. 271–299.

Wai Shing, Pamela, 'Locational and structural changes of market gardening in Pukekohe–Bombay–Patumahoe', MA Thesis, University of Auckland, Auckland, 1977.

Waldinger, Roger, H. Aldrich, R Ward and Associates, *Ethnic Entrepreneurs: Immigrant Business in Industrial Societies*, Newbury Park, CA, Sage, 1990.

Walker, Robin and Dave Roberts, *Scarcity to Surfeit: A History of Food and Nutrition in New South Wales*, Kensington, New South Wales University Press, 1988.

Wang, L. Ling Chi, 'The structure of dual domination: toward a paradigm for the study of the Chinese diaspora in the United States', in Hong Liu (ed.), *The Chinese Overseas*, Vol. 2, London, New York, Routledge, 2006, pp. 279–296.

Ward, Martin, and Shona Russell, 'Water sharing schemes: Insights from Canterbury and Otago', Report for Foundation for Research, Science and Technology, Wellington, Landcare Research, 2010.

Wass, Janice Tauer, 'Teaching history with material culture', http://www.lib.niu.edu/1998/iht529802.html, accessed 20 June 2012.

Wegars, Priscilla (ed.), *Hidden Heritage: Historical Archaeology of the Overseas Chinese*, Amityville, Baywood Publishing, 1993.

Weiping, Liu, 'Chinese newspapers in Australia from the turn of the century', Chinese Heritage of Australian Federation, http://www.chaf.lib.latrobe.edu.au/chinese_newspapers.htm, accessed 11 May 2012.

White, Colin, *Mastering Risk: Environment, Markets and Politics in Australian Economic History*, Melbourne, Oxford University Press Australia, 1992.

Whitmore, Karla, 'Willoughby's Chinese market gardens', unpublished manuscript, January, 2004.

Wickerberg, Edgar, 'Overseas Chinese adaptive organisations, past and present', in Hong Liu (ed.), *The Chinese Overseas*, Vol. 2, London, New York, Routledge, 2006, pp. 87–103.

Wikipedia, 'Hebei', http://en.wikipedia.org/wiki/Hebei, accessed 3 June 2011.

'Row cover', http://en.wikipedia.org/wiki/Row_cover, accessed 26 March 2011.

Williams, D.B. (ed.), *Agriculture in the Australian Economy*, Sydney, Sydney University Press, 1967.

Williams, Michael, 'Brief sojourn in your native land: Sydney's huaqiao and their links with south China during the first half of the twentieth century', M.Lit. Thesis, University of New England, Armidale, 1998.

———*Chinese Settlement in New South Wales: A Thematic History*, report for the NSW Heritage Office, Sydney, NSW Heritage Office, September 1999.

———'Destination qiaoxiang', PhD Thesis, University of Hong Kong, Hong Kong, 2002.

Wilton, Janis, *Hong Yuen: A Country Store and its People*, Armidale, Armidale CAE and Multicultural Education Coordinating Committee, NSW Department of Education, 1989.

———'The Chinese history and heritage of regional New South Wales', in Henry Chan, Ann Curthoys and Nora Chiang (eds.), *The Overseas Chinese in Australasia: History, Settlement and Interactions*, Canberra, National Taiwan University and Australian National University, 2001, pp. 91–101.

———*Golden Threads: The Chinese in Regional New South Wales 1850–1950*, Armidale, New England Regional Art Museum in association with Powerhouse Publishing, 2004.

Wong, Helen, *In the Mountain's Shadow: A Century of Chinese in Taranaki*, Auckland, Helen Wong, 2010.

Wordpress, 'La Perouse market gardens: sustainable and historic', http://laperousemarketgardens.wordpress.com/history/, accessed 4 February 2013.

Worrall-Smith, Roslind, 'The employment of Maori women by Asiatic market gardeners: An aspect of New Zealand race relations', *University of Auckland Historical Society Annual*, 1970, pp. 1–11.

Woudstra, Jan, 'Garden hydraulics in pre-Sistine Rome: theory and practice', in Michael Lee and Kenneth Helpland (eds.), *Technology and the Garden*, Washington D.C., Dumbarton Oaks Colloquium on the History of Landscape Architecture, 2014, pp. 111–128.

Xu, Guohua, and L. J. Peel (eds.), *The Agriculture of China*, Oxford, Oxford University Press, 1991.

Yates Australia, 'History of Yates', http://www.yates.com.au/about/history/, accessed 9 April 2012.

Yen, Doug, 'The sweet potato in historical perspective', in Ruben Villareal and T. E. Griggs, *International Symposium on Sweet Potato*, Taiwan, 23–27 March 1981, AVRDC Publication 82–172, 1982, pp. 17–30.

Yong, C. F., *The New Gold Mountain*, Richmond, S.A, Raphael Arts, 1977.

Young, Stephen, '"To be happy for the rest of your life": Chinese market gardeners in New Zealand', Address in honour of 'A Barbarous Measure; The Poll Tax and Chinese New Zealanders', National Library of New Zealand, 2 October 2003.

Yu, Henry, The Rhythm of Cantonese trans-Pacific migration' in Donna Gabaccia and Dirk Hoeder (eds.), *Connecting Seas and Connected Ocean Rims: Indian, Atlantic and Pacific Oceans and China Seas Migrations from the 1830s to the 1930s*, Leiden, Boston, Brill, 2011, pp. 393–414.

Yuan, C.M., 'Chinese in White Australia 1901–1950', in James Jupp (ed.), *The Australian People: An Encyclopedia of the Nation, its People and Their Origins*, Cambridge, England, Cambridge University Press, 2001, pp. 304–307.

Zhao, Karl, 'Chinese market gardens', *Heritage New South Wales*, Vol. 6, No. 3, October 1999, p. 15.

Index

J. Boileau, *Chinese Market Gardening in Australia and New Zealand*, Palgrave Studies in the History of Science and Technology, DOI 10.1007/978-3-319-51871-8

C

D

N

O

Kolibri 360 GmbH
Ferdinand-Jühlke-Straße 7
99095 Erfurt, Deutschland
produktsicherheit@kolibri360.de